T0073168

The Handbook of Environmental Chemistry

Founded by Otto Hutzinger

Editors-in-Chief: Damià Barceló • Andrey G. Kostianoy

Volume 58

More information about this series at http://www.springer.com/series/698

Freshwater Microplastics

Emerging Environmental Contaminants?

Volume Editors: Martin Wagner · Scott Lambert

With contributions by

E. Besseling · F.J. Biginagwa · N. Brennholt · T.B. Christensen ·
I.K. Dimzon · R. Dris · M. Eriksen · J. Eubeler · J. Gasperi ·
S.F. Hansen · J.P. Harrison · N.B. Hartmann · M. Heß · T.J. Hoellein ·
Y. Ju-Nam · F.R. Khan · T. Kiessling · S. Klein · T.P. Knepper ·
A.A. Koelmans · M. Kooi · J. Kramm · C. Kroeze · S. Lambert ·
B.S. Mayoma · J.J. Ojeda · M. Prindiville · G. Reifferscheid ·
S.E. Rist · M. Sapp · C. Scherer · K. Syberg · A.S. Tagg ·
B. Tassin · M. Thiel · C. Völker · M. Wagner · A. Weber ·
A.P. van Wenzel · C. Wu · X. Xiong · K. Zhang

Editors
Martin Wagner
Department of Biology
Norwegian University of Science
    and Technology (NTNU)
Trondheim, Norway

Scott Lambert
Department Aquatic Ecotoxicology
Goethe University Frankfurt am Main
Frankfurt, Germany

ISSN 1867-979X   ISSN 1616-864X   (electronic)
The Handbook of Environmental Chemistry
ISBN 978-3-319-61614-8   ISBN 978-3-319-61615-5   (eBook)
DOI 10.1007/978-3-319-61615-5

Library of Congress Control Number: 2017954325

Printed on acid-free paper

This Springer imprint is published by Springer Nature
The registered company is Springer International Publishing AG
The registered company address is: Gewerbestrasse 11, 6330 Cham, Switzerland

The Handbook of Environmental Chemistry
Also Available Electronically

The Handbook of Environmental Chemistry is included in Springer's eBook package *Earth and Environmental Science*. If a library does not opt for the whole package, the book series may be bought on a subscription basis.

For all customers who have a standing order to the print version of *The Handbook of Environmental Chemistry,* we offer free access to the electronic volumes of the Series published in the current year via SpringerLink. If you do not have access, you can still view the table of contents of each volume and the abstract of each article on SpringerLink (www.springerlink.com/content/110354/).

You will find information about the

– Editorial Board
– Aims and Scope
– Instructions for Authors
– Sample Contribution

at springer.com (www.springer.com/series/698).

All figures submitted in color are published in full color in the electronic version on SpringerLink.

Aims and Scope

Since 1980, *The Handbook of Environmental Chemistry* has provided sound and solid knowledge about environmental topics from a chemical perspective. Presenting a wide spectrum of viewpoints and approaches, the series now covers topics such as local and global changes of natural environment and climate; anthropogenic impact on the environment; water, air and soil pollution; remediation and waste characterization; environmental contaminants; biogeochemistry; geo-ecology; chemical reactions and processes; chemical and biological transformations as well as physical transport of chemicals in the environment; or environmental modeling. A particular focus of the series lies on methodological advances in environmental analytical chemistry.

Series Preface

With remarkable vision, Prof. Otto Hutzinger initiated *The Handbook of Environmental Chemistry* in 1980 and became the founding Editor-in-Chief. At that time, environmental chemistry was an emerging field, aiming at a complete description of the Earth's environment, encompassing the physical, chemical, biological, and geological transformations of chemical substances occurring on a local as well as a global scale. Environmental chemistry was intended to provide an account of the impact of man's activities on the natural environment by describing observed changes.

While a considerable amount of knowledge has been accumulated over the last three decades, as reflected in the more than 70 volumes of *The Handbook of Environmental Chemistry,* there are still many scientific and policy challenges ahead due to the complexity and interdisciplinary nature of the field. The series will therefore continue to provide compilations of current knowledge. Contributions are written by leading experts with practical experience in their fields. *The Handbook of Environmental Chemistry* grows with the increases in our scientific understanding, and provides a valuable source not only for scientists but also for environmental managers and decision-makers. Today, the series covers a broad range of environmental topics from a chemical perspective, including methodological advances in environmental analytical chemistry.

In recent years, there has been a growing tendency to include subject matter of societal relevance in the broad view of environmental chemistry. Topics include life cycle analysis, environmental management, sustainable development, and socio-economic, legal and even political problems, among others. While these topics are of great importance for the development and acceptance of *The Handbook of Environmental Chemistry*, the publisher and Editors-in-Chief have decided to keep the handbook essentially a source of information on "hard sciences" with a particular emphasis on chemistry, but also covering biology, geology, hydrology and engineering as applied to environmental sciences.

The volumes of the series are written at an advanced level, addressing the needs of both researchers and graduate students, as well as of people outside the field of "pure" chemistry, including those in industry, business, government, research establishments, and public interest groups. It would be very satisfying to see these volumes used as a basis for graduate courses in environmental chemistry. With its high standards of scientific quality and clarity, *The Handbook of Environmental Chemistry* provides a solid basis from which scientists can share their knowledge on the different aspects of environmental problems, presenting a wide spectrum of viewpoints and approaches.

The Handbook of Environmental Chemistry is available both in print and online via www.springerlink.com/content/110354/. Articles are published online as soon as they have been approved for publication. Authors, Volume Editors and Editors-in-Chief are rewarded by the broad acceptance of *The Handbook of Environmental Chemistry* by the scientific community, from whom suggestions for new topics to the Editors-in-Chief are always very welcome.

Damià Barceló
Andrey G. Kostianoy
Editors-in-Chief

Preface

Freshwater Microplastics as Emerging Contaminants: Much Progress, Many Questions

Historically – if one can say that given the infancy of the field – environmental plastic debris has been the baby of marine research. Driven by the rediscovery of long forgotten, 1970s studies on the occurrence of small plastic fragments (today termed microplastics) in the oceans, oceanographers and marine biologists resurrected the topic in the early 2000s. Since then, the field has rapidly expanded and established that plastics are ubiquitous in the marine system, from the Arctic to Antarctic and from the surface to the deep sea.

While obviously the sources of environmental plastics are land-based, much less research has been dedicated to investigating them in freshwater systems. At the time of writing this book, less than four percent of publications had a freshwater context, reflecting the idea that streams, rivers, and lakes are mere transport routes transferring plastics to the oceans similar to a sewer. Because this is too simplistic, this book is dedicated to the in-between. Our authors explore the state of the science, including the major advances and challenges, with regard to the sources, fate, abundance, and impacts of microplastics on freshwater ecosystems. Despite the many gaps in our knowledge, we highlight that microplastics are pollutants of emerging concern independent of the salinity of the surrounding medium.

Environmental (micro)plastics are what some call a wicked problem, i.e., there is considerable complexity involved when one tries to understand the impact of these synthetic materials on the natural world. Just as an example, there is no such thing as "the microplastic." Currently, there are in commerce more than 5,300 grades of synthetic polymers.[1] Their heterogeneous physico-chemical properties will likely result in very heterogeneous fates and effects once they enter the

---

[1]According to the plastics industry's information system CAMPUS (http://www.campusplastics.com, last visited on June 20, 2017).

environment. In the light of this, treating microplastics as a single pollutant does not make sense. Therefore, we kick off the book by giving a brief overview on what plastics are, where they come from, and where they go to in the environment. As the research on engineered nanomaterials faces similar challenges, we then look more deeply into the (dis)similarities of nanoparticles and microplastics and try to learn from past experiences.

We continue with five chapters focusing on the abundance of microplastics in freshwater systems, touching on analytical challenges, discussing case studies from Europe, Asia, and Africa as well as approaches for modeling the fate and transport of microplastics. As the biological interactions of synthetic polymers will drive their environmental impacts, we review the state of the science with regard to their toxicity in freshwater species and biofilm formation. While, admittedly, progress in this area is slow, we already learned that "It's the ecology, stupid!" to paraphrase Bill Clinton.

The last part of the book is dedicated to the question how society and microplastics interact. We take a sociological perspective on the risk perception of the issue at hand and discuss how this "vibrates" in the medial and political realm and the society at large. While the uncertainty in our understanding is still enormous, we conclude our book with an outlook on how to solve the problem of environmental plastics. We have in our hands a plethora of regulatory instruments ranging from soft to hard measures, of which some are already applied. However, because the linear economical model our societies are built on is at the heart of the problem, we critically revisit available solutions and put it into the larger context of an emerging circular economy.

Given the wickedness of the plastics problem in terms of material properties, analytical challenges, biological interactions, and resonance in society, we clearly need an inter- and transdisciplinary effort to tackle it. We hope this book promotes such view. We also hope it conveys the idea that we need to embrace the inherent complexity to solve it. We thank our authors, reviewers, the publisher, and all funders for following this path and making this book happen (and open access).

Frankfurt am Main, Germany   Martin Wagner
June 2017   Scott Lambert

Contents

Microplastics Are Contaminants of Emerging Concern in Freshwater Environments: An Overview

Scott Lambert and Martin Wagner

Abstract In recent years, interest in the environmental occurrence and effects of microplastics (MPs) has shifted towards our inland waters, and in this chapter we provide an overview of the issues that may be of concern for freshwater environments. The term 'contaminant of emerging concern' does not only apply to chemical pollutants but to MPs as well because it has been detected ubiquitously in freshwater systems. The environmental release of MPs will occur from a wide variety of sources, including emissions from wastewater treatment plants and from the degradation of larger plastic debris items. Due to the chemical makeup of plastic materials, receiving environments are potentially exposed to a mixture of micro- and nano-sized particles, leached additives, and subsequent degradation products, which will become bioavailable for a range of biota. The ingestion of MPs by aquatic organisms has been demonstrated, but the long-term effects of continuous exposures are less well understood. Technological developments and changes in demographics will influence the types of MPs and environmental concentrations in the future, and it will be important to develop approaches to mitigate the input of synthetic polymers to freshwater ecosystems.

Keywords Degradation, Ecosystem effects, Fate, Pollutants, Polymers, Sources, Toxicity

S. Lambert (✉)
Department Aquatic Ecotoxicology, Goethe University Frankfurt am Main,
Max-von-Laue-Str. 13, 60438 Frankfurt am Main, Germany
e-mail: scottl210@hotmail.co.uk

M. Wagner
Department of Biology, Norwegian University of Science and Technology (NTNU),
Trondheim, Norway

M. Wagner, S. Lambert (eds.), *Freshwater Microplastics*,
Hdb Env Chem 58, DOI 10.1007/978-3-319-61615-5_1,
© The Author(s) 2018

1 Introduction

Anthropogenic activity has resulted in the deposition of a complex combination of materials in lake sediments, including synthetic polymers (plastics) that differ greatly from the Holocene signatures. Accordingly, plastics are considered one indicator of the Anthropocene [1]. Plastic has for some time been known to be a major component of riverine pollution [2–6], and plastic degradation products have been noted as a potential issue for soil environments [7]. However, up until recently the main focus of research on plastic pollution has been the marine environment. To highlight this, a literature search on Thomson Reuters' ISI Web of Science returns 1,228 papers containing the term 'microplastic*', of which only a subset of 45 publications (3.7%) contains the term 'freshwater'. This has started to change in recent years, and attention is now also been directed towards both the terrestrial [8, 9] and freshwater environments [8, 10, 11]. These publications point out the lack of knowledge for freshwater and terrestrial environments in terms of the occurrence and impacts of plastics debris.

Monitoring studies have quantified microscopic plastics debris, so-called microplastics (MPs), in freshwater systems, including riverine beaches, surface waters and sediments of rivers, lake, and reservoirs [12–19]. Although far less data is available compared to marine systems, these studies highlight that MP is ubiquitous and concentrations are comparable [20]. Alongside the monitoring data, ecotoxicological studies have mainly explored MP ingestion by various species and their effects on life history parameters [21–24]. While the majority of studies used primary microspheres of polyethylene (PE) and polystyrene (PS) at high concentrations [25] over short-term exposures, there is some evidence that MPs may pose a risk to freshwater ecosystems [26]. In addition, there is concern that long-term exposure may lead to bioaccumulation of submicron particles with wider implications for environmental health [27–29].

This chapter provides an overview of MPs and the issues, which may be of concern to freshwater environments. The first section provides a background to the topic of discussion by describing and defining plastic materials, MPs, emerging contaminants. Subsequent sections then discuss the potential input, fate and transportation, effects, and potential risk management options for plastics and MPs in freshwater environments.

2 Plastics and Microplastics: An Overview

In this section, some context to the topic of environmental MPs is given by (1) providing a brief historical overview of the development of plastic materials, (2) describing the complex chemical composition of plastic material, and (3) defining MPs as a contaminants of emerging concern.

2.1 A Brief Overview of Plastic Development

The creation of new synthetic chemicals combined with the engineering capabilities of mass production has made plastics one of the most popular materials in modern times. Today's major usage of plastic materials can be traced back to the 1800s with the development of rubber technology. One of the key breakthroughs in this area was the discovery of vulcanisation of natural rubber by Charles Goodyear [30]. Throughout the 1800s a number of attempts were made to develop synthetic polymers including polystyrene (PS) and polyvinyl chloride (PVC), but at this time these materials were either too brittle to be commercially viable or would not keep their shape. The first synthetic polymer to enter mass production was Bakelite, a phenol-formaldehyde resin, developed by the Belgian chemist Leo Baekeland in 1909 [31]. Later, around the 1930s the modern forms of PVC, polyethylene terephthalate (PET), polyurethane (PUR), and a more processable PS were developed [32]. The early 1950s saw the development of high-density polyethylene (HDPE) and polypropylene (PP; Table 1). In the 1960s, advances in the material sciences led to the development of plastic materials produced other from natural resources [34], such as the bacterial fermentation of sugars and lipids, and include

Table 1 A brief profile of plastic development based on Lambert [33]

Year	Polymer type	Inventor/notes
1839	Natural rubber latex	Charles Goodyear
1839	Polystyrene	Discovered by Eduard Simon
1862	Parkesine	Alexander Parkes
1865	Cellulose acetate	Paul Schützenberger
1869	Celluloid	John Wesley Hyatt
1872	Polyvinyl chloride	First created by Eugen Baumann
1894	Viscose rayon	Charles Frederick Cross
1909	Bakelite	Leo Hendrik Baekeland
1926	Plasticised PVC	Walter Semon
1933	Polyvinylidene chloride	Ralph Wiley
1935	Low-density polyethylene	Reginald Gibson and Eric Fawcett
1936	Acrylic or polymethyl methacrylate	
1937	Polyurethane	Otto Bayer and co-workers
1938	Polystyrene	As a commercially viable polymer
1938	Polyethylene terephthalate	John Whinfield and James Dickson
1942	Unsaturated polyester	John Whinfield and James Dickson
1951	High-density polyethylene	Paul Hogan and Robert Banks
1951	Polypropylene	Paul Hogan and Robert Banks
1953	Polycarbonate	Hermann Schnell
1954	Styrofoam	Ray McIntire
1960	Polylactic acid	Patrick Gruber is credited with inventing a commercially viable process
1978	Linear low-density polyethylene	DuPont

polyhydroxyalkanoates (PHA), polylactides (PLA), aliphatic polyesters, and poly-saccharides [35]. PLA is on the verge of entering into bulk production, while PHA production is between pilot plant and commercial stage [36, 37].

2.2 Describing Plastic Materials

Plastics are processable materials based on polymers [38], and to make them into materials fit for purpose, they are generally processed with a range of chemical additives (Table 2). These compounds are used in order to adjust the materials properties and make them suitable for their intended purpose. Therefore, within polymer classifications plastic materials can still differ in structure and performance depending on the type and quantity of additives they are compounded with. More recently, technological advances have seen the development of new applications of elements based on nanoscales that are now producing plastic nanocomposites. The plastics industry is expected to be a major field for nanotechnology innovation. It is estimated that by 2020, the share of nanocomposites among plastics in the USA will be 7% [39]. These nanocomposites include materials that are reinforced with nano-fillers (nano-clay and nano-silica) for weight reduction, carbon nanotubes (CNTs) for improved mechanical strength, and nano-silver utilised as an antimicrobial agent in plastic food packaging materials.

2.3 Microplastics as Contaminants of Emerging Concern

The term 'microplastics' commonly refers to plastic particles whose longest dia-meter is <5 mm and is the definition used by most authors. It has been suggested that the term microplastics be redefined as items <1 mm to include only particles in the

Table 2 A selective list of additive compounds used to make plastics fit for purpose

Additive compounds	Function
Plasticisers	Renders the material pliable
Flame retardants	Reduces flammability
Cross-linking additives	Links together polymer chains
Antioxidants and other stabilisers	Increases the durability of plastics by slowing down the rate at which oxygen, heat, and light degrade the material
Sensitisers (e.g. pro-oxidant transition metal complexes)	Used to give accelerated degradation properties
Surfactants	Used to modify surface properties to allow emulsion of normally incompatible substances
Inorganic fillers	Used to reinforce the material to improve impact resistance
Pigments	For colour

micrometer size range [40, 41], and the term 'mesoplastic' introduced to account for items between 1 and 2,500 mm [42]. Lambert et al. [8] described macroplastics as >5 mm, mesoplastics as ≤5 to >1 mm, microplastics as ≤1 mm to >0.1 μm, and nanoplastics as ≤0.1 μm. However, the upper limit of 5 mm is generally accepted because this size is able to include a range of small particles that can be readily ingested by organisms [42].

Generally, MPs are divided into categories of either primary or secondary MPs. Primary MPs are manufactured as such and are used either as resin pellets to produce larger items or directly in cosmetic products such as facial scrubs and toothpastes or in abrasive blasting (e.g. to remove lacquers). Compared to this deliberate use, secondary MPs are formed from the disintegration of larger plastic debris.

MPs have undoubtedly been present in the environment for many years. For instance, Carpenter et al. [43], Colton et al. [44], and Gregory [45] reported on marine plastics in the 1970s, but they have not been extensively studied particularly in the context of freshwater systems. As research focused on the issue more intensively since the early 2000s, MPs are considered contaminants of emerging concern [8, 10, 46].

3 Sources of Plastics and Microplastics into the Freshwater Environment

Plastics will enter freshwater environments from various sources through various routes. On land littering is an important environmental and public issue [47, 48] and is a matter of increasing concern in protected areas where volumes are influenced by visitor density; consequently, measures are now needed to reduce and mitigate for damage to the environment [49]. In addition, waste management practices in different regions of the world also vary, and this may be a more important source in one geographical region compared to another [8]. As with bulk plastic items, MPs can enter the environment by a number of pathways, and an important route in one geographical region may be less important in another. For example, primary MPs used in consumer cosmetics are probably more important in affluent regions [8]. MPs have several potential environmental release pathways: (1) passage through WWTPs, either from MP use in personal care products or release of fibres from textiles during the washing of clothes, to surface waters, (2) application of biosolids from WWTPs to agricultural lands [50], (3) storm water overflow events, (4) incidental release (e.g. during tyre wear), (5) release from industrial products or processes, and (6) atmospheric deposition of fibres (discussed further in Dris et al. [51]). Plastic films used for crop production are considered an important agricultural emission, and their use is thought to be one of the most important sources of plastic contamination of agricultural soils [52–54]. There advantages include conserve of moisture, thereby reducing irrigation; reduce weed growth and increase soil temperature which reduces competition for soil nutrients and reduces fertiliser costs, thereby improving crop yields; and protect against adverse weather conditions [7, 55]. However, weathering can make them brittle and difficult to recover resulting in disintegration of the material, and when coupled with successive precipitation events, the residues and disintegrated particles can be washed into the soil where they accumulate [7, 55, 56]. Other sources exist and include emissions from manufacturing and constructions sites. Automotive tyre wear particles may also release large volumes of synthetic particles. These tyre wear particles are recognised as a source of Zn to the environment, with anthropogenic Zn concentrations that are closely correlated to traffic density [57]. The sources and emission routes of nanoplastics are also discussed in Rist and Hartmann [58].

4 Occurrence in Freshwater Systems

The isolation of MPs in environmental matrices can be highly challenging particularly when dealing with samples high in organic content such as sediments and soils. Likewise, the spectroscopic identification of synthetic polymers is complicated by high pigment contents and the weathering of particles and fibres. Accordingly, the detection and analytical confirmation of MPs require access to sophisticated equipment (e.g. micro-FTIR and micro-Raman; discussed further in Klein et al. [20]). Recent monitoring studies have established that – similar to marine environments – MPs are ubiquitously found in a variety of freshwater matrices. Reported MP concentrations in surface water samples of the Rhine river (Germany) average 892,777 particles km^{-2} with a peak concentration of 3.9 million particles km^{-2} [15]. In river shore sediments the number of particles ranged from 228 to 3,763 and 786 to 1,368 particles kg^{-1} along the rivers Rhine and Main (Germany), respectively [19]. High surface water concentrations are reported at the Three Gorges Dam, China (192–13,617 particles km^{-2}), which are attributed to a lack of wastewater treatment facilities in smaller towns, as well as infrastructure issues when dealing with recycling and waste disposal [14]. These studies may underestimate the actual MP concentrations because their separation and identification are based on visual observation methods (e.g. Reddy et al. [59]) and may exclude those in the submicron size ranges. The environmental occurrence and sources of MPs in freshwater matrices in an African, Asian, and European context are further discussed in Dris et al. [51], Wu et al. [60], Khan et al. [61], respectively.

5 Fate and Transport in Freshwater Systems

Once MPs are released or formed in the freshwater environment, they will undergo fate and transportation processes. In the following section, these processes are discussed.

5.1   Environmental Transportation

Many plastic materials that enter the environment will not remain stationary. Instead they will be transported between environmental compartments (e.g. from land to freshwater and from freshwater to marine environments), with varying residence times in each. For example, the movement from land to river systems will depend upon prevailing weather conditions, distance to a specific river site, and land cover type. The collection of plastic litter at roadside habitats is easily observed, and the regular grass cutting practices of road verges in some countries means that littered items are quickly disintegrated by mowing equipment [8]. The movement of MPs from land to water may then occur through overland run-off or dispersion (via cutting action) to roadside ditches. The movement of bulk plastics and MPs within the riverine system will be governed by its hydrology (e.g. flow conditions, daily discharge) and the morphology (e.g. vegetation pattern) at a specific river site that will have a large effect upon the propagation of litter because of stranding and other watercourse obstructions such as groynes and barrages [2]. Compared to larger plastics, MPs may also be subject to different rates of degradation as they will be transported and distributed to various environment compartments at quicker rates than macroplastics. The formation of MP-associated biofilms has been investigated for LDPE in marine setting [62]. Transport to sediments and the formation of biofilms over the surface of MPs may also limit rates of degradation as this removes exposure to light. The modelling of MP fate and transportation in freshwaters is discussed further in Kooi et al. [63], while MP-associated biofilm are discussed in Harrison et al. [64].

5.2   Environmental Persistence and Degradation

The majority of our current understanding regarding plastic degradation processes is derived from laboratory studies that often investigate a single mechanism such as photo-, thermal, or bio-degradation [65]. There is limited information on the degradation of plastics under environmentally relevant conditions where a number of degradation mechanisms occur at together. Where information is available these studies have tended to focus on weight loss, changes in tensile strength, breakdown of molecular structure, and identification of specific microbial strains to utilise specific polymer types. The degradation processes are defined in accordance with the degradation mechanism under investigation (e.g. thermal degradation) and the experimental result generated. In contrast, particle formation rates are often not investigated. This is important because polymers such as PE do not readily depoly-merise and generally decompose into smaller fragments. These fragments then further disintegrate into increasingly smaller fragments eventually forming nano-plastics [66–68].

The prediction of plastic fragmentation rates is not a simple process. Kinetic fragmentation models have been investigated in the mathematics and physics literatures, and the kinetics of polymer degradation has been researched extensively in the polymer science literature. These models describe the distribution of fragment sizes that result from breakup events. These processes can be expressed by rate equations that assume each particle is exposed to an average environment, mass is the unit used to characterise a particle, and the size distribution is taken to be spatially uniform [69, 70]. These processes can be described linearly (i.e. particle breakup is driven only by a homogeneous external agent) or nonlinearly (i.e. additional influences also play a role), and particle shape can be accounted for by averaging overall possible particle shape [69]. The models used to describe these degradation process are often frequently complicated, but as a general rule focus on chain scission in the polymer backbone through (a) random chain scission (all bonds break with equal probability) characterised by oxidative reactions; (b) scission at the chain midpoint dominated by mechanical degradation; (c) chain-end scission, a monomer-yielding depolymerisation reaction found in thermal and photodecomposition processes; and (d) in terms of inhomogeneity (different bonds have different breaking probability and dispersed throughout the system) [71–73]. The estimation of degradation half-lives has also been considered for strongly hydrolysable polymers through the use of exponential decay eqs. [65, 74, 75]. However, the applicability of modelling the exponential decay of more chemically resistant plastics requires greater investigation [74].

Important variables that will influence MP degradation and fragmentation are environmental exposure conditions, polymer properties such as density and crystallinity (Table 3), and the type and quantity of chemical additives. Molecular characteristics that generally counteract degradation are the complexity of the polymer

Table 3 Polymer type, density, and crystallinity

Polymer type	Density (g cm^{-3})	Crystallinity
Natural rubber	0.92	Low
Polyethylene–low density	0.91–0.93	45–60%
Polyethylene–high density	0.94–0.97	70–95%
Polypropylene	0.85–0.94	50–80%
Polystyrene	0.96–1.05	Low
Polyamide (PA6 and PA66)	1.12–1.14	35–45%
Polycarbonate	1.20	Low
Cellulose acetate	1.28	High
Polyvinyl chloride	1.38	High
Polylactic acid	1.21–1.43	37%
Polyethylene terephthalate	1.34–1.39	Described as high in [76] and as 30–40% in [77]
Polyoxymethylene	1.41	70–80%

Information on crystallinity was taken from [76, 77]

structure and the use of structural features that are not easy to biodegrade. Here, crystallinity is an important polymer property because the crystalline region consists of more ordered and tightly structured polymer chains. Crystallinity affects physical properties such as density and permeability. This in turn affects their hydration and swelling behaviour, which affects accessibility of sorption sites for microorganisms. Stabilisers such as antioxidants and antimicrobial agents act to prolong the life of plastics, whereas biological ingredients act to decompose the plastic in shorter time frames.

Overall, environmental degradation processes will involve MP fragmentation into increasingly smaller particles including nanoplastics, chemical transformation of the plastic fragments, degradation of the plastic fragments into non-polymer organic molecules, and the transformation/degradation of these non-polymer molecules into other compounds [65]. The environmental degradation of plastic materials is also further discussed in Klein et al. [20].

5.3   Interactions with Other Compounds

The sorption of hydrophobic pollutants to MPs is considered an important environmental process, because this will affect the mobility and bioavailability of these pollutants. It is well known that MPs in marine environments concentrate persistent organic pollutants (POPs) such as DDT, PCBs, and dioxins [78–80]. In addition, Ashton et al. [81] also found concentrations of metals in composite plastic pellet samples retrieved from the high tide line along a stretch of coastline in Southwest England. To investigate whether the metals were in fact associated with nonremovable fine organic matter associated with the pellet samples, new polypropylene pellets were suspended in a harbour for 8 weeks and were found to accumulate metals from the surrounding seawater, from low of 0.25 μg g^{-1} for Zn to a high of 17.98 μg g^{-1} for Fe [81]. So far, little data is available on freshwater and terrestrial ecosystems, which will have a pollutant makeup very different to that found in marine environments. In the freshwater environment MPs are likely to co-occur with other emerging contaminants such as pharmaceuticals, personal care products, flame retardants, and other industrial chemicals, which enter the environment as parts of complex solid and liquid waste streams.

Sorption processes will occur through physical and chemical adsorption as well as pore-filling processes. Physical adsorption is the reversible sorption to surfaces of the polymer matrix and does not involve the formation of covalent bonds. Chemical adsorption involves chemical reactions between the polymer surface and the sorbate. This type of reaction generates new chemical bonds at the polymer surface and may depend on how aged the polymer surface is. These processes can be influenced by changes in pH, temperature, and ionic strength of the localised environment [82]. Pore-filling occurs when hydrophobic pollutants enter the polymer matrix and will be dependent on the pore diameter of a particular polymer structure and the molecular size of the chemical. Here, pollutants with lower molecular weights will more easily move through a polymer matrix with larger pores.

Adsorption kinetics will depend on polymer type, polymer characteristics such as density and crystallinity, the surrounding environment, and the pollutants present. For instance, the sorption and diffusion of hydrophobic contaminants are most likely to take place in the amorphous area of a plastic material, because the crystalline region consists of more ordered and tightly structured polymer chains. Polymers that have structures with short and repeating units, a high symmetry, and strong interchain hydrogen bonding will have a lower sorption potential. A good example is low-density polyethylene (LDPE) and high-density polyethylene (HDPE; Table 3). LDPE contains substantial concentrations of branches that prevent the polymer chains from been easily stacked side by side. This results in a low crystallinity and a density of 0.90–0.94 g cm^{-3} [83]. Whereas, HDPE consists primarily of linear unbranched molecules and is chemically the closest in structure to pure polyethylene. The linearity HDPE has a high degree of crystallinity and higher density of 0.94–0.97 g cm^{-3} [83]. LDPE is often used for passive sampling devices to determine dissolved polycyclic aromatic hydrocarbons (PAH), polychlorinated biphenyls (PCB), and other hydrophobic organic compounds in aquatic environments [84–88]. Batch sorption experiments were also used to determine PAH sorption to LDPE and HDPE pellets, and LDPE was identified to exhibit higher diffusion coefficients than HDPE meaning shorter equilibrium times for low-density polymers [89]. This indicates that PE is of interest from an environmental viewpoint because of its high sorption capacity. In addition, particle size will influence the sorption parameters because the higher surface to volume ratio of smaller particles will shorten diffusion times. Isolating and quantifying the sorption mechanisms for all polymer types in use today will be challenging, because sorption behaviour may differ within polymer classification depending on the type and quantity of additive compounds the polymer is compounded with and the effects that this may have on polymer crystallinity and density. These issues are discussed in further detail in Scherer et al. [26] and Rist and Hartmann [58] in relation to MP and nanoplastics, respectively.

An interesting question is to what extent does irreversible sorption play a role? Some evidence in the pesticides literature suggests that a proportion of pesticides bind irreversibly soils [90, 91]. The study of sorption equilibrium isotherms is an important step in investigating the sorption processes that exist between different polymer types and co-occurring hydrophobic contaminants. This will make it possible to identify the sorption and diffusion relationships between case study co-occurring contaminants and MPs. Another interesting question is to what extent sorbed chemicals become bioavailable in the water column due to the continued breakdown and degradation of the MPs, or due to changes in environmental conditions, such as changes in pH, temperature, or system chemistry.

6 Effects of Plastics and Microplastics on Freshwater Ecosystems

Once in the aquatic environment, the mobility and degradation of plastics will generate a mixture of parent materials, fragmented particles of different sizes, and other non-polymer degradation products. Accordingly, biota will be exposed to a complex mixture of plastics and plastic-associated chemicals that changes in time and space.

6.1 Uptake and Biological Effects

MPs may be taken up from the water column and sediment by a range of organisms. This can occur directly through ingestion or dermal uptake most importantly through respiratory surfaces (gills). Previous investigations on freshwater zooplankton have included *Bosmina coregoni* that did not differentiate between PS beads (2 and 6 μm) and algae when exposed to combinations of both [92]. The same study also found that *Daphnia cucullata*, when exposed to PS beads (2, 6, 11, and 19 μm) in combination with algae cells of the same size, was observed to exhibit similar filtering rates for the three smaller size classes but preferred alga over the larger beads [92]. Rosenkranz et al. [93] demonstrated that *D. magna* ingests nano (20 nm) and micro (1 μm) PS beads. The authors note that both types of PS beads were excreted to some extent, but the 20 nm beads were retained to a greater degree within the organism.

The extent to which organisms are exposed to physical stress because of MP uptake depends on particle size, because particles larger than sediment or food particles may be harder to digest [94]. In addition, particle shape is also an important parameter, because particles with a more needle-like shape may attach more readily to internal and external surfaces. The indirect effects of MPs may include physical irritation, which may depend on MP size and shape. Smaller more angular particles may be more difficult to dislodge than smooth spherical particles and cause blockage of gills and digestive tract. In a recent study, the chronic effects of MP exposure to *D. magna* were evaluated [21]. Exposure to secondary MPs (mean particle size 2.6 μm) caused elevated mortality, increased inter-brood period, and decreased reproduction but only at very high MP levels (105,000 particles L^{-1}). In contract, no effects were observed in the corresponding primary MP (mean particle size 4.1 μm) [21].

There is some evidence suggesting that a trophic transfer of MP may occur, for instance, from mussels to crabs [27]. The blue mussel *Mytilus edulis* was exposed to 0.5 μm PS spheres (ca. 1 million particles mL^{-1}) and fed to crabs (*Carcinus maenas*). The concentration of microspheres in the crab haemolymph was reported to be the highest after 24 h (15,033 particles mL^{-1}) compared to 267 residual particles mL^{-1} after 21 days, which is 0.027% of the concentration fed to the mussels. Another study has demonstrated the potential of MP transfer from meso- to macro-zooplankton, using PS microspheres (10 μm) at much lower concentrations of 1,000, 2,000, and 10,000 particles mL^{-1} [28]. Because excretion rates are unavailable and MP uptake is often defined as particles present in the digestive tract (i.e. the outside and not the tissues of an organism), it is so far not clear whether the trophic transfer of MP also results in a bioaccumulation or biomagnification. However, it is clear that MP will certainly be transferred from the prey to the predator and that this can – in certain situations – be retained for longer periods in the body of the latter.

An open question is to what extent the organisms consume naturally occurring microparticles and how the effects compare to MPs (for a more in-depth discussion on this topic see Scherer et al. [26]). This is important because naturally occurring particles are an important component of aquatic ecosystems and particle properties, such as concentration, particle size distribution, shape, and chemical composition, as well as duration of exposure plays a strong role in determining their interactions with aquatic communities [95].

Overall, an understanding of the relationships between cellular level responses and population level impacts will be important in order to determine the broader implications for ecosystem functioning. Points to be assessed concern both the biological aspects (molecular target, affected endpoints) and the particle aspects such as MP physical and chemical characteristics. The bioavailability of the MPs and the penetration of submicron MPs into the cells are factors to take into consideration.

6.2 Effects of Leaching Chemicals

The environmental effects of residual starting substances and monomers, non-intentionally added substances (impurities, polymerisation byproducts, breakdown products), catalysts, solvents, and additives leaching from plastic materials are not easy to assess [96]. The mixture composition and concentration of leachable compounds depend on the physical, chemical, and biological conditions of receiving environments. The leaching of water-soluble constituents from plastic products using deionised water is considered a useful method for profiling environmental hazards posed by plastics [97, 98]. Lithner et al. used such leachates in a direct toxicity testing approach to assess their acute toxicity to *D. magna* [97, 98]. For instance, with a liquid to solid (L/S) ratio of 10 and 24 h leaching time, leachates from polyvinyl chloride (PVC), polyurethane (PUR), and polycarbonate (PC) were the most toxic with EC_{50} values of 5–69 g plastic L^{-1} [98]. Higher L/S ratios and longer leaching times resulted in leachates from plasticised PVC and epoxy resin products to be the most toxic at (EC_{50} of 2–235 g plastic L^{-1}) [99]. In a recent study, Bejgarn et al. [99] investigated the leachates from plastic that were ground to a power and had undergone artificial weathering, using a L/S of 10 and a 72 h leaching time, to the marine harpacticoid copepod *Nitocra spinipes*. Here, leachates from different PVC materials differed in their toxicity, with the toxicity of leachate from PVC packaging increasing after artificial weathering; whereas the leachate from PVC garden hose material decreased after artificial weathering [99]. This study also showed that the leachable PVC constitutes were a complex mixture of substances, and interestingly mass fragments containing chlorine were not identified. There are many challenges associated with the characterisation of such leachates owing to the potential diversity of physicochemical properties that chemical migrants and breakdown products may have. A test protocol for the identification of migration products from food contact materials has been developed that combines LC-TOF-MS and GC-MS techniques that generate accurate mass and predicted formulae to screen for volatile, semi-volatile, and non-volatile substances [100, 101].

Overall, the L/S ratio of plastic material used in these studies is higher than that typically identified during environmental monitoring studies. However, this type of screening when applied to materials manufactured from hazardous monomers and additives could facilitate the identification of compounds of interest so that they can be effectively replaced.

6.3 Biological Effects of Sub-micrometer Plastics

Depending on their use, plastic materials can contain compounds such as antimicrobial agents and nanomaterials that may be toxic to organisms such as bacteria and fungi that play a critical role in ecosystem functioning. It is possible that a combination of microscopic particles, leached additives, and other degradation products may cause subtle effects towards aquatic and terrestrial organisms that are difficult to identify in current testing methodologies. The formation of plastic particles in the submicron and nanometer size range during degradation is highly likely [8, 40, 66, 102, 103]. Engineered nanoparticles (ENPs) are able to cross cell membranes and become internalised, and the uptake of ENPs is size dependent with uptake occurring by endocytosis or phagocytosis [104]. Once inside the cell ENPs are stored inside vesicles and mitochondria and able to exert a response [104]. Cellular responses include oxidative stress, antioxidant activity, and cytotoxicity [105]. In terms of toxicity assessments, there is a need to understand the molecular and cellular pathways and the kinetics of absorption, distribution, metabolism, and excretion mechanisms that may be unique to MPs in the nano-size range. Desai et al. [106] showed that 100 nm particles of a polylactic polyglycolic acid copolymer had a tenfold higher intracellular uptake in an in vitro cell culture when compared to 10 μm particles made of the same material. ENPs have also been shown to produce cytotoxic, genotoxic, inflammatory, and oxidative stress responses in mammalian and fish systems [107]. A literature review by Handy et al. [108] highlighted the gills, gut, liver, and brain as possible target organs in fish, as well as a range of toxic effects including oxidative stress, cellular pathologies consistent with tumour formation in the liver, some organ specific ionoregulatory disturbances, and vascular injury. Taking into account the complex chemical makeup of some plastics and the ability to sorb co-occurring contaminants, experimental investigation of these endpoints for MPs seems to be merited. There are many lessons to be learned from the growing literature on the biological effects of ENPs, and these are discussed in more detail in Rist and Hartmann [58].

7 Considerations for Assessing Environmental Risks

In most countries chemical risk assessments rely on mass concentrations of substances of interest as an exposure and effect metric. In the nano-literature the mass concentrations of particles predicted to be emitted have been used to assess the risks of ENPs [109, 110]. These approaches assume particles are evenly distributed with no transfer between different environmental compartments. This approach was further developed by Gottschalk et al. [111] who used transfer coefficients to model emission flows between the different compartments used in their model, as well as the inclusion of sedimentation rates. Such modelling approaches (further discussed in Kooi et al. [63]) could be used to assess the environmental fate of primary MPs where emissions to the environment are distributed across a geographical region proportional to population density and consumption rates, assuming that the route of enter into the environment depends on the use of the MP. However, this type of approach requires extensive information on primary MP production levels, industrial applications and uses, levels in consumer products, fate in wastewater treatment, discharges to landfill, and environmental fate and distribution modelling to perform a meaningful exposure assessment. An exposure assessment for secondary MPs will require monitoring data, but this is hindered as the size ranges reported in field studies are generally constrained by the sampling techniques used [42].

The problems of using mass concentrations as an effect metric are similar to those discussed in the context of ENPs in that biological effects might not be mass dependent but dependent on physical and chemical properties of the substance in question [112, 113]. Consequently, when estimating the hazards presented by MP properties such as size, shape, polymer density, surface area, chemical composition of the parent plastic, and the chemical composition of sorbed co-occurring contaminants may need to be considered [114]. However, when considering secondary MPs information on some of these properties may be unavailable. This lack of information makes it difficult to identify the key characteristics, or combinations of characteristics, that may be responsible for hazards in the environment.

The assessment of MPs based on their chemical composition also presents a considerable challenge, because chemically MPs can be considered as a mixture. A

Table 4 A hypothetical chemical mixture risk assessment based on the chemical components of PUR flexible foam with TBBPA as a flame retardant (units are mg/L)

	Monomer 1	Monomer 2	Monomer 3	Additive 1
	Propylene oxide	Ethylene oxide	Toluene diisocyanate	TBBPA
LC_{50} algae	307	502	3.79	0.19
LC_{50} daphnid	188	278	2.61	0.02
LC_{50} fish	45	58	3.91	0.02
PNEC (AF = 1000)	0.045	0.058	0.003	0.000002
PEC (dissolved compound)	0.00067	0.00067	0.00067	0.0000032
$RQ_{PEC/PNEC}$	0.015	0.012	0.257	0.160
Mixture RQ	0.443			

LC_{50} (median lethal concentration) for this example were generated using the EPI Suite ECOSAR model; *AF* assessment factor
Monomer PECs are based on propylene oxide ECHA risk assessment [115]
TBBPA PEC based on maximum concentrations measured in UK lakes [116]

simplified example of a risk assessment for polyurethane (PUR) based on its chemical composition is provided in Table 4. PUR flexible foam is used for mattresses and car seats and is made by combining three monomers and can consist of up to 18% flame retardant content [117]. An example risk assessment based on predicted environmental concentration (PEC)/predicted no effect concentration (PNEC) ratios for all components of the mixture are then used to calculate a risk quotient (RQ; Table 4). The RQ for this particular example is less than one; however, this type of assessment does not account for potential negative effects caused by physical irritation of solid particles. In this case it becomes clear that risk assessment for MPs as with ENPs holds specific challenges (see Brennholt et al. [118] for an in-depth discussion of the regulatory challenges).

The different particles sizes of MPs in environmental systems will present different risks to organisms living in those systems. For example, small plankton feeding fish species may encounter MPs from the nanoscale through to MPs 5 mm or greater. The fish may avoid larger particles but small particles may be ingested while feeding. For filter feeding organisms the upper size boundary will depend on the size of particles that a particular organism will naturally ingest. The risk assessment of MPs could therefore be based on particle size. A simplified hypothetical case is presented in Box 1 that draws on an example given by Arvidsson [119]. This approach assumes that there is information on harm-related thresholds of MPs based on size classes and particle concentration for the most sensitive species in that particles size range. However, the use of particle size for defining environmental risk may not be that straight forward, because MPs are not monodispersed in the environment. Additionally, as described by Hansen, [120] when discussing ENPs it remains unclear whether a 'no effect threshold' can be established, what the best hazard descriptor(s) are, and what are the most relevant endpoints.

Box 1: A hypothetical case for the risk assessment for MPs based on particles size

A lake has a PEC of MPs ≤ 5 mm at 10,000 particles/L and it is assumed that the PNEC of these particles is 1,000 particles/L. Furthermore, it is assumed that 1% of the PEC consists of particles <1 mm, assuming that the lower boundary is the same; the RQ is then determined by the upper boundary of the particles size as given below:

$$RQ_{\text{upper boundary}-\leq 5mm} = \frac{PEC}{PNEC} = \frac{10,000}{1,000} = 10\ (>1).$$

$$RQ_{\text{upper boundary}-\leq 1mm} = \frac{PEC}{PNEC} = \frac{100}{1,000} = 0.1\ (<1)$$

Risk or no risk is then determined by the setting of the upper boundary.

8 Concluding Thoughts

In this chapter, we have provided a brief overview of the environmental challenges associated with MP in freshwater systems and refer the readers to the appropriate chapters of this book for more detailed information. Overall, the environmental inputs in different geographical regions may vary depending on per capita consumption of consumer plastics, population demographics [121], and the capability of infrastructure to deal with waste materials. Environmental concentrations may change in the long term (whether positivity or negative) because of urbanisation, population increase, and technological developments. A better understanding of the environmental exposure in different geographical regions will identify those areas where mitigation actions and options will be the most effective. Future work should focus on better understanding the environmental fate and ecological impacts of MPs. Such an understanding should ultimately allow the development of new modelling approaches to assess transport of MPs in soil, sediments, and the water column. Little is also known about the long-term, subtle effects of MP exposure and sensitive endpoints (e.g. oxidative stress) need to be identified that integrate particle as well as chemical toxicity. Finally, although science is far from understanding the ecological implications of freshwater MPs, technological innovation, societal action, and political interventions need to be taken to mitigate the plastics pollution, which will – in case of inaction – certainly increase over the years to come.

Acknowledgements Scott Lambert receives funding from the European Union's Horizon 2020 research and innovation programme under the Marie Skłodowska-Curie grant agreement No. 660306. Martin Wagner receives funding by the German Federal Ministry for Transportation and Digital Infrastructure and the German Federal Ministry for Education and Research (02WRS1378 and 01UU1603).

References

1. Waters CN, Zalasiewicz J, Summerhayes C, Barnosky AD, Poirier C, Galuszka A, Cearreta A, Edgeworth M, Ellis EC, Ellis M, Jeandel C, Leinfelder R, McNeill JR, Richter D, Steffen W, Syvitski J, Vidas D, Wagreich M, Williams M, Zhisheng A, Grinevald J, Odada E, Oreskes N, Wolfe AP (2016) The Anthropocene is functionally and stratigraphically distinct from the Holocene. Science 351(6269):aad2622
2. Balas CE, Williams AT, Simmons SL, Ergin A (2001) A statistical riverine litter propagation model. Mar Pollut Bull 42(11):1169–1176. doi:10.1016/S0025-326X(01)00133-3
3. Williams AT, Simmons SL (1999) Sources of riverine litter: The River Taff, South Wales, UK. Water Air Soil Pollut 112(1–2):197–216
4. Williams AT, Simmons SL (1997) Movement patterns of riverine litter. Water Air Soil Pollut 98(1):119–139. doi:10.1007/bf02128653
5. Williams AT, Simmons SL (1996) The degradation of plastic litter in rivers: implications for beaches. J Coast Conserv 2(1):63–72. doi:10.1007/bf02743038
6. Tudor DT, Williams AT (2004) Development of a 'matrix scoring technique' to determine litter sources at a Bristol channel beach. J Coast Conserv 10(1–2):119–127. doi:10.1652/1400-0350(2004)010[0119:doamst]2.0.co;2
7. Klemchuk PP (1990) Degradable plastics – a critical-review. Polym Degrad Stab 27(2):183–202
8. Lambert S, Sinclair CJ, Boxall ABA (2014) Occurrence, degradation and effects of polymer-based materials in the environment. Rev Environ Contam Toxicol 227:1–53. doi:10.1007/978-3-319-01327-5_1
9. Rillig MC (2012) Microplastic in terrestrial ecosystems and the soil? Environ Sci Technol 46(12):6453–6454. doi:10.1021/es302011r
10. Wagner M, Scherer C, Alvarez-Muñoz D, Brennholt N, Bourrain X, Buchinger S, Fries E, Grosbois C, Klasmeier J, Marti T, Rodriguez-Mozaz S, Urbatzka R, Vethaak A, Winther-Nielsen M, Reifferscheid G (2014) Microplastics in freshwater ecosystems: what we know and what we need to know. Environ Sci Eur 26(1):1–9. doi:10.1186/s12302-014-0012-7
11. Eerkes-Medrano D, Thompson RC, Aldridge DC (2015) Microplastics in freshwater systems: a review of the emerging threats, identification of knowledge gaps and prioritisation of research needs. Water Res 75:63–82. doi:10.1016/j.watres.2015.02.012
12. Zbyszewski M, Corcoran PL (2011) Distribution and degradation of fresh water plastic particles along the beaches of Lake Huron, Canada. Water Air Soil Pollut 220(1–4):365–372. doi:10.1007/s11270-011-0760-6
13. Biginagwa FJ, Mayoma BS, Shashoua Y, Syberg K, Khan FR (2016) First evidence of microplastics in the African Great Lakes: recovery from Lake Victoria Nile perch and Nile tilapia. J Great Lakes Res 42(1):146–149. doi:10.1016/j.jglr.2015.10.012
14. Zhang K, Gong W, Lv J, Xiong X, Wu C (2015) Accumulation of floating microplastics behind the three Gorges Dam. Environ Pollut 204:117–123. doi:10.1016/j.envpol.2015.04.023
15. Mani T, Hauk A, Walter U, Burkhardt-Holm P (2015) Microplastics profile along the Rhine River. Sci Rep 5:17988. doi:10.1038/srep17988
16. Bartsch WM, Axler RP, Host GE (2015) Evaluating a Great Lakes scale landscape stressor index to assess water quality in the St. Louis River area of concern. J Great Lakes Res 41(1):99–110. doi:10.1016/j.jglr.2014.11.031
17. Lechner A, Keckeis H, Lumesberger-Loisl F, Zens B, Krusch R, Tritthart M, Glas M, Schludermann E (2014) The Danube so colourful: a potpourri of plastic litter outnumbers fish larvae in Europe's second largest river. Environ Pollut 188:177–181. doi:10.1016/j.envpol.2014.02.006
18. Castañeda RA, Avlijas S, Simard MA, Ricciardi A (2014) Microplastic pollution in St. Lawrence River sediments. Can J Fish Aquat Sci 71(12):1767–1771. doi:10.1139/cjfas-2014-0281

19. Klein S, Worch E, Knepper TP (2015) Occurrence and spatial distribution of microplastics in river shore sediments of the Rhine-main area in Germany. Environ Sci Technol 49(10): 6070–6076. doi:10.1021/acs.est.5b00492

20. Klein S, Dimzon IK, Eubeler J, Knepper TP (2017) Analysis, occurrence, and degradation of microplastics in the aqueous environment. In: Wagner M, Lambert S (eds) Freshwater microplastics: emerging environmental contaminants? Springer, Heidelberg. doi:10.1007/978-3-319-61615-5_3 (in this volume)

21. Ogonowski M, Schur C, Jarsen A, Gorokhova E (2016) The effects of natural and anthropogenic microparticles on individual Fitness in *Daphnia magna*. PLoS One 11(5):e0155063

22. Booth AM, Hansen BH, Frenzel M, Johnsen H, Altin D (2016) Uptake and toxicity of methyl - methacrylate-based nanoplastic particles in aquatic organisms. Environ Toxicol Chem 35(7): 1641–1649

23. Greven AC, Merk T, Karagoz F, Mohr K, Klapper M, Jovanovic B, Palic D (2016) Polycarbonate and polystyrene nanoplastic particles act as stressors to the innate immune system of fathead minnow (*Pimephales promelas*). Environ Toxicol Chem 35(12):3093–3100

24. Green DS, Boots B, Sigwart J, Jiang S, Rocha C (2016) Effects of conventional and biodegradable microplastics on a marine ecosystem engineer (*Arenicola marina*) and sediment nutrient cycling. Environ Pollut 208:426–434. doi:10.1016/j.envpol.2015.10.010

25. Phuong NN, Zalouk-Vergnoux A, Poirier L, Kamari A, Châtel A, Mouneyrac C, Lagarde F (2016) Is there any consistency between the microplastics found in the field and those used in laboratory experiments? Environ Pollut 211:111–123. doi:10.1016/j.envpol.2015.12.035

26. Scherer C, Weber A, Lambert S, Wagner M (2017) Interactions of microplastics with freshwater biota. In: Wagner M, Lambert S (eds) Freshwater microplastics: emerging environmental contaminants? Springer Nature, Heidelberg. doi:10.1007/978-3-319-61615-5_8 (in this volume)

27. Farrell P, Nelson K (2013) Trophic level transfer of microplastic: *Mytilus edulis* (L.) to *Carcinus maenas* (L.) Environ Pollut 177:1–3

28. Setälä O, Fleming-Lehtinen V, Lehtiniemi M (2014) Ingestion and transfer of microplastics in the planktonic food web. Environ Pollut 185:77–83. doi:10.1016/j.envpol.2013.10.013

29. Murray F, Cowie PR (2011) Plastic contamination in the decapod crustacean *Nephrops norvegicus* (Linnaeus, 1758). Mar Pollut Bull 62(6):1207–1217. doi:10.1016/j.marpolbul.2011.03.032

30. Stevenson K, Stallwood B, Hart AG (2008) Tire rubber recycling and bioremediation: a review. Biorem J 12(1):1–11. doi:10.1080/10889860701866263

31. Vlachopoulos J, Strutt D (2003) Polymer processing. Mater Sci Technol 19(9):1161–1169. doi:10.1179/026708303225004738

32. Brandsch J, Piringer O (2008) Characteristics of plastic materials. In: Plastic packaging. Wiley-VCH Verlag GmbH & Co. KGaA, Weinheim, pp 15–61. doi:10.1002/9783527621422.ch2

33. Lambert S (2013) Environmental risk of polymers and their degradation products. PhD thesis, University of York

34. Lambert S (2015) Biopolymers and their application as biodegradable plastics. In: Kalia CV (ed) Microbial factories: biodiversity, biopolymers, bioactive molecules, vol 2. Springer India, New Delhi, pp 1–9. doi:10.1007/978-81-322-2595-9_1

35. Reddy CSK, Ghai R, Rashmi KVC (2003) Polyhydroxyalkanoates: an overview. Bioresour Technol 87(2):137–146. doi:10.1016/S0960-8524(02)00212-2

36. Mohan SR (2016) Strategy and design of innovation policy road mapping for a waste biorefinery. Bioresour Technol 215:76–83. doi:10.1016/j.biortech.2016.03.090

37. Amulya K, Jukuri S, Venkata Mohan S (2015) Sustainable multistage process for enhanced productivity of bioplastics from waste remediation through aerobic dynamic feeding strategy: process integration for up-scaling. Bioresour Technol 188:231–239. doi:10.1016/j.biortech.2015.01.070

38. Baner AL, Piringer O (2007) Preservation of quality through packaging. In: Plastic packaging materials for food. Wiley-VCH Verlag GmbH, Weinheim, pp 1–8. doi:10.1002/9783527613281.ch01

39. Roes L, Patel MK, Worrell E, Ludwig C (2012) Preliminary evaluation of risks related to waste incineration of polymer nanocomposites. Sci Total Environ 417:76–86. doi:10.1016/j. scitotenv.2011.12.030
40. Andrady AL (2011) Microplastics in the marine environment. Mar Pollut Bull 62(8): 1596–1605. doi:10.1016/j.marpolbul.2011.05.030
41. Browne MA, Crump P, Niven SJ, Teuten E, Tonkin A, Galloway T, Thompson R (2011) Accumulation of microplastic on shorelines woldwide: sources and sinks. Environ Sci Technol 45(21):9175–9179. doi:10.1021/es201811s
42. GESAMP (2015) Sources, fate and effects of microplastics in the marine environment: a global assessment. Joint Group of Experts on the Scientific Aspects of Marine Environmental Protection Reports and studies 90
43. Carpenter E, Anderson SJ, Miklas HP, Peck BB, Harvey GR (1972) Polystyrene spherules in coastal waters. Science 178(4062):749. doi:10.1126/science.178.4062.749
44. Colton JB, Knapp FD, Burns BR (1974) Plastic particles in surface waters of Northwestern Atlantic. Science 185(4150):491–497. doi:10.1126/science.185.4150.491
45. Gregory MR (1977) Plastic pellets on New-Zealand beaches. Mar Pollut Bull 8(4):82–84. doi:10.1016/0025-326x(77)90193-x
46. Sutherland WJ, Clout M, Cote IM, Daszak P, Depledge MH, Fellman L, Fleishman E, Garthwaite R, Gibbons DW, De Lurio J, Impey AJ, Lickorish F, Lindenmayer D, Madgwick J, Margerison C, Maynard T, Peck LS, Pretty J, Prior S, Redford KH, Scharlemann JPW, Spalding M, Watkinson AR (2010) A horizon scan of global conservation issues for 2010. Trends Ecol Evol 25(1):1–7. doi:10.1016/j.tree.2009.10.003
47. Seco Pon JP, Becherucci ME (2012) Spatial and temporal variations of urban litter in Mar del Plata, the major coastal city of Argentina. Waste Manag 32(2):343–348. doi:10.1016/j. wasman.2011.10.012
48. Njeru J (2006) The urban political ecology of plastic bag waste problem in Nairobi, Kenya. Geoforum 37(6):1046–1058
49. Cierjacks A, Behr F, Kowarik I (2012) Operational performance indicators for litter management at festivals in semi-natural landscapes. Ecol Indic 13(1):328–337. doi:10.1016/j. ecolind.2011.06.033
50. Nizzetto L, Langaas S, Futter M (2016) Pollution: Do microplastics spill on to farm soils? Nature 537(7621):488–488. doi:10.1038/537488b
51. Dris R, Gasperi J, Tassin B (2017) Sources and fate of microplastics in urban areas: a focus on Paris Megacity. In: Wagner M, Lambert S (eds) Freshwater microplastics: emerging environmental contaminants? Springer, Heidelberg. doi:10.1007/978-3-319-61615-5_4 (in this volume)
52. Xu G, Wang QH, Gu QB, Cao YZ, Du XM, Li FS (2006a) Contamination characteristics and degradation behavior of low-density polyethylene film residues in typical farmland soils of China. J Environ Sci Health Part B 41(2):189–199. doi:10.1080/03601230500365069
53. Brodhagen M, Peyron M, Miles C, Inglis DA (2015) Biodegradable plastic agricultural mulches and key features of microbial degradation. Appl Microbiol Biotechnol 99(3): 1039–1056. doi:10.1007/s00253-014-6267-5
54. Kyrikou I, Briassoulis D (2007) Biodegradation of agricultural plastic films: a critical review. J Polym Environ 15(2):125–150. doi:10.1007/s10924-007-0053-8
55. Liu EK, He WQ, Yan CR (2014) 'White revolution' to 'white pollution' – agricultural plastic film mulch in China. Environ Res Lett 9(9):091001. doi:10.1088/1748-9326/9/9/091001
56. Xu J, Wang P, Guo WF, Dong JX, Wang L, Dai SG (2006b) Seasonal and spatial distribution of nonylphenol in Lanzhou Reach of Yellow River in China. Chemosphere 65(9):1445–1451. doi:10.1016/j.chemosphere.2006.04.042
57. Councell TB, Duckenfield KU, Landa ER, Callender E (2004) Tire-wear particles as a source of zinc to the environment. Environ Sci Technol 38(15):4206–4214. doi:10.1021/es034631f
58. Rist SE, Hartmann NB (2017) Aquatic ecotoxicity of microplastics and nanoplastics - lessons learned from engineered nanomaterials. In: Wagner M, Lambert S (eds) Freshwater microplastics: emerging environmental contaminants? Springer, Heidelberg. doi:10.1007/978-3-319-61615-5_2 (in this volume)

59. Reddy MS, Basha S, Adimurthy S, Ramachandraiah G (2006) Description of the small plastics fragments in marine sediments along the Alang-Sosiya ship-breaking yard, India. Estuar Coast Shelf Sci 68(3–4):656–660. doi:10.1016/j.ecss.2006.03.018
60. Wu C, Zhang K, Xiong X (2017) Microplastic pollution in inland waters focusing on Asia. In: Wagner M, Lambert S (eds) Freshwater microplastics: emerging environmental contaminants? Springer, Heidelberg. doi:10.1007/978-3-319-61615-5_5 (in this volume)
61. Khan FR, Mayoma BS, Biginagwa FJ, Syberg K (2017) Microplastics in inland African waters: presence, sources and fate. In: Wagner M, Lambert S (eds) Freshwater microplastics: emerging environmental contaminants? Springer, Heidelberg. doi:10.1007/978-3-319-61615-5_6 (in this volume)
62. Harrison JP, Schratzberger M, Sapp M, Osborn AM (2014) Rapid bacterial colonization of low-density polyethylene microplastics in coastal sediment microcosms. BMC Microbiol 14(232):014–0232. doi:10.1186/s12866-014-0232-4
63. Kooi M, Besseling E, Kroeze C, van Wezel AP, Koelmans AA (2017) Modeling the fate and transport of plastic debris in freshwaters: review and guidance. In: Wagner M, Lambert S (eds) Freshwater microplastics: emerging environmental contaminants? Springer, Heidelberg. doi:10.1007/978-3-319-61615-5_7 (in this volume)
64. Harrison JP, Hoellein TJ, Sapp M, Tagg AS, Ju-Nam Y, Ojeda JJ (2017) Microplastic-associated biofilms: a comparison of freshwater and marine environments. In: Wagner M, Lambert S (eds) Freshwater microplastics: emerging environmental contaminants? Springer, Heidelberg. doi:10.1007/978-3-319-61615-5_9 (in this volume)
65. Lambert S, Sinclair CJ, Bradley EL, Boxall ABA (2013) Effects of environmental conditions on latex dergadation in aquatic systems. Sci Total Environ 447:225–234. doi:10.1016/j.scitotenv.2012.12.067
66. Lambert S, Wagner M (2016a) Characterisation of nanoplastics during the degradation of polystyrene. Chemosphere 145:265–268. doi:10.1016/j.chemosphere.2015.11.078
67. Lambert S, Wagner M (2016b) Formation of microscopic particles during the degradation of different polymers. Chemosphere 161:510–517. doi:10.1016/j.chemosphere.2016.07.042
68. Gigault J, Pedrono B, Maxit B, Ter Halle A (2016) Marine plastic litter: the unanalyzed nanofraction. Environ Sci Nano 3(2):346–350. doi:10.1039/c6en00008h
69. Cheng Z, Redner S (1990) Kinetics of fragmentation. J Phys A Math Gen 23(7):1233
70. Redner S (1990) Statistical theory of fragmentation. In: Charmet JC, Roux S, Guyon E (eds) Disorder and fracture, NATO ASI series, vol 204. Springer, New York, pp 31–48. doi:10.1007/978-1-4615-6864-3_3
71. Ziff RM, McGrady ED (1985) The kinetics of cluster fragmentation and depolymerisation. J Phys A Math Gen 18(15):3027
72. McCoy BJ, Madras G (1997) Degradation kinetics of polymers in solution: dynamics of molecular weight distributions. AICHE J 43(3):802–810. doi:10.1002/aic.690430325
73. Rabek JB (1975) Oxidative degradation of polymers. In: Bamford CH, Tipper CFH (eds) Comprehensive chemical kinetics: degradation of polymers, vol 14. Elsevier Scientific Publishing Company, Amsterdam
74. Bartsev SI, Gitelson JI (2016) A mathematical model of the global processes of plastic degradation in the World Ocean with account for the surface temperature distribution. Dokl Earth Sci 466(2):153–156. doi:10.1134/s1028334x16020033
75. Leejarkpai T, Suwanmanee U, Rudeekit Y, Mungcharoen T (2011) Biodegradable kinetics of plastics under controlled composting conditions. Waste Manag 31(6):1153–1161. doi:10.1016/j.wasman.2010.12.011
76. Beyler CL, Hirschler MM (2002) Thermal decomposition of polymers. In: SFPE handbook of fire protection engineering, vol 2, Sect. 1, Chap. 7. National Fire Protection Association, Quincy
77. Ehrenstein GW (2012) Polymeric materials: structure, properties, applications. Hanser Gardner Publications, Inc., München

78. Teuten EL, Saquing JM, Knappe DRU, Barlaz MA, Jonsson S, Bjorn A, Rowland SJ, Thompson RC, Galloway TS, Yamashita R, Ochi D, Watanuki Y, Moore C, Pham HV, Tana TS, Prudente M, Boonyatumanond R, Zakaria MP, Akkhavong K, Ogata Y, Hirai H, Iwasa S, Mizukawa K, Hagino Y, Imamura A, Saha M, Takada H (2009) Transport and release of chemicals from plastics to the environment and to wildlife. Philos Trans R Soc B Biol Sci 364(1526):2027–2045. doi:10.1098/rstb.2008.0284

79. Endo S, Takizawa R, Okuda K, Takada H, Chiba K, Kanehiro H, Ogi H, Yamashita R, Date T (2005) Concentration of polychlorinated biphenyls (PCBs) in beached resin pellets: variability among individual particles and regional differences. Mar Pollut Bull 50(10):1103–1114. doi:10.1016/j.marpolbul.2005.04.030

80. Mato Y, Isobe T, Takada H, Kanehiro H, Ohtake C, Kaminuma T (2001) Plastic resin pellets as a transport medium for toxic chemicals in the marine environment. Environ Sci Technol 35(2):318–324. doi:10.1021/es0010498

81. Ashton K, Holmes L, Turner A (2010) Association of metals with plastic production pellets in the marine environment. Mar Pollut Bull 60(11):2050–2055. doi:10.1016/j.marpolbul.2010. 07.014

82. Delle Site A (2001) Factors affecting sorption of organic compounds in natural sorbent/water systems and sorption coefficients for selected pollutants. A review. J Phys Chem Ref Data 30(1):187–439. doi:10.1063/1.1347984

83. Bajracharya RM, Manalo AC, Karunasena W, Lau K-T (2014) An overview of mechanical properties and durability of glass-fibre reinforced recycled mixed plastic waste composites. Mater Des 62:98–112. doi:10.1016/j.matdes.2014.04.081

84. Hale SE, Meynet P, Davenport RJ, Martin Jones D, Werner D (2010) Changes in polycyclic aromatic hydrocarbon availability in River Tyne sediment following bioremediation treatments or activated carbon amendment. Water Res 44(15):4529–4536. doi:10.1016/j.watres. 2010.06.027

85. Allan IJ, Harman C, Kringstad A, Bratsberg E (2010) Effect of sampler material on the uptake of PAHs into passive sampling devices. Chemosphere 79(4):470–475. doi:10.1016/j. chemosphere.2010.01.021

86. Fernandez LA, MacFarlane JK, Tcaciuc AP, Gschwend PM (2009) Measurement of freely dissolved PAH concentrations in sediment beds using passive sampling with low-density polyethylene strips. Environ Sci Technol 43(5):1430–1436. doi:10.1021/es802288w

87. Adams RG, Lohmann R, Fernandez LA, MacFarlane JK (2007) Polyethylene devices: passive samplers for measuring dissolved hydrophobic organic compounds in aquatic environments. Environ Sci Technol 41(4):1317–1323. doi:10.1021/es0621593

88. Booij K, Smedes F, van Weerlee EM (2002) Spiking of performance reference compounds in low density polyethylene and silicone passive water samplers. Chemosphere 46(8): 1157–1161. doi:10.1016/S0045-6535(01)00200-4

89. Fries E, Zarfl C (2012) Sorption of polycyclic aromatic hydrocarbons (PAHs) to low and high density polyethylene (PE). Environ Sci Pollut Res 19(4):1296–1304. doi:10.1007/s11356-011-0655-5

90. Wanner U, Fuhr F, DeGraaf AA, Burauel P (2005) Characterization of non-extractable residues of the fungicide dithianon in soil using 13C/14C-labelling techniques. Environ Pollut 133(1):35–41. doi:10.1016/j.envpol.2004.04.017

91. Mordaunt CJ, Gevao B, Jones KC, Semple KT (2005) Formation of non-extractable pesticide residues: observations on compound differences, measurement and regulatory issues. Environ Pollut 133(1):25–34. doi:10.1016/j.envpol.2004.04.018

92. Bern L (1990) Size-related discrimination of nutritive and inert particles by freshwater zooplankton. J Plankton Res 12(5):1059–1067. doi:10.1093/plankt/12.5.1059

93. Rosenkranz P, Chaudhry Q, Stone V, Fernandes TF (2009) A comparison of nanoparticle and fine particle uptake by Daphnia magna. Environ Toxicol Chem 28(10):2142–2149. doi:10. 1897/08-559.1

94. Besseling E, Wegner A, Foekema EM, van den Heuvel-Greve MJ, Koelmans AA (2013) Effects of microplastic on fitness and PCB bioaccumulation by the Lugworm *Arenicola marina* (L.) Environ Sci Technol 47(1):593–600. doi:10.1021/es302763x
95. Bilotta GS, Brazier RE (2008) Understanding the influence of suspended solids on water quality and aquatic biota. Water Res 42(12):2849–2861. doi:10.1016/j.watres.2008.03.018
96. Muncke J (2009) Exposure to endocrine disrupting compounds via the food chain: is packaging a relevant source? Sci Total Environ 407(16):4549–4559. doi:10.1016/j.scitotenv.2009.05.006
97. Lithner D, Nordensvan I, Dave G (2012) Comparative acute toxicity of leachates from plastic products made of polypropylene, polyethylene, PVC, acrylonitrile-butadiene-styrene, and epoxy to *Daphnia magna*. Environ Sci Pollut Res 19(5):1763–1772. doi:10.1007/s11356-011-0663-5
98. Lithner D, Damberg J, Dave G, Larsson A (2009) Leachates from plastic consumer products – screening for toxicity with *Daphnia magna*. Chemosphere 74(9):1195–1200. doi:10.1016/j.chemosphere.2008.11.022
99. Bejgarn S, MacLeod M, Bogdal C, Breitholtz M (2015) Toxicity of leachate from weathering plastics: an exploratory screening study with *Nitocra spinipes*. Chemosphere 132: 114–119. doi:10.1016/j.chemosphere.2015.03.010
100. Bradley LE, Driffield M, Guthrie J, Harmer N, Oldring TPK, Castle L (2009) Analytical approaches to identify potential migrants in polyester-polyurethane can coatings. Food Addit Contam Part A 26(12):1602–1610. doi:10.1080/19440040903252256
101. Bradley EL, Driffield M, Harmer N, Oldring PKT, Castle L (2008) Identification of potential migrants in epoxy phenolic can coatings. Int J Polym Anal Charact 13(3):200–223. doi:10.1080/10236660802070512
102. Mattsson K, Hansson LA, Cedervall T (2015) Nano-plastics in the aquatic environment. Environ Sci Process Impacts 17:1712–1721. doi:10.1039/c5em00227c
103. Syberg K, Khan FR, Selck H, Palmqvist A, Banta GT, Daley J, Sano L, Duhaime MB (2015) Microplastics: addressing ecological risk through lessons learned. Environ Toxicol Chem 34(5):945–953. doi:10.1002/etc.2914
104. Nowack B, Bucheli TD (2007) Occurrence, behavior and effects of nanoparticles in the environment. Environ Pollut 150(1):5–22. doi:10.1016/j.envpol.2007.06.006
105. Oberdörster E, Zhu S, Blickley TM, McClellan-Green P, Haasch ML (2006) Ecotoxicology of carbon-based engineered nanoparticles: effects of fullerene (C60) on aquatic organisms. Carbon 44(6):1112–1120. doi:10.1016/j.carbon.2005.11.008
106. Desai MP, Labhasetwar V, Walter E, Levy RJ, Amidon GL (1997) The mechanism of uptake of biodegradable microparticles in Caco-2 cells is size dependent. Pharm Res 14(11):1568–1573
107. Dhawan A, Pandey A, Sharma V (2011) Toxicity assessment of engineered nanomaterials: resolving the challenges. J Biomed Nanotechnol 7(1):6–7. doi:10.1166/jbn.2011.1173
108. Handy RD, Owen R, Valsami-Jones E (2008) The ecotoxicology of nanoparticles and nanomaterials: current status, knowledge gaps, challenges, and future needs. Ecotoxicology 17(5):315–325. doi:10.1007/s10646-008-0206-0
109. Boxall AB, Chaudhry MQ, Sinclair CJ, Jones A, Aitken R, Jefferson B, Watts C (2007) Current and future predicted environmental exposure to engineered nanoparticles. Report by Central Science Laboratory for the Department of Environment Food and Rural Affairs
110. Musee N (2011) Simulated environmental risk estimation of engineered nanomaterials: a case of cosmetics in Johannesburg City. Hum Exp Toxicol 30(9):1181–1195. doi:10.1177/0960327110391387
111. Gottschalk F, Sonderer T, Scholz RW, Nowack B (2009) Modeled environmental concentrations of engineered nanomaterials (TiO_2, ZnO, Ag, CNT, fullerenes) for different regions. Environ Sci Technol 43(24):9216–9222. doi:10.1021/es9015553
112. Senjen R, Hansen SF (2011) Towards a nanorisk appraisal framework. C R Phys 12(7): 637–647. doi:10.1016/j.crhy.2011.06.005

113. Foss Hansen S, Larsen BH, Olsen SI, Baun A (2007) Categorization framework to aid hazard identification of nanomaterials. Nanotoxicology 1(3):243–250. doi:10.1080/17435390701727509

114. Lambert S, Scherer C, Wagner M (2017) Ecotoxicity testing of microplastics: considering the heterogeneity of physico-chemical properties. Integr Environ Assess Manage 13(3):470–475. doi:10.1002/ieam.1901

115. Commission E (2002) Methyloxirane (propylene oxide): summary risk assessment report. Institute for Health and Consumer Protection-European Chemicals Bureau

116. Harrad S, Abdallah MA-E, Rose NL, Turner SD, Davidson TA (2009) Current-use brominated flame betardants in water, sediment, and fish from English lakes. Environ Sci Technol 43(24):9077–9083. doi:10.1021/es902185u

117. Alaee M, Arias P, Sjodin A, Bergman A (2003) An overview of commercially used brominated flame retardants, their applications, their use patterns in different countries/regions and possible modes of release. Environ Int 29(6):683–689. doi:10.1016/S0160-4120(03)00121-1

118. Brennholt N, Heß M, Reifferscheid G (2017) Freshwater Microplastics: Challenges for Regulation and Management. In: Wagner M, Lambert S (eds) Freshwater microplastics: emerging environmental contaminants? Springer, Heidelberg. doi:10.1007/978-3-319-61615-5_12 (in this volume)

119. Arvidsson R (2012) Contributions to emissions, exposure and risk assessment of nanomaterials. Doctoral Thesis, Chalmers University of Technology

120. Hansen SF (2009) Regulation and risk assessment of nanomaterials – too little, too late? PhD Thesis, Technical University of Denmark

121. Nguyen Phuc T, Matsui Y, Fujiwara T (2011) Assessment of plastic waste generation and its potential recycling of household solid waste in Can Tho City, Vietnam. Environ Monit Assess 175(1–4):23–35. doi:10.1007/s10661-010-1490-8

Aquatic Ecotoxicity of Microplastics and Nanoplastics: Lessons Learned from Engineered Nanomaterials

Sinja Rist and Nanna Bloch Hartmann

Abstract The widespread occurrence of microplastics in the aquatic environment is well documented through international surveys and scientific studies. Further degradation and fragmentation, resulting in the formation of nanosized plastic particles – nanoplastics – has been highlighted as a potentially important issue. In the environment, both microplastics and nanoplastics may have direct ecotoxicological effects, as well as vector effects through the adsorption of co-contaminants. Plastic additives and monomers may also be released from the polymer matrix and cause adverse effects on aquatic organisms. Although limited information regarding the ecotoxicological effects of nano- and microplastics is available at present, their small size gives rise to concern with respect to the adverse effects and dislocation of these particles inside organisms – similar to issues often discussed for engineered nanomaterials. In the same way, transport of co-contaminants and leaching of soluble substances are much debated issues with respect to the ecotoxicology of nanomaterials.

In this chapter, we draw on existing knowledge from the field of ecotoxicology of engineered nanomaterials to discuss potential ecotoxicological effects of nano- and microplastics. We discuss the similarities and differences between nano- and microplastics and engineered nanomaterials with regard to both potential effects and expected behaviour in aquatic media. One of the key challenges in ecotoxicology of nanomaterials has been the applicability of current test methods, originally intended for soluble chemicals, to the testing of particle suspensions. This often requires test modifications and special attention to physical chemical characterisation and data interpretation. We present an overview of lessons learned from

This chapter has been externally peer-reviewed.

S. Rist (✉) and N.B. Hartmann
Department of Environmental Engineering, Technical University of Denmark, 2800 Kongens Lyngby, Denmark
e-mail: siri@env.dtu.dk

M. Wagner, S. Lambert (eds.), *Freshwater Microplastics*,
Hdb Env Chem 58, DOI 10.1007/978-3-319-61615-5_2,
© The Author(s) 2018

nanomaterials and offer suggestions on how these can be transferred to recommendations for ecotoxicity testing of nano- and microplastics.

Keywords Biological effects, Nanoparticles, Nanotoxicology, Test methods, Vector effects

1   Engineered Nanomaterials Versus Plastic Particles: Comparing Apples and Oranges?

Over the last half century, it has become increasingly clear that environmental pollution presents a global societal challenge due to immediate and long-term hazards posed by chemicals in the environment. The focus of researchers, legislators and the population has been on chemicals such as pesticides, persistent organic pollutants, heavy metals, pharmaceuticals and endocrine-disrupting chemicals, as well as the effect of chemical mixtures. The common denominator for these groups of chemicals is that they are most often soluble in aqueous media. Ecotoxicology is a multidisciplinary field, integrating ecology and toxicology. It is the study of potentially harmful effects of chemicals on biological organisms, from the cellular to the ecosystem level. Standardised and harmonised ecotoxicological test methods have been developed within the frameworks of OECD and ISO to assess the environmental fate and effects of chemicals.

During the last decade, a new group of chemical substances has entered the limelight, namely, particles. The increasing use of nanotechnology and production of engineered nanomaterials has sharpened the public, scientific and regulatory focus on their potential consequences for the environment and human health, leading to the formation of the new scientific field of ecotoxicology of nanomaterials. The concerns apply not only to engineered nanomaterials but also to unintentionally produced anthropogenic nanomaterials such as ultrafine particles resulting from combustion processes. Similarly, it is becoming increasingly clear that microscopic plastic particles are widespread in the environment, resulting from industrial use, human activities and inadequate waste management. This plastic debris is found in the micrometre size range (i.e. microplastics) although submicron-sized plastic particles (i.e. nanoplastics) are also expected to be formed in the environment through continuous fragmentation of larger plastic particles [1, 2]. Microplastics are commonly defined as plastic particles smaller than 5 mm [3], whereas no common definition for nanoplastics has yet been established. The term has been used for particles <1 μm as well as <100 nm [2, 4]. Engineered nanomaterials, on the other hand, are more unambiguously defined as having at least one dimension in the size range of 1–100 nm [5]. Nanoparticles are a subgroup of nanomaterials possessing three dimensions within this size range. The term 'nanomaterials' is generally used here; however, 'nanoparticles' are referred to in certain places to emphasise the particulate nature of the material. To date, no established analytical methods exist for the detection of nanoplastics in the aquatic environment, and no studies have demonstrated their presence [2]. However, laboratory studies have shown the formation of nanoplastics down to sizes of 30 nm during artificial weathering of larger plastic materials, using nanoparticle tracking analysis [6]. This is a strong indication that this process can also take place in the environment. Particles as emerging environmental pollutants call for a better understanding of their environmental behaviour and potentially harmful effects on organisms. Ecotoxicity testing of particles represents a shift in test paradigm away from testing of soluble chemicals and demands reconsideration of existing test methods and procedures, including the standardised methods developed by OECD and ISO [7, 8]. On the one hand, parallels can generally be drawn between ecotoxicological testing of particles, independent of whether those particles are engineered nanomaterials or plastic particles [9]. On the other hand, it is important to understand where the similarities end, in order to avoid redundant testing, use of inappropriate test methods and generation of meaningless data. Nano- and microplastics cover a wide range in terms of particle sizes. To illustrate this: If a 1 mm particle corresponded to the size of the Earth, then a nanosized particle would correspond to the International Space Station in the orbit around it, i.e. differing in size by six orders of magnitude. Resemblances, in terms of behaviour, fate and effects, are more likely to occur for particles that are similar in size. Therefore the similarities between engineered nanomaterials and nano- and microplastic particles are more likely to apply for smaller microplastics of up to a few microns as well as the submicron-sized nanoplastic particles, which will be the main focus of this chapter. Further noteworthy differences exist in terms of their chemical properties, sources and their related methodological challenges, as described in further detail below.

2   Sources, Emissions and Regulation

The potential sources and routes by which engineered nanomaterials and nano- and microplastics enter the environment are somewhat similar (see Fig. 1). As their name suggests, engineered nanomaterials are intentionally designed and produced for specific applications, processes or products. Production can take place by synthesis (bottom-up approach) or comminution of larger materials (top-down approach). This is comparable to the production of primary nano- and microplastics, for example, microbeads intentionally produced for cosmetic products or plastic pellets used as feeding material in plastic production. Depending on the definitions applied, primary nanoplastics would actually fall under the definition of engineered nanomaterials. An estimated amount of more than 4,000 t of primary microplastic beads were used in cosmetics in Europe in 2012 [10]. Nonetheless, primary microplastics only represent a small fraction of the estimated overall environmental microplastics load [11], a fraction, however, which can relatively easily be addressed and reduced. The main sources of nano- and microplastic pollution, however, are uncontrolled processes such as abrasion and degradation of larger plastic products and fragments, i.e. secondary sources of anthropogenic origin [12]. These sources include

Fig. 1 Nano- and microplastics and engineered nanomaterials can enter the environment through different processes: intentional industrial manufacturing (as in the case of engineered nanomaterials and primary nano- and microplastics) or through uncontrolled anthropogenic processes (secondary nano- and microplastics). The different sources result in particles with different shapes, morphologies, compositions, sizes, etc. Particles manufactured under controlled industrial conditions tend to be more homogenous and uniform in their properties. *Blue*, primary sources; *red*, secondary sources

mismanaged plastic waste, either discarded in the environment directly or improperly collected and disposed of in landfills, subsequently reaching the environment by wind- or water-driven transport [13]. Also, industrial abrasion processes (e.g. air blasting), synthetic paints and car tyres are thought to contribute significantly to the generation of microplastics [11]. Wind and surface run-off water can transport these to aquatic ecosystems. Another important source is synthetic textiles, which have been shown to release large amounts of microplastic fibres into waste water during washing [14]. The relative importance of secondary sources is unique to micro- and nanoplastics, compared to engineered nanomaterials, in the sense that engineered nanomaterials are produced through controlled industrial processes and not generated from the bulk material in the environment. Their release is thereby linked to specific products or industrial applications and therefore comparable to primary microplastics.

The differences in sources between engineered/industrially produced primary particles and unintentionally produced secondary particles have consequences for risk management and regulatory options. For engineered nanomaterials, regulatory measures can ensure that risk is minimised to acceptable levels through upstream regulation of their specific production and use. Regulations addressing criteria for air emissions from various combustion processes can help to reduce the release of unintentionally produced anthropogenic nanomaterials. For micro- and nano-plastics, upstream regulation may be effective in reducing the environmental emissions of primary microplastics. Examples are the US 'Microbead-Free Waters Act of 2015' [15] prohibiting plastic microbeads in rinse-off cosmetics including toothpaste as well as the upcoming UK ban on microbeads in cosmetics by 2017 [16]. For secondary microplastics, on the other hand, reducing their environmental occurrence involves taking general action against plastics entering the environment during all steps of plastic production, use and waste management. Taxation of, or a ban on, single-use plastic shopping bags [17] and bottle return systems [18] are examples of regulatory measures aimed at reducing the general environmental plastic load. Once the plastic has entered the environment, the formation of micro-plastics is governed by the inherent properties of the plastic and the environmental conditions [19] and thereby practically impossible to mitigate through regulatory measures.

3 Material Synthesis, Chemical Composition and Consequences for Environmental Detection

A clear difference between engineered nanomaterials and nano- and microplastics relates to their chemical composition. In principle, engineered nanomaterials can be produced from any solid material. Higher production volume engineered nano-materials are typically made from metals or metal oxides (such as TiO_2, CeO_2 and Ag) or from carbon (such as carbon nanotubes (CNTs)) [20] although organic nanomaterials are also manufactured (from polymers, monomers and lipids) [21]. Nano- and microplastics, on the other hand, consist specifically of synthetic polymers, produced by polymerisation of various monomers and covering a range of materials such as polyethylene (PE), polypropylene (PP), polystyrene (PS) and polyvinylchloride (PVC) [1, 22]. Synthetic polymers differ in properties such as density, porosity and content of non-polymeric additives. Additives may constitute up to 50% of the total mass of plastics and can be composed of both organic and inorganic substances [23]. Hence, while nano- and microplastics consist of specific synthetic polymers (e.g. PE or PP), there are as many variations as there are combinations and ratios of additives. These additives may alter the properties of the material in such a way that it will behave differently in the environment and cause different environmental effects. The same is true for engineered nanomaterials: For engineered nanoparticles with a given chemical composition (e.g. TiO_2), the properties change with different crystalline structures and surface coatings. At the same time, engineered nanomaterials can be made from a range of different materials and combinations of materials. An ongoing discussion within engineered nanomaterials relates to 'sameness': When can two particles be considered the same and when are they so different that they cannot? This has consequences for categorisation and read-across for regulatory purposes [24]. For example, if data exist on the toxicity of a certain nanomaterial, can these data then be used to assess the safety of a similar nanomaterial? On what parameters should these two particles be similar: size, shape, surface chemistry? And when is 'similar' similar enough to be considered 'the same'? This discussion will be relevant for nano- and microplastics, should legislative frameworks require regulatory data on their environmental safety. According to European legislation, polymers are currently exempted from registration under REACH [25]. However, this may change in the future, making the discussion of 'sameness' also relevant for primary nano- and microplastics. For secondary microplastics, sameness is likewise relevant to categorising particles occurring in the environment, as well as to comparing observed behaviour and effects of nano- and microplastic particles between different scientific studies.

The characteristics and chemical composition of particles have consequences for the feasibility of detection and quantification of particles, especially in environmental samples and biota. It is highly challenging to detect engineered nanomaterials in the environment, especially due to their small size. Under controlled laboratory conditions, with known nanomaterials, techniques based on electron microscopy, mass spectrometry and spectroscopy can be applied to investigating the behaviour of the nanomaterials in the test system [26]. However, applying the same techniques to the detection and quantification of nanomaterials in a natural environmental matrix is not straightforward – even when looking for a known nanomaterial. For this reason, monitoring data for engineered nanomaterials are practically non-existent. One of the main problems is that the nanomaterials may be modified through sample preparation (e.g., causing dissolution or aggregation), making it difficult to 'extract' the particles from the sample in their naturally occurring state [26]. Electron microscopy, in combination with elemental ratios, has successfully been applied in detecting TiO_2 nanoparticles released from sunscreen into lake surface waters [27]. Comparing elemental ratios was necessary in order to distinguish natural Ti-bearing particles from their engineered counterparts. Even for engineered nanomaterials made of non-ubiquitous elements (e.g. Ag), detection is not straightforward due to complicated sample preparations, matrix interferences and analytical difficulties in distinguishing between different metal species [28].

Nano- and microplastics pose additional challenges due to their organic origin, affecting and limiting the analytical options when they are present in an organic matrix. While the larger-sized fractions can be collected or extracted fairly easily, for example, by filtering water samples or density-based fractionation of sand, it becomes increasingly difficult to distinguish smaller microplastics, and especially nanoplastics, from the surrounding environmental matrix. At the same time, secondary nano- and microplastics, which constitute the main environmental load of plastic particles, are irregular in shape, resulting from their formation through fragmentation rather than controlled production. Also, they are often transparent, semi-transparent or neutral in colour. A study has been carried out to compare stereomicroscopy and Fourier transform infrared spectroscopy (FT-IR) as identification methods for microplastics in environmental samples. White and transparent fragments were identified through FT-IR, but not easily detected by microscopy, leading to underestimation of the actual concentrations of microplastics [29]. In contrast, fibres, identified as cotton fibres by FT-IR, were mistaken for microplastics by stereomicroscopy, leading to overestimation of microplastic fibres using this technique [29].

The development of FT-IR combined with microspectroscopy (i.e. micro-FT-IR) greatly improved the spatial resolution, allowing the identification of particles down to a few μm [30, 31]. The technique allows measurement of transmission and reflectance. The first gives a higher-quality spectrum, but is limited to thin samples, while the latter can also be applied to thick and opaque particles [32]. However, irregular surface structures (e.g. of plastic fragments) can lead to refractive errors when using the reflectance mode [30]. In this case, attenuated total reflectance (ATR) micro-FT-IR can be used to improve the quality of the spectrum. The standard FT-IR techniques rely on a visual pre-sorting of potential plastic particles, which is time-consuming and prone to errors [30]. Therefore, the coupling of micro-FT-IR with focal plane array detectors is considered a promising method for high throughput analysis of microplastics in complex environmental samples [30, 31, 33]. Currently, however, the technique is limited to particles larger than 10–20 μm, and sample preparation is labour-intensive. As for many of the analytical techniques used for engineered nanoparticles, FT-IR is particularly useful for controlled laboratory tests with microplastics of known composition. This material can be included in the spectral library and is then detected in samples. However, it can be difficult to use FT-IR to identify unknown plastics particles from environmental samples, as the spectra of polymers change due to the weathering and chemical changes of the surface of the plastics [29]. Raman spectroscopy is another commonly used method to identify plastic particles. In combination with microscopy (i.e. micro-Raman), a resolution of less than 1 μm is achievable. However, the applicability of micro-Raman with automated spectral imaging for analysis of an entire sample is yet to be demonstrated for microplastics in environmental matrices [32].

The development of methods to detect and characterise nano- and microplastics in environmental matrices with a higher resolution, lower time consumption and high throughput is ongoing, comparable to the developments being made for engineered nanomaterials. The requirements for ideal analytical techniques are similar for both groups of particles. As previously described by Tiede et al. [26], such techniques should (a) cause minimal changes to the physical and chemical state of the particles during sample preparation; (b) provide information on several physicochemical parameters, such as chemical composition, size, shape, etc.; and (c) be able to handle complex, heterogeneous samples [26].

4 Particles as a Vector for Co-pollutants

One of the possible environmental processes, often discussed for both engineered nanomaterials and microplastics, is their ability to act as vectors for other pollutants. Through their use in, for example, consumer products and medical and industrial applications, engineered nanomaterials and primary microplastics will come into contact with other chemical substances, such as preservatives, surfactants and active ingredients in pharmaceutical drugs. Finally, through different disposal routes, the particles will come into contact with environmental contaminants present in, for example, waste water streams and landfill leachate. As a consequence, intentional and unintentional mixing of the particles with other chemical compounds takes place before, during and after their intended use. By this process, an otherwise inert and non-toxic particle potentially becomes a carrier of toxic compounds. At the same time hydrophobic pollutants with a low water solubility become more mobile when sorbed to plastic particles, which may increase their transport and consequently impact their distribution and bioavailability [34]. It has been shown that engineered nanomaterials can sorb and transport organic pollutants in the aquatic environment [35–37]. Similarly, nano- and microplastics have the potential to act as vectors for hydrophobic organic chemicals, as recently reviewed by Rochman [38].

With an increased surface area-to-volume ratio, smaller particles will generally have a larger capacity for adsorption of chemical substances (on an 'adsorption per particle mass' basis). At the same time, their small size may facilitate uptake by organisms and even potential translocation into different parts and organs. This vector function is governed by the properties of the pollutant and the particle [39]. Important particle properties include chemical composition, porosity, size and surface properties (coating, charge). Weathering processes can both increase and decrease sorption [40]. The formation of cracks and increased surface roughness leads to an increased surface area and, therefore, a potentially increased sorption capacity. Counteracting this, weathering may also change crystallinity, increase density and hardness and change surface charge. For instance, changes in surface charge as a result of weathering can increase the sorption of some substances and decrease the affinity for others [41].

Plastic to water partitioning coefficients ($\log K_{pw}$) for various organic chemicals ($\log K_{ow}$ from 0.90 to 8.76) have been collected for polydimethylsiloxane (PDMS), low density PE (LDPE), high density PE (HDPE), ultra-high molecular weight PE (UHMWPE), PP, PS and PVC [41]. Regression analysis showed generally good correlations between $\log K_{ow}$ and $\log K_{pw}$ and linear proportionality for LDPE and HDPE. This analysis suggests that the partitioning of chemicals into plastics is driven by hydrophobic interactions – similar to the partitioning of chemicals into animal lipids [41]. At the same time, pollutants may adhere to the particle surfaces. For example, it has been found that nanoplastics have a capability to adsorb hydrophobic pollutants, a process which can potentially be exploited in the removal of chemicals from contaminated soil and water [42]. Hence, for nano- and microplastics the processes of ad- and absorption may both be relevant to their potential role as pollutant vectors.

Many engineered nanomaterials are manufactured from inorganic materials – or inorganic carbon in the case of C_{60} fullerenes and CNTs. In these cases, the sorption of co-pollutants is governed by adsorption to the particle surface, rather than absorption into the particle matrix. Hence, the sorption capacity is determined by available adsorption sites on the surface of the nanomaterial. The differences in sorption processes between polymer particles and inorganic nanomaterials are illustrated in Fig. 2.

Nano- and microplastics as well as engineered nanomaterials have the potential to act as vectors for co-pollutants in the environment. The process will always depend on the specific chemical pollutant (e.g. K_{ow}), the specific particle properties (e.g. composition and size) and the properties of the surrounding media (e.g. pH), influencing the particle surface properties and the speciation and dissociation of the chemical pollutant. It has been proposed that the vector effect of particle-mediated transport of co-pollutants can be divided into three groups: (1) an environmental vector effect, whereby the co-pollutant is transported through the environment; (2) an organismal vector effect, whereby the co-pollutant is transported into organisms; and (3) a cellular vector effect whereby the co-pollutant is transported with the particle into cells [9]. Combining this with a proposed framework for different pollutant-particle interaction mechanisms, originally developed for engineered nanomaterials [37], the vector function of particle pollutants can be summarised as illustrated in Fig. 3.

Another type of vector function relates to leaching of substances that were originally part of the particle matrix. In the case of engineered nanomaterials, this is primarily metal ions (from metal and metal oxide nanomaterials) or release of coating materials. Similarly, polymer additives can leach from plastic particles. From the field of ecotoxicology of nanomaterials, the importance of properly quantifying ion release is becoming increasingly clear, as observed biological effects can often be directly linked to the concentration of free metal ions [8]. In the same way, the release of plastic additives should be examined when

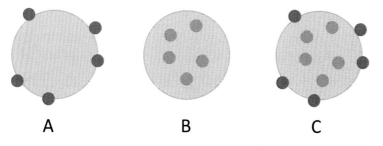

Fig. 2 Illustration of the difference between adsorption (**a**) (more pronounced for inorganic engineered nanomaterials) and absorption (**b**) (more pronounced for polymer particles). In the case of polymer particles, the sorption may also be a combination of ab- and adsorption processes (**c**)

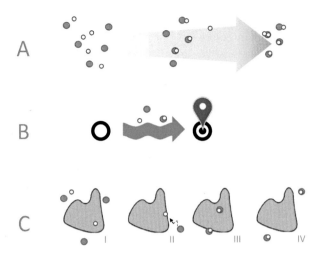

Fig. 3 Illustration of the potential vector function of particles. (**a**) The interaction between the particles (*orange, filled*) and the co-pollutants (*purple, open*) will depend on the properties of the particles, the pollutant and the surrounding medium. This will result in various degrees of absorption and/or adsorption. (**b**) The particles and pollutants are transported in the environment – individually and co-transported. This has been referred to as the 'environmental vector effect' [9]. (**c**) The particles and pollutants interact with biological organisms. This can be via 'independent action' whereby the particle and the pollutant interact with the organisms individually (*I*). It can also be via desorption of the co-pollutant (or leaching of ions/additive), which subsequently interact with the organism (*II*). The pollutant can also be co-transported into the organisms and potentially further into cells (*III*). This has been referred to as an organismal and cellular vector effect, respectively [9]. Finally, the particles can act as a 'trap' for the pollutants, thereby decreasing the interactions between the pollutant and the organisms (*IV*)

investigating the potential biological effects of nano- and microplastics. This will enable a differentiation between effects caused by the particle itself and effects caused by plastic additives.

5 Biological Effects

Engineered nanomaterials are often designed to have a certain reactivity, functionality or biological effect. As discussed, nano- and microplastics often stem from unintentional anthropogenic rather than engineered processes. Even when they are intentionally produced, they are not as such intended to be biologically active. Certain polymer additives may, however, have the purpose of, for example, preventing biotic or abiotic degradation. For both engineered nanomaterials and nano- and microplastics, it is therefore useful to consider their intended use and properties when evaluating their potential environmental risk. Engineered nanomaterials that are intended to have biocidal effects are likely to be more toxic to non-target organisms than materials intended to be inert. Similarly, plastic particles containing biocidal additives, plasticisers or flame retardants are likely to be more environmentally hazardous, as these substances may leach out of the polymer matrix.

One effect mechanism is being highlighted as important for both engineered nanomaterials and nano- and microplastics, namely, physical interactions between the particle and the organisms [43]. This includes inflammation and interference with the energy balance caused by uptake of particles into the gut, thereby limiting food uptake. Different types of engineered nanomaterials, as well as nanoplastics, have been observed to adhere to the surface of microalgae, potentially causing a physical shading effect on a cellular level [44]. Physical effects of microplastics on marine organisms have been reviewed recently [45], and mechanisms that have been described as potentially relevant include blockage of the digestive system, abrasion of tissues, blockage of feeding appendages of invertebrates, embedment in tissues, blockage of enzyme production, reduced feeding stimulus, nutrient dilution, decreased growth rates, lower steroid hormone levels and impaired reproduction. Table 1 presents an overview of effects in response to the physical particle properties that have been observed in different species.

The potential of microplastics to cause such physical effects on organisms depends on a number of factors. Particles with a high capacity to accumulate in

Table 1 Examples of biological effects observed in aquatic organisms after exposure to engineered nanoparticles or nano- and microplastics

Engineered nanoparticles	Nano- and microplastics
Molecular/cellular level	
Oxidative stress[a]	DNA damage and differential gene expression[l]
Inhibition of photosynthesis (shading)[b]	Cellular stress response and impaired metabolism[m]
Tissue level	
Histopathological changes[c]	Tissue damage[n]
Transfer into cells[d]	Transfer into tissues[o]
Organ/organismal level	
Morphological malformation[e]	Impaired respiration[p]
Decreased swimming velocities[f]	Impaired feeding[q]
Increased mucus production[g]	Impaired development and reproduction[r]
Toxic effects of released ions[h]	Decreased growth rates and biomass production[s]
Decreased growth rates and biomass production[i]	Behavioural changes[t]
Moulting inhibition[j]	Increased mortality[u]
Impaired mobility[k]	

[a]In algae [46]; [b]in algae [47]; [c]in fish [48]; [d]in algae [49]; [e]in fish embryos [50]; [f]in crustaceans [51]; [g]in fish [52]; [h]in algae [53]; [i]in algae [7]; [j]in crustaceans [54]; [k]in crustaceans [51]; [l]in echinoderms [55], bivalves [56–58] and fish [59]; [m]in polychaetes [60], echinoderms [55], bivalves [56–58, 61] and fish [62–64]; [n]in fish [59, 64, 65]; [o]in crustaceans [66], mussels [67, 68] and fish [69]; [p]in polychaetes [70], crustaceans [71] and bivalves [72]; [q]in polychaetes [60, 73], crustaceans [74, 75], bivalves [72, 76, 77] and fish [62]; [r]in crustaceans [74, 78, 79], echinoderms [80], bivalves [58] and fish [81]; [s]in crustaceans [75, 79] and bivalves [72]; [t]in fish [62, 81, 82]; [u]in crustaceans [75, 83], bivalves [72] and fish [84]

organisms and translocate into tissues are expected to have a stronger physical impact [45]. This is closely linked to particle size, as will be explained further below. Shape also plays an important role since irregular, sharp fragments are more likely to cause damage than round, smooth particles. Fibres are more likely to accumulate in the digestive system. The capacity of individual species to egest microplastics is also considered as an important factor because this process will determine how long an organism is exposed to the particles [45].

For nanomaterials, size-dependent changes in effects are of particular interest. The whole purpose of nanotechnology is to take advantage of the novel properties that come with a smaller size. For engineered nanomaterials, this involves, for example, the novel catalytic effects of some materials on the nanoscale including gold (Au), titanium dioxide (TiO_2) and cerium dioxide (CeO_2). As larger-sized (bulk) materials, these are relatively inert, but with decreasing particle size and increasing surface area, they become reactive. Therefore, as particle size decreases, there is a tendency for toxicity to increase, even if the same material is relatively inert in its corresponding bulk (micron-sized) form [85]. In addition, the small size of engineered nanomaterials may enable their uptake into tissues and cells [49]. Observed biological effects of engineered nanomaterials in aquatic organisms include oxidative stress, inhibition of photosynthesis, tissue damage, impaired growth and development, behavioural changes and increased mortality (Table 1).

Similarly, the question for nano- and microplastics is therefore: Is it likely that a decrease in size will make them more hazardous? To answer this question, we will examine the two main causes for concern: novel properties and ingestion by organisms (and potential subsequent transfer into tissues). The novel properties that would occur for smaller-sized polymer particles are linked to their increased surface-to-volume ratio. With decreasing particle size, a larger fraction of the molecules will be present on the surface of the particle. As the surface is where interactions with the surrounding environment take place, this can lead to an increase in chemical reactions and biological interactions. For example, smaller particles may (on a mass basis) have a larger adsorption capacity compared to larger particles [86], which in turn is of relevance for the vector effects. The second concern relates to the potential to cross biological barriers. Nanosized particles, such as nanoplastics, are potentially more hazardous due to their easier uptake into tissues and cells [2]. Depending on particle size, different uptake routes into organisms are also involved. For example, the freshwater crustacean *Daphnia magna* normally catches prey (mainly algae) in the size range 0.4–40 µm [87, 88]. For particles or agglomerates that are within this size range, uptake can occur through active filtration, and at the same time unwanted particles can be rejected. Particles smaller than the preferred size are not actively taken up by the animals, but may instead enter the organisms through 'drinking' of the surrounding water, resulting in non-selective, uncontrolled uptake. Depending on the feeding strategies of specific aquatic organisms and their ability to actively select their food source, they may be able to regulate their uptake of microplastics, whereas nanoplastics may enter the organisms unintentionally.

5.1 Nano- and Microplastics in Standard Ecotoxicity Tests

In the quest to determine the environmental risk posed by nano- and microplastics, laboratory-based experiments need to be carried out which analyse the effects of the particles under well-defined conditions. The number of controlled laboratory studies investigating the effects of nano- and microplastics on freshwater organisms is steadily increasing, and many different impacts have been observed – extending from the molecular and cellular to the physiological level (see Table 1). These include inflammation, disruption of lipid and amino acid metabolism, lower growth rates, decreased feeding rates, behavioural changes, impairment of reproduction and increased mortality [62, 64, 75, 79, 81, 82]. When studies involving marine organisms are also taken into account, the number and variety of biological effects of nano- and microplastics that have been found are even greater.

However, most effect studies differ greatly with respect to the parameters used, for example, particle type (different polymers, sizes, shapes, presence of chemicals), test species, exposure duration, exposure concentration and response variables. This makes it difficult to compare results between studies and hampers reproducibility. It can, therefore, be advantageous to apply standardised tests, which come with a number of benefits as they ensure controlled and reproducible test designs and inter-laboratory comparability. Another advantage of standardised ecotoxicity tests is the extensive knowledge base resulting from decades of testing the effects of chemicals on selected model organisms. For ecotoxicology of nanomaterials, this has been highlighted as a motivation for using standardised short-term tests as a starting point for gaining an insight into the fate and bioavailability of engineered nanomaterials in the environment [89]. By using a well-defined test system and a fully defined synthetic medium, other test parameters can be varied individually and in a controlled manner, thereby providing an insight into specific processes and mechanisms [90].

However, the use of standard test guidelines also comes with some potential disadvantages, especially for testing of particles. For freshwater systems, a commonly used species is the freshwater flea *Daphnia magna*, for which the OECD has developed two standard tests: an acute immobilisation test (48 h) (OECD TG 202) and a chronic reproduction test (21 days) (OECD TG 211). These tests were originally developed for soluble chemicals. Since particles show very different behaviours to soluble chemicals, it is challenging to apply the same test set-ups. Even so, some studies have used these standard tests to investigate the effects of nano- and microplastics. Casado et al. [91] conducted an acute immobilisation test with 55 and 110 nm polyethyleneimine PS beads and reported EC_{50} values of 0.8 mg/l and 0.7 mg/l, respectively. The same test with 1 μm PE beads resulted in an EC_{50} value of 57.4 mg/l [83]. This huge difference could be a consequence of the different polymer types and sizes used in the studies, but it might also indicate that mortality is not a very sensitive biological response when it comes to plastic particles. Finally, it may be indicative of a problem that has been highlighted for tests with engineered nanomaterials: That reproducibility and data interpretation in standard ecotoxicity tests with particles, rather than soluble chemicals, are challenged by the dynamic nature of particles suspended in aqueous media [90]. Particle properties and behaviour may change as a function of time or as a result of interactions with test organisms and emitted biomolecules (e.g. exudates) [7]. It has therefore been recognised as essential in the work with engineered nanomaterials to conduct a particle and exposure characterisation before and during a laboratory test [26, 92]. This includes an analysis of the size, shape, surface area and surface chemistry of the tested particle, as well as aggregation/agglomeration, sedimentation and dissolution behaviour in the test system, thereby providing information on exposure in both qualitative and quantitative terms. Furthermore, appropriate ways of dispersing the particles in aquatic media have been highlighted as an important area of future test method development [93]. The rationale behind thorough characterisation and carefully considered sample preparation methods relates to data interpretation and avoidance of the introduction of test artefacts. Such activities are currently rarely undertaken in the work with nano- and microplastics, but should be included in order to gain an insight into the behaviour of the particles in exposure media and the resulting influence on their interaction with test organisms.

Another aspect that needs to be taken into account is the leaching of molecules from particles. For engineered nanomaterials, work is ongoing within the OECD to develop test guidelines for investigating the dissolution of metal ions from metal-containing nanomaterials [94]. In the case of plastic particles, the leaching of chemicals from the polymer matrix (e.g. additives or monomers) and the release of adhered co-pollutants can influence the test results. Appropriate test methods are therefore needed to investigate the actual release of plastic additives from nano- and microplastics under relevant conditions (media, temperature, pH, etc.), and a control for the effects of chemicals and released additives or adhered pollutants needs to be included as a reference.

Transformation processes, such as oxidation/reduction, interaction with macromolecules, light exposure and biological transformation, can significantly influence the integrity, behaviour and persistence of nanomaterials in aquatic media [95–97]. Depending on the specific conditions, dissolution and degradation can be enhanced or reduced. Enhanced dissolution may result in increased toxicity of, for example, metal and metal oxide nanomaterials. At the same time it may cause a gradual decrease in particle size [97]. For nano- and microplastics, aging/weathering processes should also be accounted for as they may change particle properties (e.g. surface chemistry, polarity and density) and enhance fragmentation. It should be emphasised that a complete degradation of plastic particles under realistic environmental conditions has not yet been demonstrated [6, 98, 99]. While aging is potentially important for nanomaterials, and for nano- and microplastics, in the environment, the relevant aging processes and kinetics may differ. Based on current knowledge, nano- and microplastics may have a higher core persistence and lower release of soluble compounds than certain engineered nanomaterials (especially metal and metal oxide nanomaterials such as ZnO and Ag). However, this is clearly an area of future research – for both nanomaterials and nano- and microplastics. Aging is currently not incorporated in standard ecotoxicity test protocols, but has been proposed for engineered nanomaterials [100]. There are also indications that aging of plastic particles can influence biological effects [79]. This aspect should therefore generally be considered in the future development of ecotoxicological tests for particle testing.

For test method developments, the field of ecotoxicity testing of nanomaterials has benefitted from the availability of reference materials (e.g. NIST Standard Reference Materials) and representative industrial nanomaterials (such as those from the JRC Nanomaterials Repository). Such materials are valuable for analytical method validation and for conducting comparable inter-laboratory and inter-species studies. The field of ecotoxicity testing of nano- and microplastics would similarly benefit from the establishment of sources of well-characterised, industrially and environmentally relevant materials of various sizes and compositions.

The applicability of current standard ecotoxicity tests has been questioned for engineered nanomaterials. Development of new test guidelines and guidance is under discussion, for example, within the OECD [101]. The same concerns apply to testing of nano- and microplastics: They represent a specific challenge due to their dynamic nature in environmental media, resulting in, for example, differences in relevant exposure routes (through food or other active uptake routes, grazing on sedimented materials, etc.), as well as potentially different effect mechanisms. Soluble molecules can be taken up into aquatic organism by diffusion and then distributed within the organism based on partitioning, e.g. to lipid tissues. Cellular uptake of soluble chemicals generally relates to passage of biological membranes, mainly through passive diffusion or active uptake, such as transport through ion channels or carrier-mediated transport [102]. In the tissues, they can act non-specifically, leading to narcosis, or specifically by inhibiting or affecting certain biological processes. In comparison, particle distribution is not governed by diffusion and partitioning. Uptake of particles by organisms depends on mechanisms such as feeding rather than molecular diffusion. On a cellular level, particles may be taken up through processes such as phagocytosis. Effects will therefore most likely differ from those of soluble chemicals. An essential aspect is therefore to determine sensitive biological endpoints for the exposure to particles, potentially moving away from the current standard test organisms. A limited number of response variables and test species can be seen as a disadvantage of standardised tests. Based on the argument above, it may further be claimed that 'no effect' in a standard test does not imply a lack of ecological impact of nano- and microplastics, as these tests may not cover the most sensitive endpoints and test species for particle exposure.

As mentioned, effects of microplastics have been observed on a molecular, cellular and physiological level (see Table 1). When performing ecotoxicity testing, the aim is to establish a dose-response relationship based on the underlying assumption that effects are strongly dependent on exposure dose/concentration and time. For engineered nanomaterials, however, an inverse relationship has been observed between concentration and agglomerate size, meaning that with higher particle concentrations, particles tend to form larger agglomerates

[103]. High concentrations of engineered nanomaterials have also been linked to effects that are not due to an actual toxic response, but rather caused by an over-loading of the test organisms with engineered nanomaterials, causing physical inhibition [8]. Testing of low, environmentally relevant particle concentrations during short exposure times may, however, not be sufficient to detect effects when using endpoints on a physiological level. Before an organism shows impairment to, for example, its reproduction or survival, multiple changes must take place on a cellular level. Cellular responses may therefore be more sensitive to microplastic particle stress compared with whole-organism responses. On this level, however, we are dealing with a complex network and huge number of reactions, which makes it challenging to find and define a meaningful, reliable set of response variables. If cellular responses are to be used as indicators of the potentially hazardous properties of nano- and microplastics, more research is needed to develop suitable (standard) test methods. Another option for testing the toxicity of relatively low concentrations of particles is chronic effect studies, as chronic endpoints can prove more sensitive than acute ecotoxicity. An added benefit of testing lower concentrations is that particle agglomeration/aggregation is reduced, leading to more stable exposure.

One major criticism of current nano- and microplastic ecotoxicity studies is their lack of realism and environmental relevance when selecting test parameters [104]. Pristine particles with a clearly defined, homogenous chemical composition are most often applied in laboratory tests. This is in sharp contrast to the particles present in the environment, which undergo transformation processes, potentially influencing their morphology, and, in the case of plastic, often contain various additives. This trade-off between environmental realism and standardised test conditions is not a dilemma that is unique to testing of particles [105]. It should be kept in mind that different testing paradigms inform different scientific and regulatory questions. In standard ecotoxicity, applying simplified test systems and often synthetic media, test parameters can more easily be controlled and modified one by one in order to gain deeper insight into the mechanisms of toxicity and particle uptake [89]. They are also developed to ensure data comparability and study repeatability. For example, data generated following OECD Test Guidelines and Good Laboratory Practice are considered to satisfy the criteria for Mutual Acceptance of Data and can be used for regulatory assessment purposes in all OECD member states, ideally minimising testing efforts and use of test animals [106]. More environmentally realistic studies can, on the other hand, provide case- and site-specific information on the effects of particle pollution under specific environmental conditions. They may also provide more realistic information with regard to the combined effects of multiple environmental stressors and their interactions with plastic particles. Standard ecotoxicity tests and more environmentally realistic studies should therefore be seen as complementary tools of equal importance but potentially addressing different questions of scientific and regulatory relevance.

5.2 Detecting and Quantifying Particle Uptake as a Prerequisite for Assessing the Effects of Nano- and Microplastics

Research on the biological effects of nano- and microplastics is currently at the stage of determining possible responses and thereby investigating the interactions of organisms and plastic particles. For most organisms, there is a direct and obvious link between the uptake of nano- and microplastics by ingestion or ventilation and subsequent effects. Even so, knowledge on uptake itself is very limited, especially when it comes to quantification of this process, since the detection of small plastic particles is extremely challenging, as described earlier. Methods that have been used to quantify particle uptake include counting using a microscope and spectroscopy (Raman or FT-IR) of tissue samples. Furthermore, fluorescent particles are used for image analysis of gut sections, fluorescence microscopy and the measurement of fluorescence intensity of tissues as a proxy for the quantity of particles. All these methods have limitations and are either very difficult to use on a large scale (e.g. spectroscopy) or become increasingly challenged and even unusable with smaller particles and lower particle numbers. This is major drawback since most biological effects depend on the amount of plastic particles taken up into the organism. A possible way forward could be the use of plastic particles with a metal core which are easy to measure, even in small concentrations and sizes, by, for example, mass spectroscopy – using the same techniques as for nanoparticles. Such traceable nano- and microplastics do not reflect naturally occurring particles as found in the environment, but they could serve as model particles for investigating interactions of nano- and microplastics with biological systems. The technique could be used for precise quantification of particles as well as for localisation in tissues. Nanoparticles with a gold core and a polymer coating have previously been used in a number of studies, aimed at gaining an insight into the uptake of engineered nanomaterials in fish and daphnids [107].

6 Lessons Learned... and the Way Ahead

When the ecotoxicology of nanomaterials emerged as a scientific field around a decade ago, the already existing field of 'colloidal science' was somewhat overlooked. Over the years, it has become increasingly clear that many parallels can be drawn between the two fields. The links between particle behaviour, exposure and ecotoxicological effects, as highlighted here, demonstrate the highly interdisciplinary nature and complexity of this research field. Consequently, cooperation is required between scientists with backgrounds in biology, chemistry and colloidal science. Similarly, for studies of environmental behaviour and the effects of nano- and microplastics, it is clearly important to draw on experience from ecotoxicology of nanomaterials as well as colloidal science. This is the key to moving forwards towards an understanding of their potential environmental effects. This applies to general scientific knowledge as well as ongoing work on developing appropriate test methods that are applicable to the testing of particle pollutants rather than soluble chemicals.

Based on experience within the field of engineered nanomaterials, we recommend that the following aspects be considered in work with nano- and microplastics:

- Development of clear, common definitions for plastic particle categorisation
- Thorough particle characterisation in exposure studies (including particle intrinsic properties, aggregation, agglomeration, sedimentation, dissolution, etc.)
- Inclusion of chemical leaching controls (monomers, additives, etc.)
- Development and use of reference materials for method validation and comparison
- Development of protocols for ecotoxicity testing, sample preparation and analytical methods to minimise test artefacts
- Studies into the influence of environmental transformation processes ('aging') on nano- and microplastic behaviour and ecotoxicity
- Development of analytical techniques that introduce minimal changes to the plastic particles during sample preparation, provide information on several physicochemical parameters and can handle complex, heterogeneous samples.

While we should draw on the existing knowledge on engineered nanomaterials, it is equally important to understand where the similarities begin and where they end. In some respects, nano- and microplastics are likely to present different environmental, analytical and methodological problems compared to engineered nanomaterials, and this should be considered in the planning of experiments and in making informed decisions regarding endpoints and tests of interest.

Finally, it is very important to understand the fundamental effect mechanisms associated with nano- and microplastics: Which properties make them hazardous? This is the way forwards towards replacing problematic plastic materials with safer alternatives in consumer products and industrial applications. Such considerations are important when discussing strategies for future plastic manufacturing, minimising environmental risks and increasing the potential for plastic reuse and recycling.

References

1. Andrady AL (2011) Microplastics in the marine environment. Mar Pollut Bull 62:1596–1605. doi:10.1016/j.marpolbul.2011.05.030
2. Koelmans AA, Besseling E, Shim WJ (2015) Nanoplastics in the aquatic environment. Critical review. In: Bergmann M, Gutow L, Klages M (eds) Marine anthropogenic litter. Springer, Cham, pp 325–340. doi:10.1007/978-3-319-16510-3_12
3. Arthur C, Baker J, Bamford H (2009) Proceedings of the international research workshop on the occurrence, effects, and fate of microplastic marine debris. Group, 530

4. Browne MA, Galloway T, Thompson R (2007) Microplastic—an emerging contaminant of potential concern? Integr Environ Assess Manag 3:297–297. doi:10.1002/ieam.5630030215
5. European Commission (EC) (2011) Commission recommendation of 18 October 2011 on the definition of nanomaterial (2011/696/EU). Off J Eur Union 54, L275:38–40
6. Lambert S, Wagner M (2016) Formation of microscopic particles during the degradation of different polymers. Chemosphere 161:510–517. doi:10.1016/j.chemosphere.2016.07.042
7. Hartmann NB, Engelbrekt C, Zhang J, Ulstrup J, Kusk KO, Baun A (2013) The challenges of testing metal and metal oxide nanoparticles in algal bioassays: titanium dioxide and gold nanoparticles as case studies. Nanotoxicology 7:1082–1094. doi:10.3109/17435390.2012. 710657
8. Skjolding LM, Sørensen SN, Hartmann NB, Hjorth R, Hansen SF, Baun A (2016) A critical review of aquatic ecotoxicity testing of nanoparticles - the quest for disclosing nanoparticle effects. Angew Chemie 55:15224–15239. doi:10.1002/ange.201604964
9. Syberg K, Khan FR, Selck H, Palmqvist A, Banta GT, Daley J, Sano L, Duhaime MB (2015) Microplastics: addressing ecological risk through lessons learned. Environ Toxicol Chem 34:945–953. doi:10.1002/etc.2914
10. Gouin T, Avalos J, Brunning I, Brzuska K, de Graaf J, Kaumanns J, Koning T, Meyberg M, Rettinger K, Schlatter H, Thomas J, Van Welie R, Wolf T (2015) Use of micro-plastic beads in cosmetic products in Europe and their estimated emissions to the North Sea environment. SOFW 3:40–46
11. Lassen C, Hansen SF, Magnusson K, Norén F, Hartmann NB, Jensen PR, Nielsen TG, Brinch A (2015) Microplastics: occurrence, effects and sources of releases to the environment in Denmark, Danish Environmental Protection Agency, Copenhagen K, (2015) environmental project no. 1793
12. Kramm J, Völker C (2017) Understanding the risks of microplastics. A social-ecological risk perspective. In: Wagner M, Lambert S (eds) Freshwater microplastics: emerging environmental contaminants? Springer, Heidelberg. doi:10.1007/978-3-319-61615-5_11 (in this volume)
13. Duis K, Coors A (2016) Microplastics in the aquatic and terrestrial environment: sources (with a specific focus on personal care products), fate and effects. Environ Sci Eur 28:2. doi:10.1186/s12302-015-0069-y
14. Napper IE, Thompson RC (2016) Release of synthetic microplastic plastic fibres from domestic washing machines: effects of fabric type and washing conditions. Mar Pollut Bull 112:39–45. doi:10.1016/j.marpolbul.2016.09.025
15. US Congress (2016) Public Law 114–114, 114th Congress an act, H.R.1321 – Microbead-free waters act 2015. Public Law 3129–3130
16. UK Department for Environment, Food & Rural Affairs (2016) Microbead ban announced to protect sealife. https://www.gov.uk/government/news/microbead-ban-announced-to-protect-sealife. Accessed 1 Dec 2016
17. Convery F, McDonnell S, Ferreira S (2007) The most popular tax in Europe? Lessons from the Irish plastic bags levy. Environ Resour Econ 38:1–11. doi:10.1007/s10640-006-9059-2
18. Schneider J, Karigl B, Reisinger H, Oliva J, Süßenbacher E, Read B (2011) A European refunding scheme for drinks containers, European Parliament, Directorate General for External Policies of the Union, Brussels, Belgium
19. Lambert S, Wagner M (2017) Microplastics are contaminants of emerging concern in freshwater environments: an overview. In: Wagner M, Lambert S (eds) Freshwater microplastics: emerging environmental contaminants? Springer, Heidelberg. doi:10.1007/978-3-319-61615-5_1 (in this volume)
20. Hendren CO, Mesnard X, Dröge J, Wiesner MR (2011) Estimating production data for five engineered nanomaterials as a basis for exposure assessment. Environ Sci Technol 45: 2562–2569. doi:10.1021/es103300g
21. Lacour S (2013) Emerging questions for emerging technologies: is there a law for the nano? In: Brayner R et al (eds) Nanomaterials: a danger or a promise? A chemical and biological perspective. Springer, London, pp 357–378. doi:10.1007/978-1-4471-4213-3

22. Thompson RC, Swan SH, Moore CJ, vom Saal FS (2009) Our plastic age. Philos Trans R Soc B Biol Sci 364:1973–1976. doi:10.1098/rstb.2009.0054
23. Oehlmann J, Schulte-Oehlmann U, Kloas W, Jagnytsch O, Lutz I, Kusk KO, Wollenberger L, Santos EM, Paull GC, Van Look KJW, Tyler CR (2009) A critical analysis of the biological impacts of plasticizers on wildlife. Philos Trans R Soc B Biol Sci 364:2047–2062. doi:10.1098/rstb.2008.0242
24. The Organisation for Economic Co-operation and Development (OECD) (2016) Joint Meeting of the Chemicals Committee and the Working Party on Chemicals, Pesticides and Biotechnology Categorisation of Manufactured Nanomaterials Workshop Report. Series on the Safety of Manufactured Nanomaterials, ENV/JM/MONO(2016)9
25. European Chemicals Agency (ECHA) (2012) Guidance for monomers and polymers, version 2.0. European Chemicals Agency, Helsinki, Finland
26. Tiede K, Boxall A, Tear S, Lewis J, David H, Hassellov M (2008) Detection and characterization of engineered nanoparticles in food and the environment. Food Addit Contam Part A 25:795–821. doi:10.1080/02652030802007553
27. Gondikas AP, von der Kammer F, Reed RB, Wagner S, Ranville JF, Hofmann T (2014) Release of TiO$_2$ nanoparticles from sunscreens into surface waters: a one-year survey at the old Danube Recreational Lake. Environ Sci Technol 48:5415–5422. doi:10.1021/es405596y
28. Guo H, Xing B, Hamlet LC, Chica A, He L (2016) Surface-enhanced Raman scattering detection of silver nanoparticles in environmental and biological samples. Sci Total Environ 554–555:246–252. doi:10.1016/j.scitotenv.2016.02.084
29. Song YK, Hong SH, Jang M, Han GM, Rani M, Lee J, Shim WJ (2015) A comparison of microscopic and spectroscopic identification methods for analysis of microplastics in environmental samples. Mar Pollut Bull 93:202–209. doi:10.1016/j.marpolbul.2015.01.015
30. Harrison JP, Ojeda JJ, Romero-González ME (2012) The applicability of reflectance micro-Fourier-transform infrared spectroscopy for the detection of synthetic microplastics in marine sediments. Sci Total Environ 416:455–463. doi:10.1016/j.scitotenv.2011.11.078
31. Löder MGJ, Kuczera M, Mintenig S, Lorenz C, Gerdts G (2015) Focal plane array detector-based micro-Fourier-transform infrared imaging for the analysis of microplastics in environmental samples. Environ Chem 12:563. doi:10.1071/EN14205
32. Löder MGJ, Gerdts G (2015) Methodology used for the detection and identification of microplastics—a critical appraisal. In: Bergmann M, Gutow L, Klages M (eds) Marine anthropogenic litter. Springer, Cham, pp 201–227. doi:10.1007/978-3-319-16510-3_8
33. Tagg AS, Sapp M, Harrison JP, Ojeda JJ (2015) Identification and quantification of microplastics in wastewater using focal plane array-based reflectance micro-FT-IR imaging. Anal Chem 87:6032–6040. doi:10.1021/acs.analchem.5b00495
34. Teuten EL, Rowland SJ, Galloway TS, Thompson RC (2007) Potential for plastics to transport hydrophobic contaminants. Environ Sci Technol 41:7759–7764. doi:10.1021/es071737s
35. Hofmann T, von der Kammer F (2009) Estimating the relevance of engineered carbonaceous nanoparticle facilitated transport of hydrophobic organic contaminants in porous media. Environ Pollut 157:1117–1126. doi:10.1016/j.envpol.2008.10.022
36. Canesi L, Ciacci C, Balbi T (2015) Interactive effects of nanoparticles with other contaminants in aquatic organisms: friend or foe? Mar Environ Res 111:128–134. doi:10.1016/j.marenvres.2015.03.010
37. Hartmann NB, Baun A (2010) The nano cocktail: ecotoxicological effects of engineered nanoparticles in chemical mixtures. Integr Environ Assess Manag 6:311–313. doi:10.1002/ieam.39
38. Rochman CM (2015) The complex mixture, fate and toxicity of chemicals associated with plastic debris in the marine environment. In: Bergmann M, Gutow L, Klages M (eds) Marine anthropogenic litter. Springer, Cham, pp 117–140. doi:10.1007/978-3-319-16510-3_5

39. Ziccardi LM, Edgington A, Hentz K, Kulacki KJ, Kane Driscoll S (2016) Microplastics as vectors for bioaccumulation of hydrophobic organic chemicals in the marine environment: a state-of-the-science review. Environ Toxicol Chem 35:1667–1676. doi:10.1002/etc.3461

40. Teuten EL, Saquing JM, Knappe DRU, Barlaz MA, Jonsson S, Bjorn A, Rowland SJ, Thompson RC, Galloway TS, Yamashita R, Ochi D, Watanuki Y, Moore C, Viet PH, Tana TS, Prudente M, Boonyatumanond R, Zakaria MP, Akkhavong K, Ogata Y, Hirai H, Iwasa S, Mizukawa K, Hagino Y, Imamura A, Saha M, Takada H (2009) Transport and release of chemicals from plastics to the environment and to wildlife. Philos Trans R Soc B Biol Sci 364:2027–2045. doi:10.1098/rstb.2008.0284

41. O'Connor IA, Golsteijn L, Hendriks AJ (2016) Review of the partitioning of chemicals into different plastics: consequences for the risk assessment of marine plastic debris. Mar Pollut Bull. doi:10.1016/j.marpolbul.2016.07.021

42. Brandl F, Bertrand N, Lima EM, Langer R (2015) Nanoparticles with photoinduced precipitation for the extraction of pollutants from water and soil. Nat Commun 6:7765. doi:10.1038/ncomms8765

43. Scherer C, Weber A, Lambert S, Wagner M (2017) Interactions of microplastics with freshwater biota. In: Wagner M, Lambert S (eds) Freshwater microplastics: emerging environmental contaminants? Springer, Heidelberg. doi:10.1007/978-3-319-61615-5_8 (in this volume)

44. Hartmann NB, Legros S, Von der Kammer F, Hofmann T, Baun A (2012) The potential of TiO$_2$ nanoparticles as carriers for cadmium uptake in *Lumbriculus variegatus* and *Daphnia magna*. Aquat Toxicol 118–119:1–8. doi:10.1016/j.aquatox.2012.03.008

45. Wright SL, Thompson RC, Galloway TS (2013) The physical impacts of microplastics on marine organisms: a review. Environ Pollut 178:483–492. doi:10.1016/j.envpol.2013.02.031

46. von Moos N, Slaveykova VI (2013) Oxidative stress induced by inorganic nanoparticles in bacteria and aquatic microalgae – state of the art and knowledge gaps. Nanotoxicology 8:605–630. doi:10.3109/17435390.2013.809810

47. Sørensen SN, Engelbrekt C, Lützhøft H-CH, Jiménez-Lamana J, Noori JS, Alatraktchi FA, Delgado CG, Slaveykova VI, Baun A (2016) A multimethod approach for investigating algal toxicity of platinum nanoparticles. Environ Sci Technol. doi:10.1021/acs.est.6b01072

48. Johari SA, Kalbassi MR, Yu IJ, Lee JH (2015) Chronic effect of waterborne silver nanoparticles on rainbow trout (*Oncorhynchus mykiss*): histopathology and bioaccumulation. Comp Clin Pathol 24:995–1007. doi:10.1007/s00580-014-2019-2

49. von Moos N, Bowen P, Slaveykova VI (2014) Bioavailability of inorganic nanoparticles to planktonic bacteria and aquatic microalgae in freshwater. Environ Sci Nano 1:214–232. doi:10.1039/c3en00054k

50. Asharani PV, Lian Wu Y, Gong Z, Valiyaveettil S (2008) Toxicity of silver nanoparticles in zebrafish models. Nanotechnology 19:255102. doi:10.1088/0957-4484/19/25/255102

51. Artells E, Issartel J, Auffan M, Borschneck D, Thill A, Tella M, Brousset L, Rose J, Bottero JY, Thiéry A (2013) Exposure to cerium dioxide nanoparticles differently affect swimming performance and survival in two Daphnia species. PLoS One 8:1–11. doi:10.1371/journal.pone.0071260

52. Smith CJ, Shaw BJ, Handy RD (2007) Toxicity of single walled carbon nanotubes to rainbow trout, (*Oncorhynchus mykiss*): respiratory toxicity, organ pathologies, and other physiological effects. Aquat Toxicol 82:94–109. doi:10.1016/j.aquatox.2007.02.003

53. Heinlaan M, Ivask A, Blinova I, Dubourguier HC, Kahru A (2008) Toxicity of nanosized and bulk ZnO, CuO and TiO$_2$ to bacteria *Vibrio fischeri* and crustaceans *Daphnia magna* and *Thamnocephalus platyurus*. Chemosphere 71:1308–1316. doi:10.1016/j.chemosphere.2007.11.047

54. Dabrunz A, Duester L, Prasse C, Seitz F, Rosenfeldt R, Schilde C, Schaumann GE, Schulz R (2011) Biological surface coating and molting inhibition as mechanisms of TiO$_2$ nanoparticle toxicity in *Daphnia magna*. PLoS One 6:1–7. doi:10.1371/journal.pone.0020112

55. Della Torre C, Bergami E, Salvati A, Faleri C, Cirino P, Dawson KA, Corsi I (2014) Accumulation and embryotoxicity of polystyrene nanoparticles at early stage of development of sea urchin embryos *Paracentrotus lividus*. Environ Sci Technol 48:12302–12311. doi:10. 1021/es502569w

56. Avio CG, Gorbi S, Milan M, Benedetti M, Fattorini D, D'Errico G, Pauletto M, Bargelloni L, Regoli F (2015) Pollutants bioavailability and toxicological risk from microplastics to marine mussels. Environ Pollut 198:211–222. doi:10.1016/j.envpol.2014.12.021

57. Paul-Pont I, Lacroix C, González Fernández C, Hégaret H, Lambert C, Le Goïc N, Frère L, Cassone A-L, Sussarellu R, Fabioux C, Guyomarch J, Albentosa M, Huvet A, Soudant P (2016) Exposure of marine mussels *Mytilus spp.* to polystyrene microplastics: toxicity and influence on fluoranthene bioaccumulation. Environ Pollut 216:724–737. doi:10.1016/j. envpol.2016.06.039

58. Sussarellu R, Suquet M, Thomas Y, Lambert C, Fabioux C, Pernet MEJ, Le Goïc N, Quillien V, Mingant C, Epelboin Y, Corporeau C, Guyomarch J, Robbens J, Paul-Pont I, Soudant P, Huvet A (2016) Oyster reproduction is affected by exposure to polystyrene microplastics. Proc Natl Acad Sci 113:2430–2435. doi:10.1073/pnas.1519019113

59. Karami A, Romano N, Galloway T, Hamzah H (2016) Virgin microplastics cause toxicity and modulate the impacts of phenanthrene on biomarker responses in African catfish (*Clarias gariepinus*). Environ Res 151:58–70. doi:10.1016/j.envres.2016.07.024

60. Wright SL, Rowe D, Thompson RC, Galloway TS (2013) Microplastic ingestion decreases energy reserves in marine worms. Curr Biol 23:R1031–R1033. doi:10.1016/j.cub.2013.10. 068

61. Canesi L, Ciacci C, Bergami E, Monopoli MP, Dawson KA, Papa S, Canonico B, Corsi I (2015) Evidence for immunomodulation and apoptotic processes induced by cationic polystyrene nanoparticles in the hemocytes of the marine bivalve Mytilus. Mar Environ Res 111: 1–7. doi:10.1016/j.marenvres.2015.06.008

62. Cedervall T, Hansson L-A, Lard M, Frohm B, Linse S (2012) Food chain transport of nanoparticles affects behaviour and fat metabolism in fish. PLoS One 7:e32254. doi:10.1371/ journal.pone.0032254

63. Greven A-C, Merk T, Karagöz F, Mohr K, Klapper M, Jovanović B, Palić D (2016) Polycarbonate and polystyrene nanoplastic particles act as stressors to the innate immune system of fathead minnow (*Pimephales promelas*). Environ Toxicol Chem 35:3093–3100. doi:10.1002/etc.3501

64. Lu Y, Zhang Y, Deng Y, Jiang W, Zhao Y, Geng J, Ding L, Ren H (2016) Uptake and Accumulation of polystyrene microplastics in zebrafish (*Danio rerio*) and toxic effects in liver. Environ Sci Technol 50:4054–4060. doi:10.1021/acs.est.6b00183

65. Pedà C, Caccamo L, Fossi MC, Gai F, Andaloro F, Genovese L, Perdichizzi A, Romeo T, Maricchiolo G (2016) Intestinal alterations in European sea bass *Dicentrarchus labrax* (Linnaeus, 1758) exposed to microplastics: preliminary results. Environ Pollut 212: 251–256. doi:10.1016/j.envpol.2016.01.083

66. Rosenkranz P, Chaudhry Q, Stone V, Fernandes TF (2009) A comparison of nanoparticle and fine particle uptake by *Daphnia magna*. Environ Toxicol Chem 28:2142–2149

67. Browne MA, Dissanayake A, Galloway TS, Lowe DM, Thompson RC (2008) Ingested microscopic plastic translocates to the circulatory system of the mussel, *Mytilus edulis* (L.) Environ Sci Technol 42:5026–5031. doi:10.1021/es800249a

68. von Moos N, Burkhardt-Holm P, Köhler A (2012) Uptake and effects of microplastics on cells and tissue of the blue mussel *Mytilus edulis* L. after an experimental exposure. Environ Sci Technol 46:11327–11335. doi:10.1021/es302332w

69. Kashiwada S (2006) Distribution of nanoparticles in the see-through medaka (*Oryzias latipes*). Environ Health Perspect 114:1697–1702. doi:10.1289/ehp.9209

70. Green DS, Boots B, Sigwart J, Jiang S, Rocha C (2016) Effects of conventional and biodegradable microplastics on a marine ecosystem engineer (*Arenicola marina*) and sediment nutrient cycling. Environ Pollut 208:426–434. doi:10.1016/j.envpol.2015.10.010

71. Watts AJR, Urbina MA, Goodhead R, Moger J, Lewis C, Galloway TS (2016) Effect of microplastic on the gills of the shore crab *Carcinus maenas*. Environ Sci Technol 50: 5364–5369. doi:10.1021/acs.est.6b01187

72. Rist SE, Assidqi K, Zamani NP, Appel D, Perschke M, Huhn M, Lenz M (2016) Suspended micro-sized PVC particles impair the performance and decrease survival in the Asian green mussel *Perna viridis*. Mar Pollut Bull 111:213–220. doi:10.1016/j.marpolbul.2016.07.006

73. Besseling E, Wegner A, Foekema EM, van den Heuvel-Greve MJ, Koelmans A a (2013) Effects of microplastic on fitness and PCB bioaccumulation by the lugworm *Arenicola marina* (L.) Environ Sci Technol 47:593–600. doi:10.1021/es302763x

74. Cole M, Lindeque P, Fileman E, Halsband C, Galloway TS (2015) The impact of polystyrene microplastics on feeding, function and fecundity in the marine copepod *Calanus helgolandicus*. Environ Sci Technol 49:1130–1137. doi:10.1021/es504525u

75. Ogonowski M, Schür C, Jarsén Å, Gorokhova E (2016) The effects of natural and anthropogenic microparticles on individual fitness in *Daphnia magna*. PLoS One 11:e0155063. doi:10.1371/journal.pone.0155063

76. Wegner A, Besseling E, Foekema EM, Kamermans P, Koelmans AA (2012) Effects of nanopolystyrene on the feeding behavior of the blue mussel (*Mytilus edulis* L.) Environ Toxicol Chem 31:2490–2497. doi:10.1002/etc.1984

77. Cole M, Galloway TS (2015) Ingestion of nanoplastics and microplastics by Pacific oyster larvae. Environ Sci Technol 49:14625–14632. doi:10.1021/acs.est.5b04099

78. Lee K-W, Shim WJ, Kwon OY, Kang J-H (2013) Size-dependent effects of micro polystyrene particles in the marine copepod *Tigriopus japonicus*. Environ Sci Technol 47: 11278–11283. doi:10.1021/es401932b

79. Besseling E, Wang B, Lürling M, Koelmans AA (2014) Nanoplastic affects growth of *S. obliquus* and reproduction of *D. magna*. Environ Sci Technol 48:12336–12343. doi:10. 1021/es503001d

80. Nobre CR, Santana MFM, Maluf A, Cortez FS, Cesar A, Pereira CDS, Turra A (2015) Assessment of microplastic toxicity to embryonic development of the sea urchin *Lytechinus variegatus* (Echinodermata: Echinoidea). Mar Pollut Bull 92:99–104. doi:10.1016/j. marpolbul.2014.12.050

81. Lönnstedt OM, Eklov P (2016) Environmentally relevant concentrations of microplastic particles influence larval fish ecology. Science 352:1213–1216. doi:10.1126/science.aad8828

82. de Sá LC, Luís LG, Guilhermino L (2015) Effects of microplastics on juveniles of the common goby (*Pomatoschistus microps*): confusion with prey, reduction of the predatory performance and efficiency, and possible influence of developmental conditions. Environ Pollut 196:359–362. doi:10.1016/j.envpol.2014.10.026

83. Rehse S, Kloas W, Zarfl C (2016) Short-term exposure with high concentrations of pristine microplastic particles leads to immobilisation of *Daphnia magna*. Chemosphere 153:91–99. doi:10.1016/j.chemosphere.2016.02.133

84. Mazurais D, Ernande B, Quazuguel P, Severe A, Huelvan C, Madec L, Mouchel O, Soudant P, Robbens J, Huvet A, Zambonino-Infante J (2015) Evaluation of the impact of polyethylene microbeads ingestion in European sea bass (*Dicentrarchus labrax*) larvae. Mar Environ res 112:78–85. doi:10.1016/j.marenvres.2015.09.009

85. Nel A, Xia T, Mädler L, Li N (2006) Toxic potential of materials at the nanolevel. Science 311:622–627. doi:10.1126/science.1114397

86. Wiesner MR, Lowry GV, Casman E, Bertsch PM, Matson CW, Di Giulio RT, Liu J, Hochella MF (2011) Meditations on the ubiquity and mutability of nano-sized materials in the environment. ACS Nano 5:8466–8470. doi:10.1021/nn204118p

87. Gophen M, Geller W (1984) Filter mesh size and food particle uptake by Daphnia. Oecologia 64:408–412. doi:10.1007/BF00379140

88. Geller W, Müller H (1981) The filtration apparatus of Cladocera: filter mesh-sizes and their implications on food selectivity. Oecologia 49:316–321. doi:10.1007/BF00347591

89. Baun A, Hartmann NB, Grieger K, Kusk KO (2008) Ecotoxicity of engineered nanoparticles to aquatic invertebrates: a brief review and recommendations for future toxicity testing. Ecotoxicology 17:387–395. doi:10.1007/s10646-008-0208-y

90. Hartmann NB (2011) Ecotoxicity of engineered nanoparticles to freshwater organisms. PhD thesis, Technical University of Denmark, Kgs. Lyngby

91. Casado MP, Macken A, Byrne HJ (2013) Ecotoxicological assessment of silica and poly-styrene nanoparticles assessed by a multitrophic test battery. Environ Int 51:97–105. doi:10.1016/j.envint.2012.11.001

92. Petersen EJ, Henry TB, Zhao J, MacCuspie RI, Kirschling TL, Dobrovolskaia MA, Hackley V, Xing B, White JC (2014) Identification and avoidance of potential artifacts and misinterpretations in nanomaterial ecotoxicity measurements. Environ Sci Technol 48: 4226–4246. doi:10.1021/es4052999

93. Hartmann NB, Jensen KA, Baun A, Rasmussen K, Rauscher H, Tantra R, Cupi D, Gilliland D, Pianella F, Riego Sintes JM (2015) Techniques and protocols for dispersing nanoparticle powders in aqueous media—is there a rationale for harmonization? J Toxicol Environ Heal Part B 18:299–326. doi:10.1080/10937404.2015.1074969

94. Rasmussen K, González M, Kearns P, Sintes JR, Rossi F, Sayre P (2016) Review of achievements of the OECD working party on manufactured nanomaterials' testing and assessment programme. From exploratory testing to test guidelines. Regul Toxicol Pharma-col 74:147–160. doi:10.1016/j.yrtph.2015.11.004

95. Lin D, Tian X, Wu F, Xing B (2010) Fate and transport of engineered nanomaterials in the environment. J Environ Qual 39:1896. doi:10.2134/jeq2009.0423

96. Turco R, Bischoff M, Tong Z, Nies L (2011) Environmental implications of nanomaterials: are we studying the right thing? Curr Opin Biotechnol 22:527–532. doi:10.1016/j.copbio.2011.05.006

97. Lowry GV, Gregory KB, Apte SC, Lead JR (2012) Transformations of nanomaterials in the environment. Environ Sci Technol 46:6893–6899. doi:10.1021/es300839e

98. Brandon J, Goldstein M, Ohman MD (2016) Long-term aging and degradation of micro-plastic particles: comparing in situ oceanic and experimental weathering patterns. Mar Pollut Bull 110:299–308. doi:10.1016/j.marpolbul.2016.06.048

99. ter Halle A, Ladirat L, Gendre X, Goudouneche D, Pusineri C, Routaboul C, Tenailleau C, Duployer B, Perez E (2016) Understanding the fragmentation pattern of marine plastic debris. Environ Sci Technol 50:5668–5675. doi:10.1021/acs.est.6b00594

100. Sørensen SN, Baun A (2015) Controlling silver nanoparticle exposure in algal toxicity testing – a matter of timing. Nanotoxicology 9:201–209. doi:10.3109/17435390.2014.913728

101. Kühnel D, Nickel C (2014) The OECD expert meeting on ecotoxicology and environmental fate – towards the development of improved OECD guidelines for the testing of nano-materials. Sci Total Environ 472:347–353. doi:10.1016/j.scitotenv.2013.11.055

102. Sijm DTHM, Rikken MGJ, Rorije E, Traas TP, Mclachlan MS, Peijnenburg WJGM (2007) Transport, accumulation and transformation processes. In: van Leeuwen CJ, Vermeire TG (eds) Risk assessment of chemicals: an introduction. Springer, Dordrecht, pp 73–158. doi:10.1007/978-1-4020-6102-8_3

103. Baalousha M, Sikder M, Prasad A, Lead J, Merrifield R, Chandler GT (2016) The concentration-dependent behaviour of nanoparticles. Environ Chem 13(1). doi:10.1071/EN15142

104. Phuong NN, Zalouk-Vergnoux A, Poirier L, Kamari A, Châtel A, Mouneyrac C, Lagarde F (2016) Is there any consistency between the microplastics found in the field and those used in laboratory experiments? Environ Pollut 211:111–123. doi:10.1016/j.envpol.2015.12.035

105. Wickson F, Hartmann NB, Hjorth R, Hansen SF, Wynne B, Baun A (2014) Balancing scien-tific tensions. Nat Nanotechnol 9:870–870. doi:10.1038/nnano.2014.237

106. The Organisation for Economic Co-operation and Development (OECD) (1981) Decision of the Council concerning the mutual acceptance of data in the assessment of chemicals. 12 May 1981—C(81)30/FINAL. Amended on 26 November 1997—C(97)186/FINAL., Paris, France
107. Skjolding LM 2015 Bioaccumulation and trophic transfer of engineered nanoparticles in aquatic organisms. PhD thesis, Technical University of Denmark, Kgs. Lyngby

Analysis, Occurrence, and Degradation of Microplastics in the Aqueous Environment

Sascha Klein, Ian K. Dimzon, Jan Eubeler, and Thomas P. Knepper

Abstract Synthetic polymers are one of the most significant pollutants in the aquatic environment. Most research focused on small plastic particles, so-called microplastics (particle size, 1–5,000 μm). Compared to macroplastics, the small size complicates their determination in environmental samples and demands for more sophisticated analytical approaches. The detection methods of microplastics reported in the past are highly diverse. This chapter summarizes different strategies for the sampling of water and sediment and sample treatments, including the separation of plastic particles and removal of natural debris that are necessary prior the identification of microplastics. Moreover, the techniques used for the identification of plastics particles are presented in this chapter.

With the application of the method described in this chapter, microplastics were detected in freshwater systems, such as rivers and lakes worldwide. The abundance of microplastics reported in the studies varied in more than three orders of magnitude.

Furthermore, microplastics are not uniform, as there are many different types of synthetic polymers commercially available. Consequently, a variety of different polymer types is present in the aquatic environment. The knowledge on the type of polymer provides additional information for scientists: the type of polymer dictates its physicochemical properties and the degradation. The environmental degradation of plastics is an important factor for the formation, distribution, and accumulation of microplastics in the aquatic system. Thus, this chapter also summarizes the degradation pathways for synthetic polymers in the environment.

S. Klein, I.K. Dimzon, and T.P. Knepper (✉)
Hochschule Fresenius University of Applied Sciences, 65203 Idstein, Germany
e-mail: knepper@hs-fresenius.de

J. Eubeler
Flow Test GmbH, Katlenburg-Lindau, Germany

M. Wagner, S. Lambert (eds.), *Freshwater Microplastics*,
Hdb Env Chem 58, DOI 10.1007/978-3-319-61615-5_3,
© The Author(s) 2018

Keywords Analysis,   Degradation,   Freshwater,   Methods,   Microplastics,
Occurrence

1   Analysis of Microplastics: Sampling, Sample Preparation, and Identification

The investigation of small synthetic polymer particles (size<5 mm), so-called microplastics, strongly depends on appropriate analytical methods. These particles are present in the aquatic environment due to mechanical degradation of macroplastics (size>5 mm) or the introduction of man-made microparticles. The analysis of microplastics is a new challenge for analytical scientists. The small size of microplastics complicates their determination in environmental samples compared to macroplastics and demands for more sophisticated analytical approaches. Microplastics are heterogeneously distributed in the environment, and this impedes the representative sampling of sediments and water. The sample matrix, independent of the sampled environmental compartment, contains a high burden of particles of natural origin that strongly interfere with the visual detection of microplastics. Therefore, suitable methods for the sample preparation are needed to extract microplastics and reduce the number of natural particles. Moreover, an analytical method for the identification and confirmation of the plastic particles is mandatory to obtain reliable results. A wide range of different sampling methods, sample treatments, and detection methods were described (Fig. 1).

1.1   Sampling of Microplastics

The sampling of microplastics in the aquatic environment strongly depends on the compartment that is the subject of interest. In general, this can be differentiated between sampling of the aqueous phase (surface water, water column) and the sediment phase (shoreline sediments, riverbed, or lakebed sediments).

1.1.1   Sampling of the Aqueous Phase

The concentrations of microplastics in aqueous samples are relatively low compared to those in the sediments. Therefore, a large volume of the water samples (up to hundreds of liter) is usually filtered during the sampling process to obtain a representative sample. Sampling of the water surface is carried out in most cases with neuston or plankton nets supported by a flow meter to determine the accurate sample volume. These nets are used in different mesh sizes ranging from 50 to 3,000 μm, while 300 μm is the most commonly used mesh size along all studies [1]. This approach leads to nonquantitative sampling of microplastics with particle sizes <300 μm. The nets with smaller mesh sizes are prone to clogging. To overcome this problem, new methods are

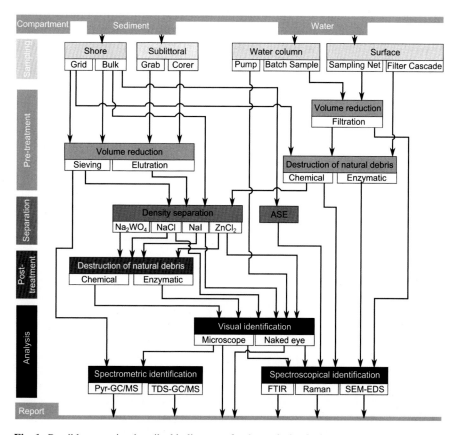

Fig. 1 Possible strategies described in literature for the analysis of microplastics in sediment and water samples starting with the sampling to the report of the results. The sample preparation is split in pretreatment, the density separation, and the posttreatment of the separated microplastics. Fourier transform infrared spectroscopy (FTIR), scanning electron microscopy energy-dispersive X-ray spectroscopy (SEM-EDS), pyrolysis- or thermal desorption-gas chromatography/mass spectrometry (Pyr-GC/MS, TDS-GC/MS) are deployed for the analysis

being developed using filter cascades that result in a size fractionation during the sampling and the reduction of the matrix burden of the small mesh sizes [2].

Less frequently, a sample of the aqueous phase is taken below the water surface. Sampling of the water column is carried out by direct filtration of the water with submersible pumps or is reported by the acquisition of batch samples [3, 4].

1.1.2 Sampling of Sediments

There is no commonly accepted sampling strategy so far for sediment samples. First, the sediments samples must be divided into samples from the shoreline and the river- or lakebed. The collection of bed sediments by sediment grabs provides relatively comparable results due to the standardized sampling instrument [5]. The applications of corers allows the determination of microplastic depth profiles but results in small sample volumes (25 cm^3) and is so far only reported for the marine environment [6]. The differences between microplastic studies for the sampling of shore sediments start with the selection of the sampled area. Shore sediments are collected parallel, perpendicular, or randomly selected in different distances to the shoreline. The majority of the studies reports the collection of grid samples with sampling depths of 2–5 cm of the upper sediment layer [7, 8]. Other studies state the sampling in relation to the lowest flotsam line of the waterbody [9, 10]. Sample collection is usually carried out with stainless steel spoons, trowels, or shovels [10, 11]. In addition, the sampling procedure used will affect how the corresponding results are reported. For example, studies that use grid samples usually report the results per surface sampled (e.g., m^2), whereas studies based on aerial bulk samples give the results referred to the volume or mass of the collected sample (e.g., m^3 or kg).

During the sampling and the sample preparation, it is important to avoid contact with plastic equipment to keep the contamination by the method low. If plastic vessels are included for transportation, blank samples must be also analyzed to quantify their contribution to the microplastic load of the sample [8, 10]. In general, blank samples need to be regarded in microplastic studies to estimate the limit of quantification of each method used, as the limit of quantification (LOQ) is mainly affected by the background contamination [12]. Especially studies dealing with fibers often neglected the analysis of blank samples; thus, the results obtained might be of limited validity. Moreover, the entire method starting from the sample preparation to the analytical detection must be critically evaluated. Therefore, a proper validation must be performed, which also allows a good comparability between different studies. This includes, for example, the determination of within-site variabilities for the sampling process or the determination of recovery rates for the separation methods used during the sample preparation [10, 13, 14].

1.2 Sample Preparation

Even large microplastics like plastic pellets, especially aged and fouled ones, are difficult to distinguish from natural matter in surface water samples with the naked eye. Various methods were developed that allow the mechanical separation of microplastics from the sediment and the removal or reduction of natural debris in the sample prior to analysis of the separated particles. A variety of techniques have been used during the sample treatment and the microplastic identification. Because not all studies conducted extensive method validation including the determination of recovery of the microplastic particles or did not provide experiments with blank samples, the resulting data can lack comparability.

1.2.1 Separation of Microplastics from Sediment Samples

In contrast to microplastics in water, which are easily filtered from the sample during the sampling process, microplastics in sediment samples must be separated in the first step of the sample preparation. A commonly used technique for the separation of plastic particles from sediment particles is the density separation. In a solution of high density, the microplastic particles float, while the very dense sediment particles settle. Numerous different techniques are described in literature, many of them based on the separation introduced by Thompson et al. [15]. Alterations to this method include the use of different salts to create the dense liquid used for separation and the development of different instrumental setups and different pretreatment and posttreatment steps of the samples (compare Fig. 1).

In addition to sodium chloride, which was used by Thompson et al. [15] and others, the application of sodium iodide and zinc chloride has also been reported [16–18]. Sodium iodide, sodium tungstate, and zinc chloride offer the possibility to produce solutions with higher densities than sodium chloride. As the density of a saturated sodium chloride solution ($\rho \approx 1.2 \, \text{g cm}^{-3}$) is rather limited and does not offer consistent separation of higher density polymers such as polyoxymethylene, polyvinyl chloride (PVC), and polyethylene terephthalate (PET), sodium iodide, sodium polytungstate ($\rho \approx 1.6 \, \text{g cm}^{-3}$), and zinc chloride are viable choices. Density separations in the microplastic research rarely use sodium polytungstate despite the possibility of solutions with high density (ρ up to $1.6 \, \text{g cm}^{-3}$), as it is too expensive for the application in large volume samples [11]. Sodium iodide ($\rho \approx 1.6–1.8 \, \text{g cm}^{-3}$) is usually combined with a pre-separation, based on elutriation that separates less dense particles from heavier particles in an upward directed stream of gas or water. This procedure is necessary to minimize the volume needed for the density separation due to the high costs of sodium iodide [14, 16]. The application of zinc chloride enables solutions with densities of $\rho > 1.6–1.7 \, \text{g cm}^{-3}$ and is suitable for the separation of most polymer types. Due to the lower costs compared to sodium tungstate and sodium iodide, zinc chloride is frequently reported in recent studies [8]. However, the ecological hazards of zinc chloride complicate the disposal of used solutions and contaminated sediments. Thus, the recycling of solutions containing zinc chloride, sodium iodide, or sodium polytungstate offers a possibility to overcome the waste management problem and reduce the material costs. To improve the effectivity, the repeatability, and the ease of handling for the density separation method, different setups were developed. The initial use of beakers or Erlenmeyer flask was substituted by the use of separation funnels, vacuum-enhanced separation of the plastic particles, or stainless steel separators with high sample volume capacity [9, 10, 13].

Recent developments focus on alternatives to density separation techniques. Elutriation seems to be a suitable and cost-effective alternative even without following density separation, yielding in good recoveries for polymers with densities of up to $\rho = 1.4$, and the versatility of this method might be improved with a pre-size fractionation of the sample [19].

A different approach includes accelerated solvent extraction (ASE) for the separation of plastics from soils. The extraction by ASE is carried out under higher pressure to increase the boiling point of the extraction solvent, which increases the extraction speed. The process usually uses metal cells of small volume that can resist the pressure. This method bypasses the need for further sample purification and benefits of a high degree of automatization and allows for a quantitative extraction of small plastic particles. However, the identification of extracts consisting of multiple polymer types is complicated, and the size of the extracted sediment sample is limited due to the small size of the extraction cell of the instrument [20].

1.2.2 Removal of Natural Debris

The identification of microplastic particles is often prevented by natural debris that is present in the sample and accompanies the microplastics during the sampling of water samples or the density separation. Thus, the destruction of natural debris or biological material is unavoidable to minimize the possibility of misidentification or underestimation of small plastic particles. The destruction of natural material can be carried out by chemical or enzymatically catalyzed reactions. Chemical destruction of natural debris is achieved through the treatment of the sample with hydrogen peroxide, mixtures of hydrogen peroxide and sulfuric acid, and Fenton-like reactions prior or after the density separation [8, 18, 21]. These harsh conditions might result in losses of plastics that are labile to oxidation or unstable in strong acidic solutions, such as poly(methyl methacrylate) or polycarbonates.

To avoid the loss of synthetic polymers, which are not resistant against acidic treatments, usage of sodium hydroxide was proposed. However, Cole et al. report that the alkaline treatment with sodium hydroxide could damage some of the synthetic polymers as well [22]. Dehaut et al. showed that the application of potassium hydroxide is preferable for the destruction of organic material, as it seems to attack the synthetic polymers less than the abovementioned methods [23].

Enzymatic treatments were developed for biota-rich marine surface water samples, which allow the detection of pH-sensitive polymers [22]. Single-enzyme approaches using proteinase K or mixtures of technical enzymes (lipase amylase, proteinase, chitinase, cellulase) were used for the removal of biological material, as the enzymatic digestion can be carried out under moderate experimental conditions in terms of pH and temperature. Unfortunately, the use of enzymes involves several disadvantages. Enzymatic treatments are, compared to chemical treatments, expensive and very time-consuming and might not result in a complete removal of the natural debris.

1.3 Identification of Microplastics

In most studies, microplastics are first identified visually, before an identification of the polymer type is undertaken. Larger particles can be identified with the naked eye, whereas small microplastics are identified using binocular microscopes or scanning electron microscopy (SEM) [6, 24, 25]. Early studies determined microplastic concentrations after visual inspection of the sample only. Depending on the efficiency of the sample treatment and particle size, the visual identification is considered not state of the art and often insufficient resulting in false-positive results. For this reason, further spectroscopic or spectrometric methods are needed to ensure the unambiguous identification of particles made from synthetic polymers.

Spectroscopic identification methods include Fourier transform infrared (FTIR) spectroscopy and Raman spectroscopy. These methods are based on the energy absorption by characteristic functional groups of the polymer particles. For larger particles (approximately >500 μm), FTIR can be carried out using an attenuated transverse reflection (ATR) unit as the particles need to be transferred on the crystal of the ATR unit manually [9, 26]. Coupling of FTIR instruments to microscopes such as reflectance or transmission micro-FTIR allows the detection of smaller microplastics [27]. The use of FTIR microscopy in transmission mode is only applicable for smaller particles or thin films that do not fully absorb the IR beam. Moreover, special filters are required in the sample treatment that are translucent to IR radiation, such as aluminum oxide membranes. Both FTIR-based and Raman-based methods are limited in the minimum particle size that can be determined by the physical diffraction of the light. FTIR measurements in transmittance mode are limited for particles between 10 and 20 μm, while Raman instruments can measure particle with sizes that are one to two orders of magnitude smaller, due to the smaller wavelengths that are applied for the excitation. Identification of the polymers by FTIR and Raman is susceptible to environmentally driven changes of the polymer surface or the additive application during polymer processing. Thus, microbial fouling, soiling, adsorption of humic acids, and colored plastics can interfere with the absorbance, reflection, or excitation of the polymer molecules and might lead to misidentification or totally prevent identification of the particles [28] (for an in-depth discussion on microplastic associated biofilms, see [29]). Besides the identification of the polymer type, visual images of particles enable the determination of particle shape.

The application of pyrolysis-gas chromatography/mass spectrometry (Pyr-GC/MS) allows the simultaneous determination of the polymer type and polymer additives by combustion of the sample and the detection of the thermal degradation products of the polymers [16, 30]. The identification of thermal degradation products serves as a marker that is specific for each polymer. The degradation products are separated by GC prior the detection of their specific mass to charge ratios in the mass spectrometer. In contrast to the spectroscopic techniques, Pyr-GC/MS is a destructive method, preventing any further analysis of the plastic particles. Results obtained through Pyr-GC/MS analysis are usually provided as the mass fraction or mass concentration of plastics. Therefore, the determination of particle counts is not possible due to the combustion of the sample. Thermal desorption GC/MS (TDS-GC/MS) in combination with thermogravimetric analysis (TGA) coupled with a solid-phase adsorber enables higher initial sample sizes compared to Pyr-GC/MS [31]. For this reason, more representative results might be obtained for inhomogeneous samples with complex matrices.

SEM can be coupled with energy-dispersive X-ray spectroscopy (SEM-EDS), which produces high-resolution images of the particles and provides an elemental analysis of the measured objects. For SEM-EDS, the particle surface of the sample is scanned by an electron beam. The contact of the electron beam with the sample surface results in the emission of secondary electrons and element-specific X-ray radiation. Thus, an image of the particle can be created and the elemental composition can be identified by using SEM-EDS. It is, therefore, possible to distinguish between microplastics and particles that are composed of inorganic elements, such as aluminum silicates [32].

Alternatively, hardness tests are reported as inspection of the separated particles. Pressure is applied to the particles by needles or tweezers. This precludes misidentifications of microplastics with fragile carbon or carbonate particles that break during the test and are not removed or formed during the sample treatment [33]. However, these tests are very time-consuming, do not provide exact polymer identification, and are less accurate as other instrumental methods.

More specialized but promising approaches for the detection of microplastics are described by Sgier et al. and Jungnickel et al. [34, 35]. The latter describe the measurement and identification of microplastics by time-of-flight secondary ion mass spectrometry. An imaging technique allows the visualization of the particles, and the ionization of the polymer molecules is carried out by a primary ion source, generating secondary ions of polymer fragments. As with Raman microscopy, this technique enables the identification of particles smaller than 10 μm. Sgier et al. detected microplastics using flow cytometry combined with visual stochastic network embedding (viSNE). viSNE is a tool for the visualization of high-dimensional cytometry data by nonlinear dimension reduction onto two dimensions [36]. This method was capable of detecting microplastic particles directly in environmental samples by using nonbiological reference data sets for the interpretation of the viSNE analysis, although the reliability of the microplastic identification needs to be proven in future studies.

2 Occurrence in the Aquatic Environment

Microplastic particles are present in surface water, sediments, and oceans all over the world, for example, at the Italian, Singapore, and Portuguese coast, at beaches of Hawaii, and islands of the equatorial Western Atlantic as well as at shores of German and Greek islands [3, 11, 17, 37–40]. First reports of smaller plastic items were primarily focused on plastic pellets that are used in the production of bulk plastic items. Plastic pellets have been quantified on numerous beaches and coastlines, for instance, in New Zealand, Lebanon, and Spain [41–43]. However, industrial plastic pellets only compose a small fraction of the numerous microscopic plastic fragments present in the ocean and other aquatic systems [15]. Monitoring studies often subdivide microplastics into categories of spheres, fibers, foams, and fragments and report range in concentrations by up to four orders of magnitude, spanning 1.3 particles kg^{-1} (German island) over 13.5 particles kg^{-1} (equatorial Western Atlantic) to 2175 particles kg^{-1} (Italy). All these studies were carried out in the marine environment, and freshwater systems have attracted less attention until 2010.

In freshwater environments of lakes and rivers, studies also report highly hetero-geneous concentrations comparable to those reported for the marine environment. High within-site variabilities, as well as different units used for microplastic quantitation, complicate the comparison of microplastic concentrations in aquatic systems. Microplastics in riverine systems were reported for large European rivers, e.g., the river Rhine and the river Danube, as well as for tributaries such as the river Main. Plastic particles in the river Danube were determined with high abundance and even exceeded the number of fish larvae. Lechner et al. stated that the river might transport high loads of plastic particles into the Black Sea [44]. Studies investigating the river Rhine showed high abundances of microplastics, especially in the German section of the river. Average concentrations amounted to approx. 900,000 particles km^{-2} for surface water and up to 4000 particles kg^{-1} for shore sediments. In the Swiss part of the rivers Rhône, Aubonne, Venoge, and Vuachière, microplastics were detected in concentrations between 0.10 and 64 particles m^3 (mean, 7 particles m^{-3}; median, 0.36 particles m^{-3}) [45]. Distinctly higher concentrations were detected in Chinese river estuaries. The estuaries of the rivers Jiaojiang, Oujiang, and Minjiang, which all are located in an urban region, contained microplastics in the range of 100–4100 particles m^{-3} [46].

A study conducted with lakeshore sediments collected from Lake Garda (Italy) showed high abundances of polyethylene (PE) and polystyrene (PS) microplastics, indicating the importance of buoyant microplastics for shore sediments. In addition, polymer particles of a higher density, such as PVC and PET, were also identified in this study underlining the variety of microplastics present in shore sediments [8]. Microplastics were detected in concentrations between 108 and 1,108 particles m^{-2} with notable spatial variation between the south and the north shore of the lake. Faure et al. reported microplastics in the Swiss parts of Lake Geneva, Lake Constance, Lake Neuchâtel, Lake Maggiore, Lake Zurich, and Lake Brienz. The concentrations of microplastics in lakeshore sediments varied between 20 and 7,200 particles m^{-2} (mean, 1,300 particles m^{-2}; median, 270 particles m^{-2}) and are comparable with concentrations reported for Lake Garda. Between 11,000 and 220,000 particles km^{-2} of microplastics were detected in the surface water of the six lakes mentioned. Microplastics were also detected in the Laurentian Great Lakes in North America with concentrations ranging between 0 and 466,305 particles m^{-2} (mean, 42,533 particles m^{-2}; median, 5,704 particles m^{-2}). Higher abundances of microplastics have been detected in proximity to urban areas [32]. The determination of microplastics in gastrointestinal tract of fish from Lake Victoria (Tanzania) shows first evidence for the microplastic pollution of African lakes [47] (microplastic uptake biological interactions are discussed in Scherer et al. [48] of this volume). A monitoring study of Lake Taihu in China detected up to 6,000,000 particles km^{-2} in surface water samples. In bulk water samples, microplastics were present in concentrations between 3 and 26 particles L^{-1}, while the sediments of Lake Taihu contained microplastics in the range of 11–235 particles kg^{-1}. The reported concentrations of microplastics in the sediments of Lake Taihu are lower compared to those detected in the sediments of the European lakes. However, it needs to be taken into account that lake-bottom sediments were sampled in Lake Taihu, whereas shore sediments were studied in the European lakes [49]. Lower concentrations of plastic particles in Asian freshwaters were detected in the surface water samples of Lake Hovsgol (Mongolia). Microplastics were quantified in the range of 997–4435 particles km^{-2}, but it should be taken into account that the catchment area of Lake Hovsgol is less populated compared to the abovementioned lakes [50]. For case study discussions on microplastic occurrence in African and Asian freshwaters, see Kahn et al. [51] and Chenxi et al. [52] of this volume.

3   Environmental Degradation of Synthetic Polymers

One of the reasons for the great versatility of many synthetic polymers is their high resistance against environmental influences. However, this fact leads to extremely low degradation and long residence times for synthetic polymers once they enter the environment. Degradation of synthetic polymers can generally be classified as biotic or abiotic, following different mechanisms, depending on a variety of physical, chemical, or biological factors. During the degradation process, polymers are converted into smaller molecular units (e.g., oligomers, monomers, or chemically modified versions) and possibly are completely mineralized [53]. The most important processes for the degradation of synthetic polymers can be divided into (Fig. 2):

- Physical degradation (abrasive forces, heating/cooling, freezing/thawing, wetting/drying)
- Photodegradation (usually by UV light)
- Chemical degradation (oxidation or hydrolysis)
- Biodegradation by organisms (bacteria, fungi, algae)

Mechanical degradation is an important factor with regard to plastics in the aquatic environment. In most cases, aging of the polymer by environmental influences, such as photodegradation or chemical degradation of additives, changes the polymer properties and leads to embrittlement of the polymer [54]. The recalcitrant material is then shredded into smaller particles by friction forces occurring during the movement through different environmental habitats (also see Kooi et al. [55] of this volume for a discussion on microplastics fate and transportation). This degradation generally leads to smaller plastic particles, which can result in particles with sizes between 1 and 5,000 µm. Such particles are classified as microplastics. However, the mechanical degradation does not stop if the particles are within the size range of microplastics. Thus, the formation of even smaller particles, so-called nanoplastics, is very likely [56]. These nanoplastics could have different properties compared to the original macroplastics or microplastics (for a discussion on nanomaterials, see Rist and Hartmann [57] of this volume). In both cases, the mechanical degradation leads to a decrease in particle size and consequently to an increase in the surface area of the polymer particles, which results in faster degradation due to higher reactivity.

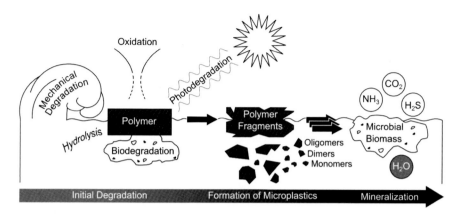

Fig. 2 Degradation pathways of synthetic polymers in the aquatic environment with degradation processes involved and intermediate steps until complete mineralization

Under normal environmental conditions in aquatic systems, the temperature is not high enough to start chemical changes of synthetic polymers; thus, thermal degradation is not significant for freshwaters [58, 59].

The degradation of synthetic polymers in the environment on a molecular basis is usually initiated by photooxidation (with UV radiation) or by hydrolysis and is eventually followed by chemical oxidation [60]. The predominant mechanisms strongly depend on the type of polymer, as there are numerous different compositions of synthetic polymers produced (i.e., polyolefins, polyesters, polyamides). After the initial reactions, the molecular weight of the polymer is decreased, and the reacted groups become available for microbial degradation. Photooxidation is usually a fast process, but the degradation rate also depends on the extent of additives in a particular polymer that could prevent oxidation processes (i.e., antioxidants). Moreover, the photodegradation of plastics floating in the aquatic environment is slower compared to degradation in terrestrial exposure [61]. Experiments on the disintegration of PE and PS showed faster degradation on the water surface compared to plastics that partially or completely submerged, likely related to the decreasing intensity of light and thus to the lower rate of photooxidation [62]. For this reason, many plastics can stay in the aquatic environment for decades or hundreds of years.

Biodegradation of synthetic polymers can occur in two different environments (aerobic and anaerobic). The extent of the degradation of polymers into CO_2, H_2O, N_2, H_2, CH_4, salts, minerals, and biomass (mineralization) can be full or partial [63]. Partial or primary degradation of the polymer chain leads to stable or temporarily stable transformation products. Biodegradation is coupled to three essential criteria:

1. Microorganisms must be present that can depolymerize the target substance and mineralize the monomeric compounds with enzymes of an appropriate metabolic pathway.
2. The environmental parameters, such as temperature, pH, moisture, and salinity must provide conditions that are necessary for biodegradation.

3. The morphology of polymer particles must render the attachment of microorganisms and the formation of a biofilm, while the structure of the polymeric substrate, e.g., chemical bonds, degree of polymerization, degree of branching, and parameter, such as hydrophobicity or crystallinity, must not hinder microbial actions.

Since the size of synthetic polymers is generally too large to penetrate the cell membranes of microorganisms, the first step of biotic degradation is the cleavage of side chains or the polymer backbone and the formation of smaller polymer units (monomers, oligomers) by extracellular enzymes [64]. In most cases, this first step of depolymerization involves an enzymatically catalyzed hydrolysis of amides, esters, or urethane bonds. These smaller molecules can then be absorbed by microorganisms and metabolized. Of course, abiotic hydrolysis can also result in intermediates that are then further metabolized by microorganisms [65].

The complete biotic degradations of poly(ε-caprolactam) and water-soluble polyethylene glycol are well described in literature [66]. However, most of the plastics occurring in the environment are water insoluble, and many of the synthetic polymers present in the aquatic environment, such as PE, polypropylene (PP), PS, and PET, degrade very slowly or not at all. The degradation of these polymers is usually a combination of abiotic and biotic degradation pathways.

Polyolefins, such as PE and PP, represent a class of substances with high industrial production volumes and are determined frequently in environmental samples. These polymers are usually not biodegradable, as the alkyl backbone is not accessible for microorganism and must undergo an abiotic transformation. The alkyl backbone of polyolefins offers a high resistivity against hydrolysis but is usually susceptible to oxidative degradation. To prevent this, additives are added during the production process, and the oxidative or photooxidative degradation of the polymer is delayed until the antioxidants are consumed. After the initial oxidation of the surface of polyolefins, the degradation could occur in several weeks but results in the formation of microplastics as possible intermediates [67]. These smaller and oxidized plastic fragments are more susceptible to microbial attack, e.g., biodegradation of PE is described for pre-oxidized fragments of the original material by *Pseudomonas* sp. [68].

4 Conclusion

The results of studies worldwide highlight the great importance of microplastics for freshwater ecosystems, as they are present in high abundance. Microplastics are emerging contaminants in the aquatic environment, and attention should be focused on a harmonized nomenclature of microplastic particles with official guidelines for microplastic studies. The definition of microplastics often remains vague, and different size classes are investigated in monitoring studies. For a thorough investigation of microplastic pollution, standardized methods, especially for the sampling of the different compartments, are a key factor for meaningful results. The quantitation of microplastics differs due to the individual sampling method and studies still lacking in comparability. Therefore, future studies should integrate additional units to describe microplastic occurrence in their investigations. Moreover, an integral step of future investigations should be a sufficient validation of the microplastic analysis, as different technological approaches might be used or future technological improvements will be implemented for the determination of microplastics. Important approaches, such as Raman-, FTIR-, and GC/MS-based techniques, seem to be very versatile for the detection of microplastics, but a better comparability of results obtained with either of these methods needs to be established. Degradation of larger plastics will result in a constant source for new microplastic particles of all size classes, and thus, as it is a slow process, microplastics will be a relevant long-term threat.

Acknowledgments We would like to thank the German Federal Ministry for Education and Research (BMBF) in the frame of the project microplastics in the water cycle (MiWa; FKZ: 02WRS1378D).

References

1. Hidalgo-Ruz V, Gutow L, Thompson RC, Thiel M (2012) Microplastics in the marine environment: a review of the methods used for identification and quantification. Environ Sci Technol 46(6):3060–3075. doi:10.1021/es2031505
2. Löder MGJ, Gerdts G (2015) Methodology used for the detection and identification of microplastics—a critical appraisal. In: Bergmann M, Gutow L, Klages M (eds) Marine anthropogenic litter. Springer International Publishing, Cham, pp 201–227. doi:10.1007/978-3-319-16510-3_8
3. Ng KL, Obbard JP (2006) Prevalence of microplastics in Singapore's coastal marine environment. Mar Pollut Bull 52(7):761–767. doi:10.1016/j.marpolbul.2005.11.017
4. Norén F, Naustvoll L-J (2010) Survey of microscopic anthropogenic particles in Skagerrak. Commissioned by KLIMA – Og Forurensningsdirektoratet
5. Castañeda RA, Avlijas S, Simard MA, Ricciardi A (2014) Microplastic pollution in St. Lawrence River sediments. Can J Fish Aquat Sci 71(12):1767–1771. doi:10.1139/cjfas-2014-0281
6. Van Cauwenberghe L, Vanreusel A, Mees J, Janssen CR (2013) Microplastic pollution in deep-sea sediments. Environ Pollut 182:495–499. doi:10.1016/j.envpol.2013.08.013
7. Zbyszewski M, Corcoran PL, Hockin A (2014) Comparison of the distribution and degradation of plastic debris along shorelines of the Great Lakes, North America. J Great Lakes Res 40(2):288–299. doi:10.1016/j.jglr.2014.02.012
8. Imhof HK, Ivleva NP, Schmid J, Niessner R, Laforsch C (2013) Contamination of beach sediments of a subalpine lake with microplastic particles. Curr Biol 23(19):R867–R868. doi:10.1016/j.cub.2013.09.001
9. Browne MA, Galloway TS, Thompson RC (2010) Spatial patterns of plastic debris along Estuarine shorelines. Environ Sci Technol 44(9):3404–3409. doi:10.1021/es903784e
10. Klein S, Worch E, Knepper TP (2015) Occurrence and spatial distribution of microplastics in river shore sediments of the Rhine-Main area in Germany. Environ Sci Technol 49(10):6070–6076. doi:10.1021/acs.est.5b00492

11. Corcoran PL, Biesinger MC, Grifi M (2009) Plastics and beaches: a degrading relationship. Mar Pollut Bull 58(1):80–84. doi:10.1016/j.marpolbul.2008.08.022

12. Leslie H, Van Velzen M, Vethaak A (2013) Microplastic survey of the Dutch environment. Novel data set of microplastics in North Sea sediments, treated wastewater effluents and marine biota Amsterdam. Institute for Environmental Studies, VU University, Amsterdam

13. Imhof HK, Schmid J, Niessner R, Ivleva NP, Laforsch C (2012) A novel, highly efficient method for the separation and quantification of plastic particles in sediments of aquatic environments. Limnol Oceanogr Methods 10:524–537. doi:10.4319/lom.2012.10.524

14. Claessens M, Van Cauwenberghe L, Vandegehuchte MB, Janssen CR (2013) New techniques for the detection of microplastics in sediments and field collected organisms. Mar Pollut Bull 70(1–2):227–233. doi:10.1016/j.marpolbul.2013.03.009

15. Thompson RC, Olsen Y, Mitchell RP, Davis A, Rowland SJ, John AWG, McGonigle D, Russell AE (2004) Lost at sea: where is all the plastic? Science 304(5672):838–838. doi:10.1126/science.1094559

16. Nuelle MT, Dekiff JH, Remy D, Fries E (2014) A new analytical approach for monitoring microplastics in marine sediments. Environ Pollut 184:161–169. doi:10.1016/j.envpol.2013.07.027

17. Vianello A, Boldrin A, Guerriero P, Moschino V, Rella R, Sturaro A, Da Ros L (2013) Microplastic particles in sediments of Lagoon of Venice, Italy: first observations on occurrence, spatial patterns and identification. Estuar Coast Shelf Sci 130:54–61. doi:10.1016/j.ecss.2013.03.022

18. Liebezeit G, Dubaish F (2012) Microplastics in beaches of the East Frisian islands Spiekeroog and Kachelotplate. Bull Environ Contam Toxicol 89(1):213–217. doi:10.1007/s00128-012-0642-7

19. Kedzierski M, Le Tilly V, Bourseau P, Bellegou H, Cesar G, Sire O, Bruzaud S (2016) Microplastics elutriation from sandy sediments: a granulometric approach. Mar Pollut Bull 107(1):315–323. doi:10.1016/j.marpolbul.2016.03.041

20. Fuller S, Gautam A (2016) A procedure for measuring microplastics using pressurized fluid extraction. Environ Sci Technol 50(11):5774–5780. doi:10.1021/acs.est.6b00816

21. Yonkos LT, Friedel EA, Perez-Reyes AC, Ghosal S, Arthur CD (2014) Microplastics in four estuarine rivers in the Chesapeake Bay, U.S.A. Environ Sci Technol 48(24):14195–14202. doi:10.1021/es5036317

22. Cole M, Webb H, Lindeque PK, Fileman ES, Halsband C, Galloway TS (2014) Isolation of microplastics in biota-rich seawater samples and marine organisms. Sci Rep 4:4528. doi:10.1038/srep04528

23. Dehaut A, Cassone A-L, Frère L, Hermabessiere L, Himber C, Rinnert E, Rivière G, Lambert C, Soudant P, Huvet A, Duflos G, Paul-Pont I (2016) Microplastics in seafood: benchmark protocol for their extraction and characterization. Environ Pollut 215:223–233. doi:10.1016/j.envpol.2016.05.018

24. Karapanagioti HK, Klontza I (2007) Investigating the properties of plastic resin pellets found in the coastal areas of lesvos island. Global Nest J 9(1):71–76

25. Gilfillan LR, Ohman MD, Doyle MJ, Watson W (2009) Occurrence of plastic micro-debris in the Southern California current system. Cal Coop Ocean Fish 50:123–133

26. Doyle MJ, Watson W, Bowlin NM, Sheavly SB (2011) Plastic particles in coastal pelagic ecosystems of the Northeast Pacific ocean. Mar Environ Res 71(1):41–52. doi:10.1016/j.marenvres.2010.10.001

27. Harrison JP, Ojeda JJ, Romero-Gonzalez ME (2012) The applicability of reflectance micro-Fourier-transform infrared spectroscopy for the detection of synthetic microplastics in marine sediments. Sci Total Environ 416:455–463. doi:10.1016/j.scitotenv.2011.11.078

28. Rocha-Santos T, Duarte AC (2015) A critical overview of the analytical approaches to the occurrence, the fate and the behavior of microplastics in the environment. TrAC Trends Anal Chem 65:47–53. doi:10.1016/j.trac.2014.10.011

29. Harrison JP, Hoellein TJ, Sapp M, Tagg AS, Ju-Nam Y, Ojeda JJ (2017) Microplastic-associated biofilms: a comparison of freshwater and marine environments. In: Wagner M,

Lambert S (eds) Freshwater microplastics: emerging environmental contaminants? Springer, Heidelberg. doi:10.1007/978-3-319-61615-5_9 (in this volume)

30. Trimpin S, Wijerathne K, McEwen CN (2009) Rapid methods of polymer and polymer additives identification: multi-sample solvent-free MALDI, pyrolysis at atmospheric pressure, and atmospheric solids analysis probe mass spectrometry. Anal Chim Acta 654(1):20–25. doi:10.1016/j.aca.2009.06.050

31. Dümichen E, Barthel A-K, Braun U, Bannick CG, Brand K, Jekel M, Senz R (2015) Analysis of polyethylene microplastics in environmental samples, using a thermal decomposition method. Water Res 85:451–457. doi:10.1016/j.watres.2015.09.002

32. Eriksen M, Mason S, Wilson S, Box C, Zellers A, Edwards W, Farley H, Amato S (2013) Microplastic pollution in the surface waters of the Laurentian Great Lakes. Mar Pollut Bull 77 (1–2):177–182. doi:10.1016/j.marpolbul.2013.10.007

33. Eriksen M, Lebreton LCM, Carson HS, Thiel M, Moore CJ, Borerro JC, Galgani F, Ryan PG, Reisser J (2014) Plastic pollution in the world's oceans: more than 5 trillion plastic pieces weighing over 250,000 tons afloat at sea. PLoS One 9(12):e111913. doi:10.1371/journal.pone.0111913

34. Sgier L, Freimann R, Zupanic A, Kroll A (2016) Flow cytometry combined with viSNE for the analysis of microbial biofilms and detection of microplastics. Nat Commun 7:11587. doi:10.1038/Ncomms11587

35. Jungnickel H, Pund R, Tentschert J, Reichardt P, Laux P, Harbach H, Luch A (2016) Time-of-flight secondary ion mass spectrometry (ToF-SIMS)-based analysis and imaging of polyethylene microplastics formation during sea surf simulation. Sci Total Environ 563:261–266. doi:10.1016/j.scitotenv.2016.04.025

36. Amir E-AD, Davis KL, Tadmor MD, Simonds EF, Levine JH, Bendall SC, Shenfeld DK, Krishnaswamy S, Nolan GP, Pe'er D (2013) viSNE enables visualization of high dimensional single-cell data and reveals phenotypic heterogeneity of leukemia. Nat Biotechnol 31 (6):545–552. doi:10.1038/nbt.2594

37. Martins J, Sobral P (2011) Plastic marine debris on the Portuguese coastline: a matter of size? Mar Pollut Bull 62(12):2649–2653. doi:10.1016/j.marpolbul.2011.09.028

38. Ivar do Sul JA, Spengler Â, Costa MF (2009) Here, there and everywhere. Small plastic fragments and pellets on beaches of Fernando de Noronha (Equatorial Western Atlantic). Mar Pollut Bull 58(8):1236–1238. doi:10.1016/j.marpolbul.2009.05.004

39. Dekiff JH, Remy D, Klasmeier J, Fries E (2014) Occurrence and spatial distribution of microplastics in sediments from Norderney. Environ Pollut 186:248–256. doi:10.1016/j.envpol.2013.11.019

40. Karapanagioti HK, Endo S, Ogata Y, Takada H (2011) Diffuse pollution by persistent organic pollutants as measured in plastic pellets sampled from various beaches in Greece. Mar Pollut Bull 62(2):312–317. doi:10.1016/j.marpolbul.2010.10.009

41. Gregory MR (1978) Accumulation and distribution of virgin plastic granules on New Zealand beaches. N Z J Mar Freshw Res 12(4):399–414. doi:10.1080/00288330.1978.9515768

42. Shiber JG (1979) Plastic pellets on the coast of Lebanon. Mar Pollut Bull 10(1):28–30. doi:10.1016/0025-326X(79)90321-7

43. Shiber JG (1987) Plastic pellets and tar on Spain's Mediterranean beaches. Mar Pollut Bull 18 (2):84–86. doi:10.1016/0025-326X(87)90573-X

44. Lechner A, Keckeis H, Lumesberger-Loisl F, Zens B, Krusch R, Tritthart M, Glas M, Schludermann E (2014) The Danube so colourful: a potpourri of plastic litter outnumbers fish larvae in Europe's second largest river. Environ Pollut 188:177–181. doi:10.1016/j.envpol.2014.02.006

45. Faure F, Demars C, Wieser O, Kunz M, de Alencastro LF (2015) Plastic pollution in Swiss surface waters: nature and concentrations, interaction with pollutants. Environ Chem 12 (5):582–591. doi:10.1071/EN14218

46. Zhao SY, Zhu LX, Li DJ (2015) Microplastic in three urban estuaries, China. Environ Pollut 206:597–604. doi:10.1016/j.envpol.2015.08.027

47. Biginagwa FJ, Mayoma BS, Shashoua Y, Syberg K, Khan FR (2016) First evidence of microplastics in the African Great Lakes: recovery from Lake Victoria Nile perch and Nile tilapia. J Great Lakes Res 42(1):146–149. doi:10.1016/j.jglr.2015.10.012

48. Scherer C, Weber A, Lambert S, Wagner M (2017) Interactions of microplastics with freshwater biota. In: Wagner M, Lambert S (eds) Freshwater microplastics: emerging environmental contaminants? Springer Nature, Heidelberg. doi:10.1007/978-3-319-61615-5_8 (in this volume)

49. Su L, Xue Y, Li L, Yang D, Kolandhasamy P, Li D, Shi H (2016) Microplastics in Taihu Lake, China. Environ Pollut 216:711–719. doi:10.1016/j.envpol.2016.06.036

50. Free CM, Jensen OP, Mason SA, Eriksen M, Williamson NJ, Boldgiv B (2014) High-levels of microplastic pollution in a large, remote, mountain lake. Mar Pollut Bull 85(1):156–163. doi:10.1016/j.marpolbul.2014.06.001

51. Khan FR, Mayoma BS, Biginagwa FJ, Syberg K (2017) Microplastics in inland African waters: presence, sources and fate. In: Wagner M, Lambert S (eds) Freshwater microplastics: emerging environmental contaminants? Springer, Heidelberg. doi:10.1007/978-3-319-61615-5_6 (in this volume)

52. Chenxi W, Kai Z, Xiong X (2017) Microplastic pollution in inland waters focusing on Asia. In: Wagner M, Lambert S (eds) Freshwater Microplastics: Emerging Environmental Contaminants? Springer, Heidelberg. doi:10.1007/978-3-319-61615-5_5 (in this volume)

53. Eubeler JP, Zok S, Bernhard M, Knepper TP (2009) Environmental biodegradation of synthetic polymers I. Test methodologies and procedures. TrAC Trends Anal Chem 28 (9):1057–1072. doi:10.1016/j.trac.2009.06.007

54. Duwez AS, Nysten B (2001) Mapping aging effects on polymer surfaces: specific detection of additives by chemical force microscopy. Langmuir 17(26):8287–8292. doi:10.1021/la0113623

55. Kooi M, Besseling E, Kroeze C, van Wezel AP, Koelmans AA (2017) Modeling the fate and transport of plastic debris in freshwaters: review and guidance. In: Wagner M, Lambert S (eds) Freshwater microplastics: emerging environmental contaminants? Springer, Heidelberg. doi:10.1007/978-3-319-61615-5_7 (in this volume)

56. Lambert S, Wagner M (2016) Formation of microscopic particles during the degradation of different polymers. Chemosphere 161:510–517. doi:10.1016/j.chemosphere.2016.07.042

57. Rist SE, Hartmann NB (2017) Aquatic ecotoxicity of microplastics and nanoplastics - lessons learned from engineered nanomaterials. In: Wagner M, Lambert S (eds) Freshwater microplastics: emerging environmental contaminants? Springer, Heidelberg. doi:10.1007/978-3-319-61615-5_2 (in this volume)

58. Anderson DA, Freeman ES (1961) The kinetics of the thermal degradation of polystyrene and polyethylene. J Polym Sci 54(159):253–260. doi:10.1002/pol.1961.1205415920

59. McNeill IC, Leiper HA (1985) Degradation studies of some polyesters and polycarbonates—2. Polylactide: degradation under isothermal conditions, thermal degradation mechanism and photolysis of the polymer. Polym Degrad Stab 11(4):309–326. doi:10.1016/0141-3910(85)90035-7

60. Andrady AL (2011) Microplastics in the marine environment. Mar Pollut Bull 62 (8):1596–1605. doi:10.1016/j.marpolbul.2011.05.030

61. Andrady AL, Pegram JE, Song Y (1993) Studies on enhanced degradable plastics. II. Weathering of enhanced photodegradable polyethylenes under marine and freshwater floating exposure. J Environ Polym Degrad 1(2):117–126. doi:10.1007/BF01418205

62. Leonas KK, Gorden RW (1993) An accelerated laboratory study evaluating the disintegration rates of plastic films in simulated aquatic environments. J Environ Polym Degrad 1(1):45–51. doi:10.1007/bf01457652

63. Grima S, Bellon-Maurel V, Feuilloley P, Silvestre F (2000) Aerobic biodegradation of polymers in solid-state conditions: a review of environmental and physicochemical parameter settings in laboratory simulations. J Polym Environ 8(4):183–195. Unsp 1566-2543/00/1000-0183/0. doi:10.1023/A:1015297727244

64. Gu J-G, Gu J-D (2005) Methods currently used in testing microbiological degradation and deterioration of a wide range of polymeric materials with various degree of degradability: a review. J Polym Environ 13(1):65–74. doi:10.1007/s10924-004-1230-7
65. Müller R-J, Kleeberg I, Deckwer W-D (2001) Biodegradation of polyesters containing aromatic constituents. J Biotechnol 86(2):87–95. doi:10.1016/S0168-1656(00)00407-7
66. Eubeler JP, Bernhard M, Knepper TP (2010) Environmental biodegradation of synthetic polymers II. Biodegradation of different polymer groups. Trac, Trends Anal Chem 29 (1):84–100. doi:10.1016/j.trac.2009.09.005
67. Weinstein JE, Crocker BK, Gray AD (2016) From macroplastic to microplastic: degradation of high-density polyethylene, polypropylene, and polystyrene in a salt marsh habitat. Environ Toxicol Chem 35(7):1632–1640. doi:10.1002/etc.3432
68. Reddy MM, Deighton M, Gupta RK, Bhattacharya SN, Parthasarathy R (2009) Biodegradation of oxo-biodegradable polyethylene. J Appl Polym Sci 111(3):1426–1432. doi:10.1002/app.29073

Sources and Fate of Microplastics in Urban Areas: A Focus on Paris Megacity

Rachid Dris, Johnny Gasperi, and Bruno Tassin

Abstract Since the beginning of the 2010s, the number of investigations on microplastics in freshwater increased dramatically. However, almost no study aims at investigating the various sources and fate of microplastics in a catchment. This chapter aims at analyzing the various sources and fate of microplastics for an urban catchment and its hydrosystem (sewage, runoff, etc.). It presents the results obtained during a 3-year study of the Paris Megacity. Such a study required the development of appropriate sampling strategies for each compartment. It was highlighted that fibers are highly concentrated in the studied area, and therefore a focus in this category of microplastics was carried out. The atmospheric fallout exhibited important levels of fibers. However, at the scale of the Parisian agglomeration, wastewater treatment plant disposals and combined sewer overflows represent the major sources (number of fibers introduced per year) among the studied ones.

Keywords Fibers, Freshwater, Microplastics, Plastic pollution, Urban areas, Urban impact

1 Introduction

Although the first scientific articles that identified plastic in the environment as an issue are rather old [1, 2], efforts of the scientific community on this subject actually started at the beginning of the twenty-first century with the 2004 paper of Thompson et al. [3]. Studies highlighted the issue of microplastics (MPs) and raise two main questions: (1) the interaction between plastic items and the

R. Dris (✉), J. Gasperi, and B. Tassin
LEESU, UMR MA 102, École des Ponts, UPEC, AgroParisTech, UPE, Paris, France
e-mail: rachid.dris@enpc.fr

M. Wagner, S. Lambert (eds.), *Freshwater Microplastics*,
Hdb Env Chem 58, DOI 10.1007/978-3-319-61615-5_4,

physical-chemical environment (fragmentation, micro-pollutant exchanges) and (2) their interaction with the biological compartment including ecotoxicological effects but also biodegradation. As underlined by Dris et al. [4], the term "microplastics" was used first in the Thompson article to describe mainly plastic particles that are "fibrous, 20 μm in diameter, and brightly colored." However, in 2008 another definition of microplastics with a much broader scope was proposed that included all the particles with a size smaller than 5 mm [5]. Although, on the basis of the usual scientific meaning of "micro," microplastics will describe "micrometric" particles (i.e., 1–1,000 μm), for the scientific community, the name "micro" has to be understood following its Greek word μικρός, meaning small (i.e., all particles <5 mm).

This issue of microplastics in the environment has received increasing attention from the scientists, and the number of articles on this subject increased from less than 10 in 2011 to more than 150 in 2016 (request on Scopus with the words "microplastic" and "environment"). However, microplastics in freshwater environments represent only a small fraction of this amount. The first articles on microplastics in freshwater were published in 2011 and focus mainly on Lake Huron [6] and Los Angeles rivers [7] in the USA. Since that period, numerous studies have been published covering all continents, with the exception of Antarctica, and all the potentially impacted environments (water and sediment for the aquatic environment, banks for the terrestrial environment). While at the beginning mainly lakes have been investigated, it can be considered that now both lentic and lotic ecosystems are investigated. Nevertheless, the dynamics of fibers and plastics in urban catchments and hydrosystems are practically unknown. Their fate, transfer routes, and processes in continental water have yet to be determined.

Cities can be considered as one of the major sources of MPs as they gather at an especially high density all the activities that involve plastics and MPs including textile uses, packaging, transportation, electronics, buildings, and constructions. The aim of this work is to provide a comprehensive analysis of the sources and fate of microplastics in an urban environment with a focus on the Paris Megacity. Often referred to as the Paris agglomeration, this megacity is one the world's 40 largest with a population of over 10 million. The Paris agglomeration is crossed by the Seine River, whose catchment drains an area of approximately 32,000 km² from the river's headwaters to Paris. This catchment combines intense anthropogenic pressures with a very limited dilution factor due to the low average flow of the Seine River (350 m³ s⁻¹ in Paris). As a consequence, the Paris agglomeration exerts a dramatic pressure regarding pollutants on the river and provides a good case study in order to understand the consequences of urban environments on continental water. Moreover, to our knowledge, there is no case study dealing with MPs and covering all the compartments of the urban system and fluxes that occur between them. This study on Paris Megacity provides, from our point of view, the most comprehensive overview of MP sources in an urban area and its interaction with continental water.

2 Types and Shapes of Microplastics: A Focus on Fibers in Urban Areas?

MPs observed in freshwater not only cover several orders of magnitude in size, they also cover a wide spectrum of shapes. This includes fibers (length $\gg$ diameter) and fragments (diameter $\gg$ thickness) composed of different irregular shapes and spheres. These can be considered as either primary or secondary MPs depending on their origin.

Primary MPs are already manufactured in a size smaller than 5 mm. Two different forms of primary MPs exist: preproduction pellets and microbeads. Preproduction pellets are used in plastic industry. These virgin resins are melted and then formed into consumer products [8]. Microbeads were first present in hand cleaners that are used on rare occasion by the average consumer [9]. Microbeads have also come to replace natural exfoliating materials in facial cleansers, which are often used on a daily or at least weekly basis [10]. Secondary MPs stem from the degradation and fragmentation of large debris. Thermal, mechanical, and photo-degradation are important factors during the fragmentation process [11, 12].

We consider textile fibers to represent a special case. They can be considered as secondary MPs as they come from the breakdown of large items (clothes). This breakdown does not primarily occur in the environment but in the washing machines during the laundry [13]. As a consequence, fibers are found in the disposal of washing machines and, like primary MPs, enter the environment in a micro-scopic size. The same study showed the presence of fibers at the disposal of waste-water treatment plants (WWTPs). As a consequence, we would expect a predominance of fibers in urban areas with large WWTPs. In addition, because of the complexity of studying fibers, they are often overlooked. It is here decided as consequence to give a specific attention to fibers.

Fibers are often not included on the key figures concerning plastic materials [14]. However, a great proportion of the produced fibers are derived from petro-chemical polymers. The international organization for standardization (ISO/TR 11827:2012 Textiles – Composition testing – Identification of fibers) proposed a classification of the fibers according to their nature and origin.

Fibers that are used and commercialized can be either natural or man-made. The natural fibers are categorized according to their origin into animal, vegetal, or mineral fibers. For instance, cotton is a vegetal natural fiber and is very widely used. Man-made fibers are obtained by a manufacturing process. The artificial ones are made by the transformation of natural polymers. For example, rayon is artifi-cially manufactured but is made from cellulose, which is a natural polymer. On the other side, synthetic fibers are made from polymers that were chemically synthetized. In this latter category, we can find plastic polymers (polypropylene, PP; polyamide, PA; polyether sulfone, PES; etc.). The latter are most often the only fibers that are included in the microplastics definition in the different studies. Bicomponent fibers also exist and are composed of two fibers forming polymer components, which are chemically and/or physically different.

In 2014, a total of 91 million metric tons of textile fibers were produced [15], of which 63 million metric tons were man-made fibers and 58 million metric tons were plastic fibers (≈20% of the world's plastic production).

3 Source and Fate of Microplastics on the Paris Megacity

3.1 Overview of the Approach

Between 2014 and 2016, an investigation of several sources and fluxes of MPs has been carried out on the Paris Megacity. The following sources have been investigated:

1. Atmospheric fallout
2. Runoff water
3. Gray water
4. Wastewater and WWTP outlets
5. Combined sewer overflows (CSOs)

A map of Paris Megacity as well as the location of the various sampling sites is presented (Fig. 1). More details on the methodology and the approach can be found in Dris [16–18].

One of the challenges of a holistic study is that the different sample types will require different sampling methods. For instance, atmospheric fallout requires the use of a funnel for sampling, while automatic samplers including pumping devices are needed for the inaccessible canals of wastewater and CSOs. The other difficulty for such a systemic approach is the interpretation and comparison of the results. The atmospheric fallout is intrinsically expressed differently than the other compartments (fibers per surface and time unit rather than a concentration). Moreover, the interpretation has to take into consideration the time and space scale differences among the compartments. For instance, atmospheric fallout is a diffusive permanent source, whereas WWTP outlets represent a punctual but permanent source, while CSOs are punctual in time and space. The best way to work around this and be able to understand the sources of MPs is to consider the fluxes between the various compartments, for instance, on an annual basis. In this chapter, a first attempt to describe this approach is presented (Table 1).

3.2 Microplastics Encountered in the Different Compartments

The various concentrations of fibers and fragments encountered in each compartment are synthetized in Fig. 2. The total atmospheric fallout was investigated at two

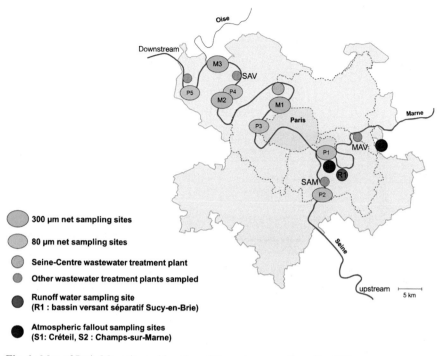

Fig. 1 Map of Paris Megacity and location of the various sampling sites [23]

Table 1 Used methods and sampled volumes for the various compartments

	Sampling method	Sampled volume
Atmospheric fallout	A collection funnel	Continuous monitoring
Urban runoff, WWTP effluents, CSOs	Automatic samplers	200–1,500 mL
Fibers in freshwater	80 μm mesh size net	0.2–4 m^3
Fragments in freshwater	300 μm mesh size net	50–200 m^3

sampling sites: one in a dense urban environment and one in suburban environment [17]. Only fibers were encountered and fragments were not detected. Throughout the year of monitoring (site 1), the atmospheric fallout ranged from 2 to 355 fibers m^{-2} day^{-1} (110 ± 96 fibers m^{-2} day^{-1}, mean ± SD), indicating a high annual variability. On site 2, the atmospheric fallout was estimated around 53 ± 38 fibers m^{-2} day^{-1}. When the levels on both sites are compared, the suburban site systematically showed fewer fibers than the urban one. We hypothesize that this difference can be explained by the density of the surrounding population, which is considered as a proxy of local activity. However, this has to be confirmed through investigations at other sites.

A suburban catchment with a separate sewer (wastewater and runoff water are collected separately) was considered for the urban runoff (unpublished data). The Sucy-en-Brie (R1, Fig. 1) catchment area reaches 261 ha, and its impervious

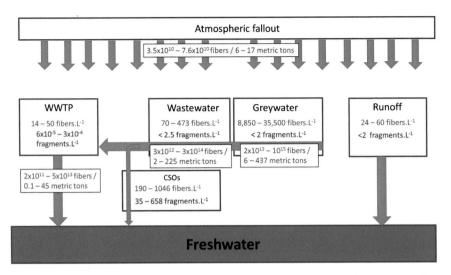

Fig. 2 A synthesis of the total fibers and fragment concentrations in the investigated urban compartments as well as the fluxes of synthetic fibers between different compartments. Only data about total fibers (natural fiber like cotton + artificial and synthetic fibers) are available for CSOs and runoff, and therefore the latter is not included in the synthetic fiber flux estimations [24]

surface coefficient is about 27%. At the outlet of this catchment, stormwater was collected during five different precipitation events between October 2014 and March 2016. Runoff concentrations at the catchment outlet ranged between 24 and 60 fibers L^{-1} with an average of 35 fibers L^{-1}. By considering the rainwater volumes collected during the atmospheric fallout experiments, a median concentration of 40 fibers L^{-1} of rainwater can be estimated. Fragments of plastic were also recovered in the runoff on three of the five sampled events and consisted mainly of foamy and irregular shapes. Concentrations ranged from <2 to 16 fragments L^{-1}. The number of fragments was systematically lower than the number of fibers, and there is no correlation between both types of MPs.

MPs were also analyzed at various scales of the sewer system: from the outlet of washing machines to the outlet of WWTPs (unpublished data except for few samples of one WWTP outlet [18]). Washing machine effluents have been collected at four volunteers' homes. The washings were carried out in their own washing machines with their own items. To have a more realistic vision, participants were asked to not change anything from their habitual washing process. Therefore, clothes made from natural, artificial, and synthetic fibers were all included, without any preselection. Concentrations between 8,850 and 35,500 fibers L^{-1} were encountered compared to 120 and 810 fibers L^{-1} for control washes (without clothes), confirming the large amount of the fibers disposed of in washing machines' effluents.

The wastewater entering four WWTPs of the Paris agglomeration was considered. Generally, wastewater is highly contaminated with fibers with the average concentration being of 248 ± 109 fibers L^{-1} (mean $\pm$ SD, $n = 20$) reaching a maximum of 473 fibers L^{-1}. In comparison, another study found similar values about 180 textile fibers L^{-1} [19].

Unlike fibers, fragments were only rarely collected, probably because of the low volumes of the samples (100–400 mL). In the 20 samples studied, 16 did not contain any fragments. For the four remaining samples, three contained one fragment, while the other contained two fragments. These small values indicate a potential upper limit for the concentration of these particles. No fragments in a 400 mL sample, for instance, imply a concentration <2.5 particles L^{-1}. Larger values were observed in other studies: 430 particles L^{-1} [19] or 13 particles L^{-1} [20]. If this variation in fragment concentrations is site dependent or is related to different methodologies, it remains unknown at the moment. Fragments could, in an oversized sewage network, settle inside the sewer during dry weather periods and reach the WWTP only at low concentrations. It is known that Paris wastewater sewerage system is oversized leading to a very low flow velocity and therefore a high sedimentation of particles inside the sewer.

Two WWTPs were selected in order to estimate the treatment efficiency. Seine Centre WWTP was considered for the case of a treatment by biofilters, while Seine Aval WWTP is a conventional activated sludge plant. For Seine Centre, three sampling campaigns on three consecutive working days were carried out. Raw wastewater after pretreatment, settled wastewater (after the primary clarifier), and treated water (after the biofilters treatment) were all considered. Only limited volumes were considered here. As the disposal channel of Seine Aval is large and long enough to be navigable, a manta trawl with a 330 μm mesh size was towed by a motor boat in the upstream direction in order to sample high volumes (68 m^3) and count fragments.

In the case of Seine Centre, only fibers were encountered. The results suggested a removal from 83 to 95% of the fiber contamination. Other works showed higher removal efficiencies of MPs by WWTPs, i.e., between 95 and 99% [20, 21]. However, all these results are hardly comparable since methods used vary from one study to another.

Fragments were detected at the outlet of the WWTP Seine Aval because larger volumes were integrated. Irregular fragments were observed primarily, and foams, films, and spheres were found punctually. The number of fragments observed in the treated water varies between 6×10^{-5} and 3×10^{-4} fragments L^{-1}. The concentrations of fragments are much lower than the levels of fibers observed in the automatic sampler samples (10^5 more fibers). Being able to observe fragments in the manta trawl samples is also an additional indication that fragments are present in WWTP effluents but just too scarce to be observed in small sample volumes.

A combined sewer system collects rainwater runoff and domestic sewage. However, during heavy rainfall events, the volume of water may exceed the capacity of the sewer and can cause flood events. In this situation, untreated water is discharged directly into the water bodies. In this context, 500 mL samples were collected at La Briche CSO outlet during three events.

Both fragments and fibers were encountered in the CSO samples. For fibers, levels of 190,898 and 1,046 fibers L^{-1} were detected. The presence of fibers is most likely due to the combined contribution of both wastewater and runoff. The levels observed in CSOs are higher than the levels in strict separated runoff. For two of the three samples, concentrations were higher than the concentrations in wastewater. A settlement of particles during dry weather periods and their re-suspension when the flow increases was described on the literature [22]. A similar behavior could be expected for the fibers. The level of fibers also depends on the previous rain events: the first sampling was conducted after a long period of heavy rainfall, which might have induced a decrease in the amount of fibers in the sewer system.

Levels of fragments vary between 35 and 3,100 fragments L^{-1}. The levels are especially high even if they vary by two orders of magnitude. Lower concentrations of fragments in comparison to fibers can be observed except for the event presenting the highest runoff contribution and volume.

3.3   Fiber Fluxes in Different Compartments

As fibers are of utmost importance concerning MP pollution in freshwater, it is especially interesting to assess the relative contribution of the various sources. In this section, the number of fiber fluxes is estimated at the scale of the Paris Megacity (surface area around 2,500 km^2–10 million inhabitants). The number of MP fibers is used to estimate the mass fluxes. Because the fibers were measured, the cumulated total length of the fibers was calculated. The length was coupled with their approximated diameter to evaluate the volume. It was estimated that diameters ranged between 5 and 100 μm with an average diameter of 25 μm. Therefore, with the total volume and specific densities of the plastic polymers (1 g cm^{-3} for the PA and 1.45 g cm^{-3} for the PET, corresponding to polymers widely used in the textile industry), total masses can be estimated. The results are summarized in Fig. 2.

- *Atmospheric fallout.* According to the average atmospheric flux of total fibers at the urban and suburban sites (110 and 53 fibers m^{-2} day^{-1}), we can estimate that at the scale of the Paris agglomeration between 1.2 and 2.5 $\times$ 10^{11} fibers could originate annually from the atmosphere. According to our analyses, we calculate that 30% of these fibers are plastic polymers. Therefore, between 3.5 and 7.6 $\times$ 10^{10} MPs would fall per year from the atmosphere on the Paris agglomeration. The masses of plastic fibers are likely between 6 and 17 metric tons.
- *Gray water.* Based on the water consumption for washing machine in France (14.4 L $inhab^{-1}$ day^{-1}) and 10 millions of inhabitants, we assess that between 4 $\times$ 10^{14} and 2 $\times$ 10^{15} fibers are discharged annually into the wastewater. However, the washing machine fibers were not chemically characterized in this work. Two hypotheses can be assumed. For the first one, we can consider that, at a global scale, 60% of these fibers are synthetic, according to the Europeans' uses and supposing also that both categories of fibers tear off similarly from clothes. The second hypothesis is based on a talk at the SETAC Europe 2016 conference indicating that MPs account for 5% of the fibers at the

disposal of washing machines [23]. Therefore, based on both scenarios (5 and 60% of synthetic fibers), between 2×10^{13} and 1×10^{15} annual MP fibers are discharged into the wastewater at the Paris agglomeration scale. This corresponds to a mass between 6 and 437 metric tons/year.

- *Wastewater.* As 2.3 million m^3 wastewater are treated daily for the Paris agglomeration, between 6×10^{13} and 4×10^{14} fibers flow annually in the wastewater. Assuming that the proportion of synthetic fibers remains constant between the washing machine disposal and the entry of the wastewater treatment plant, it was considered that between 5 and 60% of the fibers in wastewater are synthetic. Therefore, between 3×10^{12} and 3×10^{14} synthetic fibers, i.e., between 2 and 225 metric tons of fibrous MPs, flow annually on wastewater. By applying the removal rates of WWTP Seine Centre (between 80 and 95%) to the global estimation made above for the wastewater, we estimate that the Paris agglomeration releases annually between 2×10^{11} and 5×10^{13} plastic fibers into the surface waters, corresponding to a mass between 0.1 and 45 metric tons.
- *Combined sewer overflows.* CSO discharges in the Paris combined sewage system are approximated about 21 million m^3 year^{-1}, corresponding to a potential introduction into the freshwater of between 4 and 5×10^{12} fibers annually. It is hard to provide an accurate estimation of the proportion of plastic polymers among those fibers.

3.4   Comparison of Microplastic Sources in Freshwater

Among the various sources investigated, fibers were always present, while fragments were mainly detected in the urban runoff and the CSOs. The atmospheric compartment was confirmed as a source of fibers including MPs. These fibers could have different sources including synthetic fibers from clothes and houses, degradation of macroplastics, landfills, or waste incineration. The characterization indicates that the hypothesis of the clothing being the main source of these fibers is the most plausible (proportion of polymers close to the uses on the textile industry). These fibers in the atmosphere, including MPs, could be transported by wind to the aquatic environment or deposited on surfaces of cities or agrosystems. After deposition, they could impact terrestrial organisms or be transported into the aquatic systems through runoff. Future work is needed in order to investigate these atmospheric fibers and understand where they come from, where they end up, and which mechanisms and factors lead to their transport and their fallout. The distance over which a fiber could be transported is also still unknown. In this study, it was not possible to assess whether the observed fibers come from very close sources in the proximity or from distant places. MPs found in isolated lakes suggest that the transport could occur over long distances [24]. It seems that atmospheric fallout is a significant source of MPs and should be considered when investigating MPs in freshwater. In addition to atmospheric fallout, other sources have to be considered like fibers that can deposit directly from the clothes of people walking on streets. In addition, fibers coming from dry weather atmospheric fallout could be re-suspended by wind or washed during the roads and streets cleaning, limiting the accumulation between two rain events.

While only fibers were detected in atmospheric fallout, runoff contained also fragments. It suggests that fibers can be transported by air, while the larger fragments seem to fall directly on surfaces and wait to be transported by the urban runoff, if not cleaned by road cleaning services. We can suppose that the fragments stem from the degradation of larger debris, but this hypothesis still needs verification.

WWTPs were also studied as potential sources of MPs. Fibers were found in washing machine effluents and consequently in wastewater. Mechanisms and dynamics that fibers undergo inside of a sewage system are up to now not reported in the literature. The roughly estimated flux of fibers entering WWTPs lies in the same range as the amounts supposedly discharged by washing machines. The estimation range is however large and fluxes could be potentially lower. Because the transport duration in the sewer systems is probably short (max. 48–72 h), it is considered that no fragmentation occurs. On the other hand, a sedimentation process during dry weather periods in the sewage network is possible. WWTP effluents are perhaps the most investigated sources to the receiving systems. For the different Parisian WWTPs, the estimated number of MP fibers even with a removal of fibers between 80 and 95% is higher than the fibers coming from atmospheric fallout. The WWTP effluents seem to be the major source of fibers in comparison to other MPs.

In contrast, CSOs contain both high fragments and fiber concentrations. It appears that CSOs are the main and major input of MP fragments into the freshwater. Moreover, the fact that it presents concentrations of fibers sometimes higher than wastewater tends to confirm a re-suspension of sewer deposits during wet weather periods.

The knowledge on the various dynamics and mechanisms of MPs in urban catchments is still very coarse. For instance, conditions driving the fibers suspension, the aerial transport, and the fallout processes are unknown. If some sources and fluxes have been identified, it is necessary to compare the results obtained on the Paris Megacity on other case studies all over the world.

4 Monitoring Microplastics in the River Seine

4.1 Overview of the Approach

In a preliminary study published at early stages of this work, we tested two different mesh size nets to sample the river Seine [18]. It highlighted the differences between a small (80 μm) and a larger mesh size (330 μm). Fibers are highly concentrated and the use of the 80 µm mesh size is preferred. Fragments on the other hand are less abundant, and sampling higher volumes is mandatory, requiring large mesh size (330 µm). In this chapter, both methods were employed. Moreover, because fibers seem to characterize the area of the case study and are less investigated by the previously published investigations on freshwater, a long-term monitoring from April 2014 to December 2015 was carried out on four stations (P2–P5) on the Seine River from upstream to downstream Paris plus one station on the Marne River (P1). For the fragments, only five different campaigns were carried out on the various sites (Fig. 1).

4.2   Fibers in the Seine and Marne Rivers

Concentrations through the year in the Marne River (P1) range between 5.7 and 398.0 fibers m^{-3} with a mean concentration of 100.6 ± 99.9 fibers m^{-3} (mean ± SD). From the upstream to the downstream points, the concentrations are 48.5 ± 98.5 fibers m^{-3} (P2), 27.9 ± 26.3 fibers m^{-3} (P3), 27.9 ± 40.3 fibers m^{-3} (P4), and 22.1 ± 25.3 fibers m^{-3}. Variations occurred in a parallel way on the different sites. This could indicate that global factors that vary equally for all sites are more likely to affect the concentrations than local factors. The variations in diffusive inputs or seasonal changes could be at cause. We could also suspect a relation with the river flow variations, but no clear correlation was found. Nonetheless, a tendency to always have low fiber levels during high-water-flow periods was observed. During low-water-flow periods, levels are much more variable and could be influenced by different parameters such as the input of fibers, either from punctual sources (WWTP, CSO), diffusive sources (atmospheric fallout), or a possible re-suspension of fibers from the sediments.

It is possible to assess the annual fluxes of fibers in the Seine River using the 19 punctual fluxes calculated at each site. The increase between the most upstream and most downstream point (P2 and P5) is only 6%, which is much smaller than the uncertainty induced by the short-term variability (unpublished data). As a consequence, it seems regarding the fibers that the impact of the Paris agglomeration is rather small. Current state of knowledge does not allow to understand and explain this non-increasing pattern. In fact, between P2 and P5, two tributaries (Marne and Oise Rivers), three WWTP disposals (Seine Amont, Seine Centre, and Seine Aval), and numerous CSOs join the Seine River. Sinks that counterbalance these inputs could explain the fact that similar fluxes are found from upstream and downstream Paris. We suspect an important role related to the sedimentation and deposition on the banks of the fibers. Further research on the fate of the fibers is still required.

The minimum and maximum estimated fluxes for the site P5 are 2.8 × 10^{10} and 6.1 × 10^{11} fiber/year with a mean of 1.8 × 10^{11}. With the hypothesis that 65% of the fibers are synthetic, we approximate that between 1.8 × 10^{10} and 4.0 × 10^{11} MP fibers flow in 1 year in this site corresponding to an estimated mass of synthetic fibers between 0.01 and 0.34 metric tons flow per year. The evolution of this flux toward the estuary has to be determined in order to be able to determine the input from freshwater to the marine environment in terms of MPs.

4.3 Comparison with the Fragments

There is a large difference in the concentration levels between fibers and fragments. The mean fiber concentration is around 45 fibers m^{-3} ($n = 95$ samples), while the mean fragment concentrations considering both methods is around 0.54 fragments m^{-3} ($n = 17$ samples). As a consequence, using two different sampling methods for fibers and fragments seems really pertinent. While analyzing fibers needs the use of a small mesh size, sampling higher volumes is mandatory to collect other shapes of MPs.

By assuming a mean fiber length (973 μm) and diameter (25 μm) and for the fragments the mean area (168,000 μm [2]) and roughly estimated thickness (35 μm), the volumes of a typical particle for each shape can be approximated. Combining MP proportions and polymer densities hypotheses (1 and 1.45 g cm^{-3}), the mass concentrations were approached.

It was estimated that the mean concentration for synthetic fibers is of 2×10^{-5} g m^{-3}, while it is of 3×10^{-6} g m^{-3} for fragments. Because of the small amount of data, the fragment mass flux was not estimated. However, it seems with this result that even if a fragment is bigger than a fiber on average, the fragment mass fluxes would be one order of magnitude smaller than fiber mass fluxes.

5 Conclusions and Perspectives

Although information on MPs in freshwater increased dramatically over the very recent years, there is, until now, neither a systematic overview of the sources, fate, and sinks on a catchment scale nor a link between the catchment characteristics and the concentration of MPs in receiving systems.

A first attempt was made on the urban catchment of Paris Megacity and its main drainage system: the Seine and the Marne Rivers. During almost 3 years, samples have been collected from atmospheric fallouts and urban runoff, from upstream to downstream of the sewage system, and in the rivers. The key results are:

• The importance of the fiber category (near urban areas at least), which includes not only plastic fibers but also other synthetic fibers like rayon, which might also have an environmental impact.

- The necessity to split MPs in two categories: fibers and fragments. Sampling protocols often do not allow the sampling of fibers. This study shows that microplastic fibers are numerous in the catchment and in the receiving system and cannot be neglected. Of course the impact of such fibers is still unknown and must be investigated. Fragments are two orders of magnitude less abundant than fibers, but with different environmental impacts and must also be analyzed.
- Atmospheric fallouts constitute a significant flux of fibers at the scale of an urban catchment. Up to now, the dynamics of these fibers in the atmosphere, its interaction with the catchment surface, the alternating mechanisms of fallout/re-suspension, and the length scale of the movement of a fiber in the atmosphere remain unknown. Moreover, while studies point out at the rivers as a major introduction way of the fibers from the continental into the marine environment, this study suggests that the atmospheric compartment and the wind should be further investigated as a potentially major contributor.
- Washing machines seem to be a major source of fibers, including MP fibers. However, wastewater treatment plants play a major role in the reduction of fibers and microplastic fragments, which are probably transferred to sludge. Investigation of sludges and, when sludge spreading takes place, of agricultural land has still to be reinforced.
- Sampling protocol of fragments in rivers has to be improved, in order to decrease the uncertainty in downstream flow estimates.
- Concerning fibers, surprisingly no obvious increase of the upstream-downstream flow has been observed. Various in-river phenomena may explain this, but both in situ measurements and lab experiments have to be conducted to clarify this point.

Actually, this study is only the starting point of a more comprehensive work related to the dynamics of (micro)plastics at a catchment and especially urban catchment scale. Other places have to be investigated and consistent interdisciplinary research programs conducted. This study showed mainly that the identification of the sources and the fate of microplastic in freshwater is complicated and that a systematic and holistic approach is required.

Acknowledgments The authors acknowledge all the partners who collaborated during this work. We thank the technical teams of SIAAP; those of the Val-de-Marne General Council; Anne Chabas of the Interuniversity Laboratory of Atmospheric Systems (LISA); Emmanuel Rinner and Kada Boukerma of the IFREMER Detection, Sensors, and Measurements Laboratory; and the Bruker demonstrating laboratory in Champs-sur-Marne and its staff. This work is part of a thesis funded by the Paris research network on sustainable development (R2DS Ile-de-France).

References

1. Carpenter EJ, Anderson SJ, Harvey GR, Miklas HP, Peck BB (1972) Polystyrene spherules in coastal waters. Science 178:749–750
2. Carpenter EJ, Smith KL (1972) Plastics on the Sargasso Sea surface. Science 175:1240–1241
3. Thompson RC, Olsen Y, Mitchell RP, Davis A, Rowland SJ, John AWG et al (2004) Lost at sea: where is all the plastic? Science 304:838–838
4. Dris R, Imhof H, Sanchez W, Gasperi J, Galgani F, Tassin B et al (2015) Beyond the ocean: contamination of freshwater ecosystems with (micro-)plastic particles. Environ Chem 12:539
5. Arthur C, Baker J, Bamford H (2008) In: Arthur C, Baker J, Bamford H (eds) Proceedings of the international research workshop on the occurrence, effects, and fate of microplastic marine Debris. National Oceanic and Atmospheric Administration: University of Washington Tacoma campus in Tacoma, Washington. http://marinedebris.noaa.gov/sites/default/files/Microplastics.pdf
6. Zbyszewski M, Corcoran PL (2011) Distribution and degradation of fresh water plastic particles along the beaches of Lake Huron. Can Water Air Soil Pollut 220:365–372
7. Moore CJ, Lattin GL, Zellers AF (2011) Quantity and type of plastic debris flowing from two urban rivers to coastal waters and beaches of southern. J Integr Coastal Zone Manage 11:65–73
8. Wilber RJ (1987) Plastic in the North Atlantic. Oceanus 30:61–68
9. Gregory MR (1996) Plastic 'scrubbers' in hand cleansers: a further (and minor) source for marine pollution identified. Mar Pollut Bull 32:867–871
10. Fendall LS, Sewell MA (2009) Contributing to marine pollution by washing your face: microplastics in facial cleansers. Mar Pollut Bull 58:1225–1228
11. Cooper DA, Corcoran PL (2010) Effects of mechanical and chemical processes on the degradation of plastic beach debris on the island of Kauai. Hawaii Mar Pollut Bull 60:650–654
12. Gregory MR, Andrady AL (2003) Plastics in the marine environment. In: Andrady AL (ed) Plastics and the environment. Wiley, Hoboken, pp 379–401. doi:10.1002/0471721557.ch10/summary
13. Browne MA, Crump P, Niven SJ, Teuten E, Tonkin A, Galloway T et al (2011) Accumulation of microplastic on shorelines worldwide: sources and sinks. Environ Sci Technol 45:9175–9179
14. Plastics Europe (2015) Plastics – the Facts 2015, an analysis of European latest plastics production, demand and waste data. Plastics Europe, Association of Plastic Manufacturers, Brussels, p 40
15. Industrievereinigung Chemiefaser, Chemical fiber global production by type 2000–2014 statistic [internet]. Statista 2015 [cited 20 Jun 2016]. http://www.statista.com/statistics/271651/global-production-of-the-chemical-fiber-industry/
16. Dris R (2016) First assessment of sources and fate of macro- and micro-plastics in urban hydrosystems: case of Paris megacity. PhD thesis, LEESU, Paris
17. Dris R, Gasperi J, Saad M, Mirande C, Tassin B (2016) Synthetic fibers in atmospheric fallout: a source of microplastics in the environment? Mar Pollut Bull 104:290–293
18. Dris R, Gasperi J, Rocher V, Saad M, Renault N, Tassin B (2015) Microplastic contamination in an urban area: a case study in Greater Paris. Environ Chem 12:592–599
19. Talvitie J, Heinonen M, Paakkonen J-P, Vahtera E, Mikola A, Setala O et al (2015) Do wastewater treatment plants act as a potential point source of microplastics? Preliminary study in the coastal Gulf of Finland, Baltic Sea. Water Sci Technol 72:1495–1504
20. Murphy F, Ewins C, Carbonnier F, Quinn B (2016) Wastewater treatment works (WwTW) as a source of microplastics in the aquatic environment. Environ Sci Technol 50:5800–5808
21. Carr SA, Liu J, Tesoro AG (2016) Transport and fate of microplastic particles in wastewater treatment plants. Water Res 91:174–182

22. Gasperi J, Garnaud S, Rocher V, Moilleron R (2009) Priority pollutants in surface waters and settleable particles within a densely urbanised area: case study of Paris (France). Sci Total Environ 407:2900–2908
23. Llorca M, Schirinzi G, Sanchis J, Barcelo D, Marinella F (2016) Evaluation of the analysis of microplastics by mass spectrometry and assessment of their adsorption capacity for organic contaminants. SETAC Europe Nantes
24. Free CM, Jensen OP, Mason SA, Eriksen M, Williamson NJ, Boldgiv B (2014) High-levels of microplastic pollution in a large, remote, mountain lake. Mar Pollut Bull 85:156–163

Microplastic Pollution in Inland Waters Focusing on Asia

Chenxi Wu, Kai Zhang, and Xiong Xiong

Abstract The presence of microplastics in marine environment is increasingly reported and has been recognized as an issue of emerging concern that might adversely affect wildlife and cause potential risk to the health of marine ecosystems. In addition, preliminary works demonstrated that microplastics are ubiquitously present in many inland waters with concentrations comparable or higher than those observed in marine environments. Asia is the most populous continent in the world, and most Asian countries are under rapid development while facing serious environmental problems. In this chapter, we review the available literature reporting on the occurrence of microplastics in inland waters in Asia. Limited works have provided basic information on the occurrence, distribution, and properties of microplastics in lakes, reservoirs, and estuaries in Asia. Comparison with data from other regions worldwide suggests that microplastic pollution in inland waters in Asia can be more serious. These preliminary results call for more research efforts to better characterize the sources, fate, effects, and risks of microplastics in inland waters. Extensive and in-depth studies are urgently needed to bridge the knowledge gaps to enable a more comprehensive risk assessment of microplastics in inland waters and to support the development of policy addressing this issue.

Keywords Asia, Lakes, Microplastics, Occurrence, Rivers, Source

C. Wu (✉), K. Zhang, and X. Xiong
Institute of Hydrobiology, Chinese Academy of Sciences, Wuhan 430072, China
e-mail: chenxi.wu@ihb.ac.cn

M. Wagner, S. Lambert (eds.), *Freshwater Microplastics*,
Hdb Env Chem 58, DOI 10.1007/978-3-319-61615-5_5,
© The Author(s) 2018

1 Introduction

Plastics are the most versatile materials invented by man. The use of plastic materials has brought great convenience to our daily lives but not without downsides [1]. Inappropriate disposal of wasted plastics has caused serious environmental problems. The presence of plastic debris in the environment not only affects the aesthetical and recreational values of ecosystems but may also present a persistent pollution problem that will continue to accumulate into future generations [2–4]. Once entering the environment, plastics are subject to physical, chemical, and biological weathering processes, which act to slowly break large pieces of plastic into smaller fragments. Plastics less than 5 mm are considered as "microplastics" [5]. However, no universally accepted definition in terms of the size range for microplastics is currently available [6]. Microplastics can be ingested by aquatic organisms, which might cause potential adverse effects and arouse food safety concerns [7–10]. As a result, microplastic pollution has become an issue of emerging concern and is drawing increasing attention from both the public and scientific community.

Microplastic pollution in the marine environment has received widespread attention. Microplastics are found ubiquitously in benthic and pelagic environments in the oceans [11, 12]. In oceans, the high abundance of microplastics observed in the large-scale subtropical convergence zones is attributed to the circulation of ocean currents [13–15]. Accumulation of microplastics in shoreline sediments has also been observed worldwide [16–19]. The majority of plastic debris in oceans originates from land, although discharges from ocean vessels, military operations, and general shipping activities cannot be discounted [20]. It was estimated that 275 million metric tons of plastic wastes were generated in 192 coastal countries in 2010, and about 4.8–12.7 million metric tons are estimated to end up in the ocean [20]. Based on this estimation, over 95% of the plastic wastes will remain on continents to be either recycled, disposed of in landfills, go for incineration (with or without energy recovery), or otherwise be discarded and stay on continents [21].

Only a few studies have addressed the issue of microplastic pollution in terrestrial environments and inland waters in contrast to the vast amount of research in marine environments. These studies suggest inland waters are facing similar microplastic accumulation problems as found in the oceans [22, 23]. Many inland waters are habitats for aquatic species that have important ecological and economic value and provide services for recreation, aquatic products, and water resources. Therefore, it is important to understand the occurrence, fate, and effects of microplastics in inland waters [24–26].

Asia is the largest and the most populous continent in the world. Asia covers about 30% of Earth's total land area and supports about 60% of the world's population. There are about a 1,000 ethnic groups with diverse languages and cultures. Nearly all countries in Asia are developing countries, which are under rapid development while facing growing environmental problems at the same time. In this chapter, we reviewed the available literature on microplastic pollution in inland waters with a specific focus on Asia. A considerable lack of data for inland waters was found. Extensive and in-depth studies are urgently needed to bridge the knowledge gaps to enable a more comprehensive risk assessment of microplastics in inland waters and to support the development of policy addressing this issue.

2 Production and Use of Plastics in Asia

According to the data from Plastics Europe [27], world production of plastics reached 311 million metric tons in 2014, an increase of 38% from 2004. China and Japan are the two leading countries with the highest plastic production in Asia accounting for 26% and 4% of the world's total production in 2014, respectively. Plastic production in the rest of Asia accounted for 16% of the world production. All together Asia produced nearly a half of the world's plastic materials in 2014. These plastic materials are used in a wide variety of markets, including packaging, building and construction, automotive, electrical and electronic, agriculture, consumer and household appliance, etc. Polypropylene (PP), polyethylene (PE), and polyvinyl chloride (PVC) are the most-used polymer types and account for 19.2%, 29.3%, and 10.3% of the plastics demand in Europe, respectively [27]. Although the production of plastics is the highest in Asia, the per capita consumption is low compared to developed regions. For example, the per capita plastic consumption is 9.7 and 45 kg per person in India and China, comparing to 65 and 109 kg per person in Europe and the USA [28].

Plastic wastes are recycled at a much lower ratio in developing countries than in developed countries. In China, less than 10% of the plastic wastes are recycled, while about 30% of the plastic wastes are recycled in Europe [27, 29]. What's more, developing countries usually have a high percentage of mismanaged plastic waste. Among the top 20 countries ranked by the mass of mismanaged plastic waste, all of them are developing countries except the USA which has the highest waste generation rate but the lowest percentage of mismanaged plastic waste [20]. Twelve Asian countries were on the list with China, Indonesia, and the Philippines ranked top three. The percentage of mismanaged plastic waste among these Asian countries varied from 1.0 to 27.7%. As a result the environmental release of plastic wastes is more likely in these Asian countries.

3 Microplastics in Inland Waters in Asia

3.1 Occurrence of Microplastics

Although inland waters in Asia have a high potential to be polluted by microplastics, relevant studies are scarce among the literature. Only seven studies were found from the databases accessible to us and are summarized in Table 1. Among these studies, one was carried out in Mongolia, while all others were performed in

Table 1 Studies reporting microplastic pollution in inland waters in Asia and other regions

Study area	Samples	Size (mm)	Abundance (items/km^2)	Main types	Source
Asia					
Lake Hovsgol, Mongolia	Surface water	>0.333	Maximum: 44,435 Mean: 20,264	Not identified	[30]
Three Gorges Reservoir, China	Surface water	0.112–5.0	Maximum: 13.6175×10^6 Mean: 6.395×10^6	PE, PP, PS	[22]
Siling Co basin, China	Beach sediment	<5.0	Maximum: 563×10^6 Mean: 113×10^6	PE, PET, PP, PS, PVC	[23]
Taihu Lake, China	Surface water	0.333–5.0	Range: 0.01×10^6–6.8×10^6	CP, PA, PET, PP	[31]
Taihu Lake, China	Lake sediment	0.005–5.0	Range: 11.0–234.6 items/kg	CP, PA, PE, PET, PP	[31]
Urban waters of Wuhan, China	Surface water	0.05–5.0	Range: 1660 ± 639.1–8925 ± 1591 items/m^3	PA, PE, PET, PP, PS	[32]
Yangtze River Estuary, China	Surface water	>0.5	Maximum: 10,200 items/m^3 Mean: 4,137 items/m^3	Not identified	[33]
Three urban estuaries, China	Surface water	>0.5	Range: 100–4,100 items/m^3	PE, PP, PVC, PTFE	[34]
Europe					
Swiss lakes, Switzerland	Surface water	0.3–5.0	Mean: 91,000 Median: 48,000	PE, PP, PS, PVC	[35]
Swiss lakes, Switzerland	Beach sediment	0.3–5.0	Mean: $1,300 \times 10^6$ Range: 20×10^6–7200×10^6	PE, PP, PS, PVC	
Lake Garda, Italy	Beach sediment	<5.0	Range: 108–1108	PA, PE, PP, PS, PVC	[36]
Lake Bolsena and Lake Chiusi, Italy	Surface water	0.3–5.0	Range: 0.21–4.08 items/m^3	Not identified	[37]
	Beach sediment	0.3–5.0	Range: $1,772 \times 10^6$–$2,462 \times 10^6$	Not identified	[37]
Rhine-Main area, Germany	Beach sediment	<5.0	Range: 228–3,763 item/kg	PA, PE, PP, PS	[38]
Seine River, France	Surface water	–	0.8–5.1% of total debris by weight	PE, PET, PP, PS, PVC	[39]
Danube River, Austria	Surface water	0.5–2.0	Mean: 316.8 items/1,000 m^3 Range: 0–141647.7 items/1,000 m^3	Not identified	[40]

				PE, PP, PS, PVC	[41]
Tamar Estuary, England	Surface water	0.3–5.0	Mean: 0.028/m^3	PE, PP, PS, PVC	[41]
North America					
Lake Superior	Surface water	>0.355	Mean: 5390.8 Range: 6,875–12,645	Not identified	[42]
Lake Huron	Surface water	>0.355	Mean: 2779.4 Range: 0–6,541	Not identified	[42]
Lake Huron	Beach sediment	Not specified	Mean: 5.43×10^6 Range: 0–34×10^6	PE, PP	[43]
Lake Erie	Surface water	>0.355	Mean: 105,502 Range: 4,686–466,305	Not identified	[42]
Lake Erie	Beach sediment	Not specified	Mean: 1.537×10^6 Range: 0.36×10^6–3.7×10^6	PE, PP	[43]
Lake St. Clair	Beach sediment	Not specified	Mean: 1.726×10^6 Range: 0.18×10^6–8.38×10^6	PE, PP	[43]
St. Lawrence River, Canada	River sediment	>0.5	Range: 0–$136,926 \times 10^6$	PE	[44]
Four estuaries in the Chesapeake Bay, USA	Surface water	0.3–5.0	Range: 5,534–259,803	PE	[45]
North Shore Channel, Chicago, USA	Surface water	>0.333	Range: 0.73×10^6–6.698×10^6	Not identified	[46]
Africa					
Lake Victoria, Mwanza Region of Tanzania	Fish	>0.25	20% of fish	PE, PES, PUR, PE/PP copolymer, silicone rubber	[47]
Five urban estuaries of KwaZulu-Natal, South Africa	Surface water	>0.25	Range: 2–487 item/10,000 L	PS	[48]
Five urban estuaries of KwaZulu-Natal, South Africa	River sediment	>0.02	Range: 13.7–745.4 item/500 mL	PS	[48]

CP cellophane, PA polyamide, PE polyethylene, PES polyester, PET polyethylene terephthalate, PP polypropylene, PS polystyrene, PTFE polytetrafluoroethylene, PUR polyurethane, PVC polyvinyl chloride

China. Lake Hovsgol, a remote mountain lake in Mongolia, was surveyed for pelagic microplastics [30]. Results showed that microplastic abundance ranged from 997 to 44,435 items/km^2. Microplastic abundance decreased with distance from the southwestern shore which had the highest human impact and was distributed by the prevailing winds. Zhang et al. [22] investigated microplastic occurrence in the surface waters of the Three Gorges Reservoir in China and found microplastic abundance up to $136,175 \times 10^6$ items/km^2, which is the highest microplastic abundance ever reported in the literature. The authors suggested that the high accumulation of microplastics is related to the damming, and reservoirs can act as potential hot spots for microplastics. In another study by Zhang et al. [23], microplastics were sampled from the shorelines of four lakes within the Siling Co basin in northern Tibet. Microplastics were detected in six out of seven sampling sites, and the site with the highest microplastic abundance was related to the riverine input. Su et al. [31] reported microplastic pollution in Taihu Lake, which is the third largest freshwater lake in China located in a well-developed area under extensive human influence. Microplastics were detected in plankton net, surface water, sediment, and Asian clams samples. More recently, microplastic pollution was studied in inland freshwaters in Wuhan, the largest city in Central China [32]. Microplastics were detected with concentrations ranged from $1,660 \pm 639.1$ to $8,925 \pm 1,591$ items/m^3 in surface water, and microplastic abundance was negatively correlated with the distance from the city center. Another two studies have also investigated the occurrence of microplastics in surface water from the estuaries of Yangtze, Jiaojiang, Oujiang, and Minjiang in China [33, 34]. Results demonstrated that microplastics were present in high abundance in these transitional zones between rivers and the sea and suggested that rivers are important sources of microplastics to marine environment.

The literature reporting the occurrence of microplastics in inland waters from other geographical regions is also summarized in Table 1. A comparison of data from different regions can be challenging due to the difference in sampling methods used, size ranges investigated, and the reporting units that are employed. Therefore, it is urgently needed to adopt universal criteria for sampling and reporting microplastics occurrence data to facilitate a comparison [49]. Additionally, the abundance of microplastics from different regions differs by several orders of magnitude. Even within the same region, the abundance of microplastics varies considerably. This uneven distribution pattern can be related to their relatively low density, which means that they can be transported easily with the current and accumulation in areas with weaker hydrodynamic conditions. In addition, the loading rate of plastic waste can differ significantly in different regions. Previously, Yonkos et al. [45] demonstrated that the abundance of microplastics was positively correlated with population density and proportion of urban/suburban development within the watersheds. However, researches also demonstrated that microplastics were also found at relatively high concentrations in inland waters from remote areas with limited human activities [23, 30]. This is likely due to a lack of proper waste management measures in those areas. In many Asian countries, high population density and unsound waste management systems lead to a high risk of inland water pollution by microplastics as well as many other pollutants. This might explain the very high abundance of microplastics observed in Taihu Lake and Three Gorges Reservoirs in China. Therefore, inland waters in Asia deserve more attention in the future.

3.2 Characteristics of the Microplastics

3.2.1 Particle Shape

After sample collection, potential microplastics are usually examined using stereo microscopes. According to their shapes, microplastics are typically categorized as follows: sheet, film, line/fiber, fragment, pellet/granule, and foam (Fig. 1). However, there is no set protocol, and different classifications might be used by different researchers. This morphological information from the microplastic samples can be used to indicate their potential origins. For example, line/fiber usually originates from fishing lines, clothing, or other textiles, while film mainly originates from bags or wrapping materials. In Lake Hovsgol, fragments and films were found to be the most abundant, together accounting for 78% of the total microplastics [31]. In the Three Gorges Reservoir, sheet particles and miscellaneous fragments were dominant in most sites [22], whereas fibers and fragments were predominant in samples from Taihu Lake [31]. In inland freshwaters of Wuhan, fiber, granule, film, and pellet were commonly detected, and fibers were most frequently detected accounting for 52.9–95.6% of the total plastics [32]. For microplastic samples from the four estuaries in China, fibers and granules were the more abundant [33, 34]. Different patterns observed in these study areas suggest that the sources of the microplastics

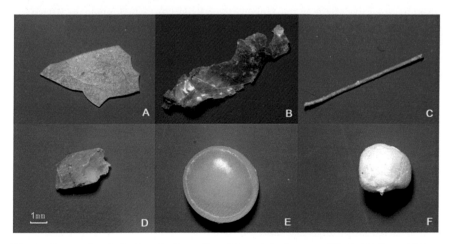

Fig. 1 Shapes of typical microplastics collected from inland waters (Qinghai Lake and Three Gorges Reservoir) in China (**a**, sheet; **b**, film; **c**, line/fiber; **d**, fragment; **e**, pellet/granule; **f**, foam)

might differ considerably in different regions. Fibers appeared to be more abundant in more populated areas.

3.2.2 Particle Size

Size is another parameter usually measured for microplastics, but no unified criteria are currently available. Different size classes were reported by different authors, which make it difficult to compare the data from different works [50]. Due to the restriction of the sampling methods used, usually only microplastics >0.333 mm (mesh size of the manta trawl net) are assessed in neustonic samples collected by trawling. Smaller microplastics can be examined for sediment and biota samples as density separation combined with filtration is used. Whereas the examination of microplastics <0.05 mm will get increasingly difficult, advanced instruments such as Raman microscopy, micro-Fourier transform infrared spectroscopy (μ-FTIR), or scanning electron microscope (SEM) with energy dispersive spectroscopy (EDS) should be used [36]. Generally, microplastic abundance increases with decreasing size [51–53]. In lake Hovsgol, 0.355–0.999, 1.00–4.749, and >4.75 mm size classes accounted for 41, 40, and 19% of the total plastics, respectively [30]. In the freshwaters of Wuhan, 0.05–0.5, 0.5–1, and 1–2 mm size classes together accounted for over 80% of the total microplastics, and 0.05–0.5 mm microplastics were the most abundant in most of the studied waters [32]. In Yangtze Estuary, 0.5–1, 1–2.5, 2.5–5, and >5 mm size classes made up 67, 28.4, 4.4, and 0.2% of the total plastics, respectively [33]. In the estuary of Minjiang, Oujiang, and Jiaojiang, the smallest size class (0.5–1.0 mm) was also found the most abundant followed by the 1.0–2.0 mm size class, and these two size classes together accounted for over 70% of the total plastics [34]. However, among the four size classes (0.112–0.3, 0.3–0.5, 0.5–1.6, 1.6–5 mm), 0.5–1.6 mm microplastics were the most abundant from the majority of site in the Three Gorges Reservoir, which made up 30–57% of the total microplastics [22]. While for microplastic samples from the lakeshore sediment of the Siling Co basin, different size distribution patterns were observed from different sampling sites [23]. The patterns of microplastic size distribution can be related to the sources of microplastics and might also reflect the degree of weathering. A higher degree of weathering might result in a higher abundance of smaller particles. Biofouling and hydrodynamic conditions were also believed to affect the size distribution of microplastics [54–56].

3.2.3 Color

In some studies, colors of the microplastics were described. Microplastics can inherit their colors from their parent plastic products, but their colors can change due to weathering. Previous research infers that predators may preferably ingest microplastics with colors resembling their prey [57–59]. Therefore, color information of microplastics may be used to indicate their potential to be ingested by aquatic animals. In Taihu Lake, recovered microplastics were found in a variety of colors including transparent, black, white, red, yellow, green, and blue [31]. In addition, blue was the most dominant color in plankton net and surface water samples, while white microplastics were the most abundant in sediments [31]. In the freshwaters of Wuhan, microplastics were found to be transparent or in blue, purple, red, or other colors, and colored microplastics, accounting for 50.4–86.9% of the total microplastics, were more abundant than transparent ones [32]. From the estuaries of Jiaojiang, Oujiang, and Minjiang, microplastics were divided into transparent, white, black, and colored groups, and colored microplastics were identified as the most dominant [34]. It may be interesting to investigate further how color affects the environmental fate and ecological effects of microplastics. As an example, colorants can often influence the final thermal and UV stability of a plastic material [60, 61].

3.2.4   Surface Texture

Once entering the environment, plastics are subject to weathering processes, and these processes will influence the surface of the microplastics (Fig. 2). Featured surface textures on microplastics can be used to indicate the processes of mechanical and oxidative weathering [62, 63]. Surface textures are usually examined using SEM. Features such as grooves, fractures, and mechanical pits are believed to result from mechanical weathering, while flakes, granules, and solution pits are considered as oxidative weathering features [43]. The surface oxidation of plastics can be confirmed using FTIR as indicated by the appearance of peaks for carbonyl groups [31, 43]. Zhang et al. [23] examined the surface textures of microplastics from the lakeshore sediments of the Siling Co basin, and mechanical weathering features were more often observed, which were attributed to the windy weather condition in the study area. This result agrees with the overall trend of microplastics recovered from the Great Lakes, but differs from those recovered from beach sands in Hawaii [43, 63]. Hawaii has a warmer and more humid climate than northern Tibet and the Great Lakes region and might therefore favor the oxidative weathering of the plastics.

3.3   Polymer Types Found

A variety of polymers were used in the production of plastics. Properties and performances of the plastic materials are largely determined by the polymer types they are made of. Thus, polymer types can have a great impact on the longevity and buoyancy of microplastics, thus affecting their fate in the environment. Polymer types are typically identified using FTIR and Raman spectrometry, and less often pyrolysis-gas chromatography/mass spectrometry (Pyr-GC/MS) is used [64, 65]. Detailed reviews of the typically used techniques for the

Fig. 2 Surface texture of typical microplastics collected from inland waters (Siling Co Basin) in China (**a**, grooves; **b**, fractures; **c**, mechanical pits; **d**, flakes; **e**, granular; **f**, solution pits)

identification of microplastics were published recently [66–68]. An overview of the advantages and limitations of these techniques is summarized in Table 2.

In the Three Gorges Reservoir, only PE, PP, and PS were identified from the recovered microplastics, which may be related to their lower density. The weak hydraulic conditions of the Three Gorges Reservoir will generally favor the sedimentation of microplastics originating from denser polymer types [22]. In lakeshore sediments from Siling Co basin, PE and PP were predominant, while PVC, PET, and PS were only identified from one sampling site [23]. Whereas cellophane (CP) which is a transparent material made of regenerated cellulose was found to

Table 2 Advantages and limitations of commonly used techniques for the identification of microplastics from environmental samples

Technique	Size	Advantages	Limitations
ATR-FTIR	>1 mm	Easy simple preparation; low cost	Unable to identify small samples; unsuitable for convex particles or severely aged or contaminated samples
μ-FTIR	>10 μm	Suitable for small samples	Sample preparation is complex; unable to analyze polyamides; high cost
FPA-based μ-FTIR	>20 μm	Suitable for small samples; no need for visual sorting	Time consuming; high cost
Raman spectroscopy	>1 μm	Suitable for very small samples; lower water interference	Sensitive to fluorescence interference; laser-induced degradation; high cost
CARS[c]	>1 μm	Applicable to living organisms; minimal or no sample preparation; strong signal; less interference	Very high cost
Pyr-GC/MS	>0.1 mg	Organic plastic additive can be analyzed	Destructive; time consuming

ATR attenuated total reflectance, *FPA* focal plane array, *CARS* coherent anti-Stokes Raman scattering

be the most abundant in Taihu Lake likely due to a low biodegradability of the material [31]. In the freshwaters of Wuhan, PET, PP, PE, PA, and PS were identified, and PET and PP were more abundant [32]. From the estuaries of Jiaojiang, Oujiang, and Minjiang, selected microplastics were found mostly PP and PE [33], which agrees with the result from the Three Gorges Reservoir. Frequent detection of PP and PE was also reported in many other works [36, 45, 69, 70]. This can be related to their low density, meaning these polymer types are buoyant and readily transported with water. In addition, a larger global demand for PE and PP makes them more prevalent in the environment.

4 Conclusions

Inland waters are facing similar issues as marine environments with regard to microplastics. Analysis suggests that inland waters in many Asian countries have a high risk to be polluted by microplastics. However, only limited works have been performed in investigating microplastic occurrence in inland waters in Asia currently. Available researches have demonstrated the presence of microplastics in lakes, reservoirs, and river estuaries. The abundances of microplastics in the Three Gorges Reservoir and Taihu Lake in China are among the highest reported data in inland waters worldwide. Relatively high abundance of microplastic was also observed in lakes from remote areas in Mongolia and in central Tibet. Waste management systems need to be improved in these developing Asian countries to mitigate the microplastic pollution problems. Along with abundance, features of microplastic samples such as shapes, sizes, colors, surface textures, and polymer types were measured in these studies, which can be used to interpret the origins and experiences of the microplastics. These preliminary results call for further research efforts to better understand the sources and fate of microplastics in inland waters. The biological and ecological risk of microplastic exposure should be assessed especially at environmentally relevant circumstances. Sampling, pretreatment, and reporting of microplastics should be standardized for the future monitoring programs.

References

1. Andrady AL, Neal MA (2009) Applications and societal benefits of plastics. Philos Trans R Soc B Biol Sci 364(1526):1977–1984
2. Wright SL, Thompson RC, Galloway TS (2013) The physical impacts of microplastics on marine organisms: a review. Environ Pollut 178:483–492
3. Bakir A, Rowland SJ, Thompson RC (2014) Enhanced desorption of persistent organic pollutants from microplastics under simulated physiological conditions. Environ Pollut 185: 16–23
4. Yang CZ, Yaniger SI, Jordan VC, Klein DJ, Bittner GD (2011) Most plastic products release estrogenic chemicals: A potential health problem that can be solved. Environ Health Perspect 119(7):989–996
5. Arthur C, Baker J, Bamford H (eds) (2008) Proceedings of the international research workshop on the occurrence, effects and fate of microplastic marine debris. NOAA technical memorandum NOS-OR&R-30, 9–11 Sept 2008
6. Hidalgo-Ruz V, Gutow L, Thompson RC, Thiel M (2012) Microplastics in the marine environment: a review of the methods used for identification and quantification. Environ Sci Technol 46(6):3060–3075
7. Rochman CM, Hoh E, Kurobe T, Teh SJ (2013) Ingested plastic transfers hazardous chemicals to fish and induces hepatic stress. Sci Rep 3:3263
8. Browne MA, Dissanayake A, Galloway TS, Lowe DM, Thompson RC (2008) Ingested microscopic plastic translocates to the circulatory system of the mussel, *Mytilus edulis* (L.) Environ Sci Technol 42(13):5026–5031
9. Seltenrich N (2015) New link in the food chain? Marine plastic pollution and seafood safety. Environ Health Perspect 123(2):A34–A41
10. da Costa JP, Santos PSM, Duarte AC, Rocha-Santos T (2016) (Nano)plastics in the environment – sources, fates and effects. Sci Total Environ 566–567:15–26
11. Ivar do Sul JA, Costa MF (2014) The present and future of microplastic pollution in the marine environment. Environ Pollut 185:352–364
12. Van Cauwenberghe L, Devriese L, Galgani F, Robbens J, Janssen CR (2015) Microplastics in sediments: a review of techniques, occurrence and effects. Mar Environ Res 111:5–17
13. Law KL, Morét-Ferguson S, Maximenko NA, Proskurowski G, Peacock EE, Hafner J, Reddy CM (2010) Plastic accumulation in the North Atlantic subtropical gyre. Science 329(5996): 1185–1188

14. Eriksen M, Maximenko N, Thiel M, Cummins A, Lattin G, Wilson S, Hafner J, Zellers A, Rifman S (2013) Plastic pollution in the South Pacific subtropical gyre. Mar Pollut Bull 68(1–2):71–76
15. Cózar A, Echevarría F, González-Gordillo JI, Irigoien X, Úbeda B, Hernández-León S, Palma AT, Navarro S, García-de-Lomas J, Ruiz A, Fernández-de-Puelles ML, Duarte CM (2014) Plastic debris in the open ocean. Proc Natl Acad Sci 111(28):10239–10244
16. Browne MA, Crump P, Niven SJ, Teuten E, Tonkin A, Galloway T, Thompson R (2011) Accumulation of microplastic on shorelines worldwide: sources and sinks. Environ Sci Technol 45(21):9175–9179
17. Jayasiri HB, Purushothaman CS, Vennila A (2013) Quantitative analysis of plastic debris on recreational beaches in Mumbai, India. Mar Pollut Bull 77(1–2):107–112
18. Smith SDA, Gillies CL, Shortland-Jones H (2014) Patterns of marine debris distribution on the beaches of Rottnest Island, Western Australia. Mar Pollut Bull 88(1–2):188–193
19. Lee J, Hong S, Song YK, Hong SH, Jang YC, Jang M, Heo NW, Han GM, Lee MJ, Kang D, Shim WJ (2013) Relationships among the abundances of plastic debris in different size classes on beaches in South Korea. Mar Pollut Bull 77(1–2):349–354
20. Jambeck JR, Geyer R, Wilcox C, Siegler TR, Perryman M, Andrady A, Narayan R, Law KL (2015) Plastic waste inputs from land into the ocean. Science 347(6223):768–771
21. Rochman CM, Browne MA, Halpern BS, Hentschel BT, Hoh E, Karapanagioti HK, Rios-Mendoza LM, Takada H, Teh S, Thompson RC (2013) Policy: classify plastic waste as hazardous. Nature 494(7436):169–171
22. Zhang K, Gong W, Lv J, Xiong X, Wu C (2015) Accumulation of floating microplastics behind the three Gorges dam. Environ Pollut 204:117–123
23. Zhang K, Su J, Xiong X, Wu X, Wu C, Liu J (2016) Microplastic pollution of lakeshore sediments from remote lakes in Tibet plateau, China. Environ Pollut 219:450–455
24. Wagner M, Scherer C, Alvarez-Muñoz D, Brennholt N, Bourrain X, Buchinger S, Elke F, Grosbois C, Klasmeier J, Marti T, Rodriguez-Mozaz S, Urbatzka R, Vethaak AD, Winther-Nielsen M, Reifferscheid G (2014) Microplastics in freshwater ecosystems: what we know and what we need to know. Environ Sci Eur 26(1):1–9
25. Eerkes-Medrano D, Thompson RC, Aldridge DC (2015) Microplastics in freshwater systems: a review of the emerging threats, identification of knowledge gaps and prioritisation of research needs. Water Res 75:63–82
26. Lambert S, Sinclair C, Boxall A (2014) Occurrence, degradation, and effect of polymer-based materials in the environment. In: Whitacre DM Reviews of environmental contamination and toxicology, vol 227. Springer, Cham, pp 1–53
27. PlasticsEurope (2015) Plastics-the facts 2015: an analysis of European plastics production, demand and waste data. http://www.plasticseurope.org
28. TataStrategic (2014) Potential of plastics industry in Northern India with special focus on plasticulture and food processing-2014. A report on plastics industry. Federation of Indian Chambers of Commerce and Industry. New Delhi
29. Tang G, Hu B, Kang Z, Meng C, Zhang X, Zhang L, Feng HY, Sun WP (2013) Current status and problems on waste plastic recycling. Recy Resour Circ Econ 6(1):31–35. (In Chinese with English Abstract)
30. Free CM, Jensen OP, Mason SA, Eriksen M, Williamson NJ, Boldgiv B (2014) High-levels of microplastic pollution in a large, remote, mountain lake. Mar Pollut Bull 85(1):156–163
31. Su L, Xue Y, Li L, Yang D, Kolandhasamy P, Li D, Shi H (2016) Microplastics in Taihu Lake, China. Environ Pollut. doi:10.1016/j.envpol.2016.06.036
32. Wang W, Ndungu AW, Li Z, Wang J (2016) Microplastics pollution in inland freshwaters of China: a case study in urban surface waters of Wuhan, China. Sci Total Environ. doi:10.1016/j.scitotenv.2016.09.213
33. Zhao S, Zhu L, Wang T, Li D (2014) Suspended microplastics in the surface water of the Yangtze estuary system, China: first observations on occurrence, distribution. Mar Pollut Bull 86(1–2):562–568

34. Zhao S, Zhu L, Li D (2015) Microplastic in three urban estuaries, China. Environ Pollut 206:597–604
35. Faure F, Corbaz M, Baecher H, de Alencastro L (2012) Pollution due to plastics and microplastics in Lake Geneva and in the Mediterranean Sea. Arch Sci 65:157–164
36. Imhof HK, Ivleva NP, Schmid J, Niessner R, Laforsch C (2013) Contamination of beach sediments of a subalpine lake with microplastic particles. Curr Biol 23(19):R867–R868
37. Fischer EK, Paglialonga L, Czech E, Tamminga M (2016) Microplastic pollution in lakes and lake shoreline sediments – a case study on Lake Bolsena and lake Chiusi (central Italy). Environ Pollut 213:648–657
38. Klein S, Worch E, Knepper TP (2015) Occurrence and spatial distribution of microplastics in river shore sediments of the Rhine-Main area in Germany. Environ Sci Technol 49(10): 6070–6076
39. Gasperi J, Dris R, Bonin T, Rocher V, Tassin B (2014) Assessment of floating plastic debris in surface water along the Seine River. Environ Pollut 195:163–166
40. Lechner A, Keckeis H, Lumesberger-Loisl F, Zens B, Krusch R, Tritthart M, Glas M, Schludemann E (2014) The Danube so colourful: a potpourri of plastic litter outnumbers fish larvae in Europe's second largest river. Environ Pollut 188:177–181
41. Sadri SS, Thompson RC (2014) On the quantity and composition of floating plastic debris entering and leaving the Tamar estuary, Southwest England. Mar Pollut Bull 81(1):55–60
42. Eriksen M, Mason S, Wilson S, Box C, Zellers A, Edwards W, Farley H, Amato S (2013) Microplastic pollution in the surface waters of the Laurentian Great Lakes. Mar Pollut Bull 77(1–2):177–182
43. Zbyszewski M, Corcoran PL, Hockin A (2014) Comparison of the distribution and degradation of plastic debris along shorelines of the Great Lakes, North America. J Great Lakes Res 40(2): 288–299
44. Castañeda RA, Avlijas S, Simard MA, Ricciardi A (2014) Microplastic pollution in St. Lawrence River sediments. Can J Fish Aquat Sci 71(12):1767–1771
45. Yonkos LT, Friedel EA, Perez-Reyes AC, Ghosal S, Arthur CD (2014) Microplastics in four estuarine rivers in the Chesapeake bay, USA. Environ Sci Technol 48(24):14195–14202
46. McCormick A, Hoellein TJ, Mason SA, Schluep J, Kelly JJ (2014) Microplastic is an abundant and distinct microbial habitat in an urban river. Environ Sci Technol 48(20):11863–11871
47. Biginagwa FJ, Mayoma BS, Shashoua Y, Syberg K, Khan FR (2016) First evidence of microplastics in the African Great Lakes: recovery from Lake Victoria Nile perch and Nile tilapia. J Great Lakes Res 42(1):146–149
48. Naidoo T, Glassom D, Smit AJ (2015) Plastic pollution in five urban estuaries of KwaZulu-Natal, South Africa. Mar Pollut Bull 101(1):473–480
49. Phuong NN, Zalouk-Vergnoux A, Poirier L, Kamari A, Châtel A, Mouneyrac C, Lagarde F (2016) Is there any consistency between the microplastics found in the field and those used in laboratory experiments? Environ Pollut 211:111–123
50. Filella M (2015) Questions of size and numbers in environmental research on microplastics: methodological and conceptual aspects. Environ Chem 12(5):527–538
51. Collignon A, Hecq JH, Galgani F, Voisin P, Collard F, Goffart A (2012) Neustonic micro-plastic and zooplankton in the North Western Mediterranean Sea. Mar Pollut Bull 64(4): 861–864
52. Isobe A, Uchida K, Tokai T, Iwasaki S (2015) East Asian seas: a hot spot of pelagic micro-plastics. Mar Pollut Bull 101(2):618–623
53. Imhof HK, Laforsch C, Wiesheu AC, Schmid J, Anger PM, Niessner R, Ivleva NP (2016) Pigments and plastic in limnetic ecosystems: a qualitative and quantitative study on micro-particles of different size classes. Water Res 98:64–74
54. Isobe A, Kubo K, Tamura Y, Si K, Nakashima E, Fujii N (2014) Selective transport of micro-plastics and mesoplastics by drifting in coastal waters. Mar Pollut Bull 89(1–2):324–330
55. Fazey FMC, Ryan PG (2016) Biofouling on buoyant marine plastics: an experimental study into the effect of size on surface longevity. Environ Pollut 210:354–360

56. Peter GR (2015) Does size and buoyancy affect the long-distance transport of floating debris? Environ Res Lett 10(8):084019
57. Boerger CM, Lattin GL, Moore SL, Moore CJ (2010) Plastic ingestion by planktivorous fishes in the North Pacific central gyre. Mar Pollut Bull 60(12):2275–2278
58. Moser ML, Lee DS (1992) A fourteen-year survey of plastic ingestion by western North Atlantic seabirds. Colon Waterbird 15(1):83–94
59. Moore CJ (2008) Synthetic polymers in the marine environment: a rapidly increasing, long-term threat. Environ Res 108(2):131–139
60. Saron C, Felisberti MI (2006) Influence of colorants on the degradation and stabilization of polymers. Quím Nova 29(1):124–128
61. Russell S (2003) Color compounding. In: Charvat RA (ed) Coloring of plastics: fundamentals. Wiley, Hoboken
62. Zbyszewski M, Corcoran PL (2011) Distribution and degradation of fresh water plastic particles along the beaches of Lake Huron, Canada. Water Air Soil Pollut 220(1):365–372
63. Cooper DA, Corcoran PL (2010) Effects of mechanical and chemical processes on the degradation of plastic beach debris on the island of Kauai, Hawaii. Mar Pollut Bull 60(5):650–654
64. Rocha-Santos T, Duarte AC (2015) A critical overview of the analytical approaches to the occurrence, the fate and the behavior of microplastics in the environment. TrAC Trends Anal Chem 65:47–53
65. Käppler A, Fisher D, Oberbechmann S, Schernewski G, Labrenz M, Eichhorn KJ, Voit B (2016) Analysis of environmental microplastics by vibrational microspectroscopy: FTIR, Raman or both? Anal Bioanal Chem 408:8377–8391
66. Qiu Q, Tan Z, Wang J, Peng J, Li M, Zhan Z (2016) Extraction, enumeration and identification methods for monitoring microplastics in the environment. Estuar Coast Shelf Sci 176:102–109
67. Ribeiro-Claro P, Nolasco MM, Araújo C (2017) Characterization of microplastics by Raman spectroscopy. Compr Anal Chem 75:119–151
68. Renner G, Schmid TC, Schram J (2017) Characterization and quantification of microplastics by infrared spectroscopy. Compr Anal Chem 75:67–118
69. Veerasingam S, Saha M, Suneel V, Vethamony P, Rodrigues AC, Bhattacharyya S, Naik BG (2016) Characteristics, seasonal distribution and surface degradation features of microplastic pellets along the Goa coast, India. Chemosphere 159:496–505
70. Chae DH, Kim IS, Kim SK, Song YK, Shim WJ (2015) Abundance and distribution characteristics of microplastics in surface seawaters of the Incheon/Kyeonggi coastal region. Arch Environ Con Tox 69(3):269–278

Microplastics in Inland African Waters: Presence, Sources, and Fate

Farhan R. Khan, Bahati Sosthenes Mayoma, Fares John Biginagwa, and Kristian Syberg

Abstract As the birthplace of our species, the African continent holds a unique place in human history. Upon entering a new epoch, the Anthropocene defined by human-driven influences on earth systems, and with the recognition that plastic pollution is one of the hallmarks of this new age, remarkably little is known about the presence, sources, and fate of plastics (and microplastics (MPs)) within African waters. Research in marine regions, most notably around the coast of South Africa, describes the occurrence of MPs in seabirds and fish species. More recently environmental sampling studies in the same area have quantified plastics in both the water column and sediments. However, despite Africa containing some of the largest and deepest of the world's freshwater lakes, including Lakes Victoria and Tanganyika as part of the African Great Lakes system, and notable freshwater rivers, such as the River Congo and the Nile, the extent of MPs within the inland waters remains largely unreported. In the only study to date to describe MP pollution in the African Great Lakes, a variety of polymers, including polyethylene, polypropylene, and silicone rubber, were recovered from the gastrointestinal tracts of Nile perch (*Lates niloticus*) and Nile tilapia (*Oreochromis niloticus*) fished from Lake Victoria. The likely sources of these plastics were considered to be human activities linked to fishing and tourism, and urban waste. In this chapter we discuss the need for research focus on MPs in Africa and how what has been described in

F.R. Khan (✉) and K. Syberg
Department of Science and Environment, Roskilde University, Universitetsvej 1, P.O. Box 260, DK-4000 Roskilde, Denmark
e-mail: frkhan@ruc.dk

B.S. Mayoma
Department of Livestock and Fisheries Development, Mtwara District Council, P.O. Box 528, Mtwara, Tanzania

F.J. Biginagwa
Department of Biological Sciences, Faculty of Science, Sokoine University of Agriculture, P.O. Box 3038, Morogoro, Tanzania

M. Wagner, S. Lambert (eds.), *Freshwater Microplastics*,
Hdb Env Chem 58, DOI 10.1007/978-3-319-61615-5_6,
© The Author(s) 2018

the coastal regions and other freshwater environments can be applied to inland African waters. The aforementioned study in Lake Victoria is used to exemplify how small-scale investigations can provide early indications of MP pollution. Lastly we discuss the current challenges and future needs of MP research in African freshwaters.

Keywords Africa, African Great Lakes, Freshwater, Microplastics, MP sampling

1 Introduction

1.1 Africa, the Anthropocene, and Plastic Pollution

As the birthplace of our species, the African continent holds a unique place in human history. Current scientific consensus places the evolution of modern humans in East Africa approximately 200,000 years ago from where they successfully dispersed approximately 72,000 years ago during the late Pleistocene [1, 2]. From here our species continued to spread and over the next 50,000 years or so colonized the majority of the Earth's land surface. Fast-forward through the following epoch, the Holocene, which is regarded as being relatively stable in terms of climate, and we arrive at a point in time in which humankind have established themselves as the dominant force and major driver for environmental change. Accordingly a new era is said to have now dawned – the Anthropocene [3, 4]. While the exact start date of the Anthropocene is subject to much current debate, the advent of the industrial age (ca. 1800s) changed the dynamics between humans and the environment. The Anthropocene is thus defined by human actions which perturb the Earth's land, oceans, and biosphere [5]. These dramatic effects include climate change, ocean acidification, deforestation, and plastic pollution.

Plastics (and microplastics, MPs, defined as <5 mm in size) are considered a hallmark of this new anthropogenic age, having become widely used in the last 60 years [6], and are now a ubiquitous pollutant found worldwide and in all aquatic compartments (surface waters, water column, and sediments) and numerous animals (invertebrates, fish, seabirds, and marine mammals) [7]. Up until recently MP pollution had been viewed solely as a marine issue, but there is now an increasing amount of information regarding the presence of MPs in freshwaters [8, 9]. MPs have been sampled from both freshwater lakes, such as Lakes Erie, Huron, and Superior in Canada [10], Lake Geneva in Switzerland [11], and Lake Garda in Italy [12], and rivers, such as the River Thames in London (UK, [13]), River Seine in Paris (France, [14]), and the Danube [15], to name but a few. In this last study, the mass and abundance of drifting plastic items in the Austrian Danube were found to be higher than those of larval fish [15], which is an indication of the magnitude of the problem. However, there is remarkably little information on the presence of MPs in the freshwaters of Africa – the place where it all started for humans!

In this chapter we begin by outlining the scope for plastic pollution in African inland waters, both through the nature of the water bodies and the human pressures they face. We then focus briefly on the marine and estuarine MP research that has been conducted in Africa. There are only two studies that have investigated the prevalence of plastics in African freshwaters, specifically the Tanzanian waters of Lake Victoria, Ngupula et al. [16] and Biginagwa et al. [17], and only the latter is focused directly on MPs. They are exemplified as case studies, which, in addition to providing useful data, may also be a template for similar research in other African freshwater bodies. Lastly, we discuss the current challenges and knowledge gaps and future research needs that require attention in order to gain a better understanding of the presence, sources, and fate of MPs in inland African waters.

1.2 African Freshwaters and the Potential for MP Pollution

The African continent contains some of the most famous and notable freshwater bodies in the world. The River Nile, which is the second longest river, has been described as the "donor of life to Egypt" [18], and the River Congo is the second largest by river discharge (in both cases the Amazon is number one) and also the world's deepest river. Lakes Victoria, Tanganyika, and Turkana are perhaps the three most well-known of the African Great Lakes that are located in East Africa. Lake Nasser is a vast man-made reservoir that was created by the construction of the Aswan Dam across the River Nile. Each of these freshwater bodies, identified in Fig. 1, supports significantly sized populations (see Table 1a, b). The city of Cairo, through which the Nile flows, will have a population of over 20 million inhabitants by the year 2020 according to United Nations Sources [19]. The River Congo flows through the capital city of the Democratic Republic of the Congo, Kinshasa, with a population of over 14 million, and the Lakes Victoria and Tanganyika both have on their banks urban centers of >1 million people. Many of Africa's cities have undergone rapid urban expansion [20], with sub-Saharan urban growth averaging 140% between the 1960s and the 1990s, at a rate 10 times faster than OECD countries and 2.5 times than the rest of the developing world [21].

Inevitably, the pace of increase has placed pressure on urban services and not least in the management of solid waste, where it is common in the developing world for municipalities to be short of funds, deficient in institutional organization and interest, have poor equipment for waste collection, and lack urban planning [22, 23]. Among this waste are plastics. Plastics, as we know, are used in a variety of products including packaging, bags, bottles, and many other short-lived products that are discarded within a year of production [6]. In some parts of the world, plastic recycling procedures are well established, but in African countries, even when reuse and/or recycling practices are present, they often lack legal foundation and are therefore conducted on an ad hoc basis [24]. Thus much solid waste ends up in landfills or is subject to illegal dumping. In proximity to freshwater systems, plastic waste then has the potential to enter the aquatic environment where subsequent degradation can form MPs. The link between urban waste and MPs has been established in the freshwater MP literature [10, 11, 13, 15], but to date there is

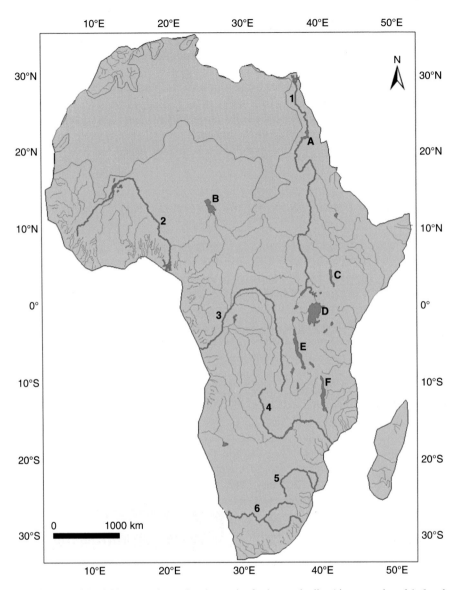

Fig. 1 Map of the African continent showing major freshwater bodies (rivers numbered 1–6 and lakes lettered *A–F*). Water bodies and their characteristics including highly populated neighboring urban centers can be found in Table 1

little information on this specific to African freshwaters. Contributions to MP concentrations may also be from fishing and tourism activities [25] which are commonly linked to freshwaters. Given the magnitude of Africa's freshwater bodies and the populations and activities they support, the likelihood for MP pollution in these waters is substantial.

Table 1 Major African freshwater bodies, their characteristics, and neighboring urban centers (>0.3 million people) with estimated populations (in 2020). (a) Major rivers depicted on Fig. 1 (numbered 1–6) are described by the location of river mouth, length, and average discharge. (b) Major lakes depicted on Fig. 1 (lettered A-F) are described by the counties in the basin, surface area and water volume

River[a]		Flows through	Length (km)	Average discharge (m^3/s)	Major urban centers and estimated population size in 2020 (million)[a]
1	Nile	Egypt, Sudan, South Sudan, Uganda	6,853	2,830	Cairo (20.57), Alexandria (5.23), Aswan (0.34) (all Egypt); Khartoum (5.91) (Sudan); Juba (0.40) (South Sudan)
2	Niger	Guinea, Mali, Niger, Benin, Nigeria	4,180	5,589	Bamako (3.27) (Mali); Niamey (1.32) (Niger); Lokoja (0.66) (Nigeria)
3	Congo	DR Congo, Congo	4,700	41,000	Kinshasa (14.12), Kisangani (1.25), Mbandaka (0.44) (all DR Congo); Brazzaville (2.21) (Congo)
4	Zambezi	Angola, Zambia, Zimbabwe, Mozambique	2,574	3,400	No urban centers >0.3 m people
5	Limpopo	South Africa, Botswana, Zimbabwe, Mozambique	1,750	170	No urban centers >0.3 m people
6	Orange	South Africa	2,200	365	No urban centers >0.3 m people
Lakes[b]		Basin countries	Surface area (km^2)	Water volume (km^3)	Major urban centers and estimated population size in 2020 (million)[a]
A	Nasser	Egypt, Sudan	5,250	132	Aswan (0.34) (Egypt)
B	Chad	Chad, Cameroon, Niger, Nigeria	1,350	72	N'Djaména (1.54) (Chad)
C	Turkana	Kenya, Ethiopia	6,405	203.6	No urban centers >0.3 m people
D	Victoria	Tanzania, Uganda, Kenya	68,800	2,750	Kampala (2.39) (Uganda); Mwanza (1.12) (Tanzania)
E	Tanganyika	Burundi, DR Congo, Tanzania, Zambia	32,900	18,900	Bujumbura (1.01) (Burundi); Uriva (0.57) (DR Congo)
F	Malawi	Malawi, Mozambique, Tanzania	29,600	8,400	No urban centers >0.3 m people

[a]Source: United Nations, Department of Economic and Social Affairs, Population Division (2014) [18]

Despite the lack of scientific confirmation of MPs in Africa's freshwaters, it would be unfair to say that there is a lack of recognition of the plastic issue. On the contrary, there has been a great deal of research conducted on the presence of MPs within the marine and estuarine environment (described in the following section), and also, there has been much progress made on reducing and banning the use of plastic bags in some countries. This progress has not been made solely to reduce the plastic waste but also on the grounds of environmental and public health. Improperly discarded plastic bags have been shown to block gutters and drains which create storm water problems and collect water which provides a breeding ground for mosquitos that spread malaria, and the use of bags as toilets has been linked to the spread of disease [26, 27]. The government of South Africa introduced levies on the use of plastic bags in 2003 [28], in 2005 Rwanda imposed a ban on the use and importation of plastic bags of <100 microns thick, and Tanzania similarly imposed a ban based on thickness in 2006 [27]. Such measures may not always be successful as in South Africa levies were not predicted to reduce the plastic bag litter stream [28]. Subsequently the actions taken, while positive, may have little impact in terms of the potential for MP pollution in African freshwaters. However, the scale of the problem first needs to be assessed, and in this regard, studies conducted in marine and estuarine waters may show the way forward.

2   Presence of MPs in African Marine and Estuarine Environments

In comparison to the rest of Africa, significant knowledge has been gathered about the presence, sources, and fate of plastics and MPs in the coastal regions around South Africa and their biota. The earliest documented reports of plastics are from the mid- to late 1980s with Ryan [29] having sampled the sea surface water off the southwestern Cape province between 1977 and 1978 with a total of 1,224 neuston trawls that found a mean plastic density of 3,640 particles km^{-2} with the majority of the particles in the MP range. Commonly found types were fragments, fibers, and foamed plastic particles with polyethylene being a predominant polymer [29]. A follow-up study [30] conducted at 50 South African beaches in 1984 and 1989 found a significant increase in the mean MP density from 491 m^{-1} in 1984 to 678 m $^{-1}$ 5 years later. Analysis of the distribution of MPs found that inshore currents rather than local sources were responsible for the variation in abundances between beaches. Conversely, in the case of macroplastics, it was the local sources that had the greater influence. More recent research conducted by Nel and Froneman [31] reached the same conclusion regarding the primary influence on the distribution of MPs in both sediment and water. Across 21 sampling locations along South Africa's southeastern coastline, comprising both bay and open coast areas, with both sediment and water samples analyzed for MP abundance, the authors found that MP densities in general did not significantly vary between sites in either matrix. As with the study conducted 25 years prior, the conclusion was that water circulation rather than proximity to land-based sources was main driver to MP abundances in coastal regions [31].

Biological sampling in this region has also revealed a number of interesting details regarding the fate of marine plastics. Plastic particles were found in more than half the seabirds predominantly sampled off Southern Africa and African sector of the Southern Ocean [32]. The size of the ingested particles was related to the body size of the bird, and smaller species exhibited a higher incidence of plastic ingestion. Dark-colored particles were more abundant suggesting a selection for easily visible particles rather than transparent ones. Omnivorous species were the most likely to confuse plastics with prey items, whereas feeding specialists were less likely to mistake plastics for food, unless they shared a resemblance [32]. A comparison of this historic dataset with a more recent sampling period (1999–2006) revealed a decrease in virgin pellet ingestion, but no overall change in total plastic ingestion [33]. This decrease suggested a change in the make-up of small plastic debris at sea in the intervening period.

Studies in the estuarine environment are less common than marine studies and, like freshwater research on MPs, have only recently started to gain momentum. However, estuaries provide pathways for the transport of MPs from catchments to the oceans, notably in urban areas where estuarine waters serve as industrial out-flows or fishing grounds [34, 35]. The characterization of MPs in five urban estuaries of Durban (KwaZulu-Natal, South Africa) found the highest concentrations in sediments collected from Durban harbor, which included cosmetic microbeads and fibers [35]. Possible sources were thought to include the several rivers that flow through Durban's industrial suburbs and enter the harbor, the industrial companies that use plastic powders and pellets around the harbor, and the closeness of dry docks where ship repairs take place. The fate of these plastics was revealed in a follow-up study by the same authors looking at plastic ingestion by the estuarine mullet (*Mugil cephalus*) in Durban harbor [36]. Plastic particles were found in the digestive tracts of 73% of the sampled fish, with more than half of the recovered plastics in the form of fibers and approximately one-third as frag-ments. Plastic concentrations found in the mullet were higher than those reported elsewhere for other species, and it appears that, as with omnivorous seabirds, the nonselective feeding mode of *M. cephalus* (i.e., ingestion of sediments) was a contributing factor.

Studies into South Africa's plastic and MP pollution are particularly pertinent as the country is ranked within the top 20 counties with the highest mass of mismanaged plastic waste [37]. Other African countries are also on the list, and although focused on marine debris, the relevance to freshwaters should not be ignored.

3   Presence, Sources, and Fate of MPs in Inland African Freshwaters

3.1   Presence of MPs in Freshwaters

The presence of MPs has been extensively reported in the marine environment [38–40], including that of South Africa's coast (as described in the previous section). In comparison, describing MPs in freshwaters is still in its infancy with the majority of research only arriving in the last 5 years [9]. Thus, only a few studies have investigated the occurrence of MPs in freshwaters with research conducted in the vicinity of urbanization and industrialization, such as Laurentian Great Lakes in North America [10] and Lake Geneva in Switzerland [11], as well as in more remote locations, such as Lake Hovsgol in Mongolia [25] and Lake Garda in Italy [12]. Not only do these studies show that MPs are present in freshwaters, but also relate the type of plastics found to their likely sources.

In the Laurentian Lakes (Lakes Superior, Huron, and Erie), MPs were found in 20 out of 21 surface samples, and in many of the tows, the most notable MPs were multicolored spherical beads that were determined to be polyethylene in composition. Shape, size, and composition were comparable to the microbeads used in exfoliating facial cleansers and cosmetic products and were likely to originate from nearby urban effluents [10]. Although there have been efforts to raise scientific, regulatory, and public awareness to ban the use of microbeads [41–43], successfully in some countries, the Canadian Great Lake study demonstrated that they are already abundant in the environment, and in the Laurentian lakes, "hot spots" were found where lake currents converge. Logically, it would be expected that remote lakes with lower population densities would have less plastic pollution than freshwaters near urban centers, but in case of Lake Hovsgol, the remote mountain lake in northwest Mongolia near the Russian border, the opposite was true. An MP density of 20,264 particles km^2 was averaged from nine transects making the lake more polluted than Lakes Huron and Superior. No microbeads were found with fragments and films instead being the most abundant MP shapes. The shoreline was dominated by discarded household waste (bottles, plastic bags) and fishing gear, and the likely source of the pelagic MPs was the degradation of this shoreline debris [25]. Thus even low population densities can cause significant levels of MP pollution in the absence of waste management infrastructures. Taken together it appears that plastics recovered from freshwaters in different parts of the world closely reflect the anthropogenic activities and waste generated by the local populations. Although this would seem obvious, further research is required to verify this link with the aim of more specific waste management relating to the nature of plastic pollution within a given location.

To date only two studies have attempted to document the presence of plastic debris in African freshwaters [16, 17], and only one specifically focused on MPs [17]. Both studies were conducted in the Tanzanian waters of Lake Victoria, and in the following sections, we describe them as case studies. In addition to providing valuable baseline data for MPs in Lake Victoria, these investigations may serve to inform how research could be conducted in other African freshwater bodies.

3.2 Plastics in the Tanzanian Waters of Lake Victoria

Lake Victoria is the world's second largest freshwater lake by area (the largest being Lake Superior in North America) and has been described as eutrophic and polluted due to human influences within the catchment area [44]. The area surrounding the lake is among the most densely populated in the world, and this population growth is set to continue – by the year 2020, an estimated 53 million people will inhabit the lake basin [45]. The majority of economic activities in the region are associated with the lake with one of the most important being fishing.

Case Study I details the work of Ngupula et al. [16] in which the authors documented presence and distribution of solid waste including plastic bags and fishing gear at six depth strata reaching 80 m below the surface. Thus, while they did not specifically look for MPs in the waters of Lake Victoria, the work of these authors greatly increases our understanding of where MPs originate from in the lake system. In the second case study by Biginagwa et al. [17], the ingestion of MPs by resident fish species in Lake Victoria was used in place of environmental sampling. The recovery of MPs from the gastrointestinal tracts of Lake Victoria Nile perch (*Lates niloticus*) and Nile tilapia (*Oreochromis niloticus*), and their subsequent characterization, provided the first evidence of MPs within African inland freshwaters.

3.2.1 Case Study I: Abundance, Composition, and Distribution of Solid Wastes in Lake Victoria

To determine the vertical distribution of solid wastes in Lake Victoria, the waters were categorized into three main ecological zones: (1) the nearshore, which is described as highly influenced by anthropogenic input and was sampled at depths of <10 m and 10.1–20 m; (2) the intermediate zone which is moderately influenced by the catchment and was sampled at depths of 20.1–30 m and 30.1–40 m; and (3) the deep offshore waters which are the most isolated from the human activities and were sampled at depths of 40.1–50 m and then >50.1. The maximum depth of Lake Victoria is 80 m; thus, this last depth stratum extended to bottom trawls. Across these three zones and six strata, 68 samples were taken in total during two periods, May and late September to early October 2013. Trawls were conducted at three knots and debris collected by 4 mm mesh trawl net.

Plastic debris was found at all depths and all sampling locations. Across all trawls, the dominant waste types originated from fishing activities; multifilament gillnets compromised 44% of all debris, monofilament gillnets (42%), longlines and hooks (7%), and floats (1%). Plastic bags (4%) and clothing (2%) accounted for the remaining solid waste. Gillnets, which compromised more than 80% of all the debris found and 96% of waste in the fourth depth strata, are constructed using synthetic fibers, and although nylon was used in the 1960s, newer materials, such as ultrahigh-molecular-weight polyethylene (UHMWPE) or polyethylene terephthalate (PET), are now commonplace as they are cheaper, are more durable, and require less maintenance. Multifilament gillnets are used in the fishing of Nile perch, while monofilaments are used for catching tilapiine species, including Nile tilapia. Both species were found to contain MPs in their intestinal tracts, as described in Case Study II [17].

There were only minor differences in the abundance by weight of debris sampled from the different depths, and of the six waste types identified, the proportion found at each depth did not vary to any great degree, with the exception of the bottom strata in which longlines and hooks (67%) were most abundant. Of the three ecological zones, the intermediate zone (20.1–40 m) contained most waste and is also known to have the highest levels of fishing activities; thus, within this zone, there was a reduced abundance of clothing and plastic bags. Fishing activity appears to be the major source of solid (plastic) waste in Lake Victoria, but land-based waste was not accounted for due to the inability to trawl at shallow depths (<4 m) in the nearshore. Land-based waste is often an important component of marine waste and through tidal action is transported to the lower depths of the sea [46]. However, without strong currents, this mechanism of circulating waste is ineffective within the lake environment. Nevertheless, within shallow waters, land-based waste would undoubtedly be important.

While not specifically focusing on the abundance of MPs, this study demonstrates that plastic waste is present at all levels of Lake Victoria and is strongly linked to fishing activities and discarded fishing gear. Though authors do not discount other sources including land runoff and transportation of cargo, the limitations of the study do not allow these to be investigated further.

3.2.2 Case Study II: Recovery of MPs from Lake Victoria Nile Perch and Nile Tilapia

A number of studies have used the ingestion of MPs by resident fish species as a marker of MP pollution. Lusher et al. [47] found that marine pelagic and demersal fish sampled from the English Channel readily consume plastics, and Sanchez et al. [48] similarly reported, for the first time in freshwaters, that wild gudgeons (*Gobio gobio*) inhabiting French rivers ingest MPs. Using Nile perch (*Lates niloticus*) and Nile tilapia (*Oreochromis niloticus*) as proxies for environmental MP contamination in Lake Victoria, a small-scale study was conducted in the Mwanza region of Tanzania, located on the Lake's southern shore (Fig. 2). Both species are economically and ecologically important and were introduced to Lake Victoria in the 1950s and 1960s with the aim of supplementing native fish populations that had declined due, in part, to overfishing [49]. However, this introduction was detrimental to the native species, particularly the native tilapiine species such as the Victoria tilapia

Fig. 2 Map of the Lake Victoria study area showing the Mwanza region. (**a**) Nile perch and Nile tilapia were purchased from the harbor market at Mwanza (regional capital). The local fishing area extends across the Mwanza Gulf and to Ukerewe Island. Inset Lake Victoria (LV) bordered by Uganda, Kenya, and Tanzania. The Mwanza region located on the southern shore of Lake Victoria is highlighted. (**b**) Urban waste in Mwanza, including plastic debris, collects in drainage ditches which are a potential source of plastic pollution in Lake Victoria. (**c**) Nile Tilapia used in this study (photographed prior to dissection) were purchased from the market

(*Oreochromis variabilis*) and Singidia tilapia (*Oreochromis esculentus*), which subsequently disappeared from parts of the lake [49, 50]. Thus Nile perch and Nile tilapia have established themselves as dominant commercial and ecological species and therefore represent logical choices by which to monitor MP pollution in the area. Moreover, their differing feeding habits could provide additional information by which to contextualize plastic ingestion. Nile perch are predatory fish feeding on *haplochromine* cichlids and gastropod snails, whereas Nile tilapia are omnivorous with a diet consisting of plankton and fish.

In March 2015, 20 fish of each species were purchased from Mwanza harbor market, where fish are caught and sold daily. The fishing territory for both species extends to Ukerewe Island (the largest island in Lake Victoria) to the North of Mwanza and across the Mwanza Gulf to the neighboring district of Sengerema (Fig. 2). Nile perch and Nile tilapia were 46–50 cm and 25–30 cm in length and 500–800 g and 500–700 g in weight, respectively. For each fish, the dissection of the entire gastrointestinal tract (buccal cavity to anus) was conducted on site. All efforts were made to eliminate sample contamination with separate clean dishes used for each fish and thorough cleaning of dissection utensils between samples. A preliminary examination was made of each gastrointestinal tract, and in the case of Nile perch, undigested gastropods and cichlids were removed. Gastrointestinal tracts and their contents were then individually preserved in 96% ethanol and transported to laboratory facilities at the University of Dar es Salaam (Dar es Salaam, Tanzania), a journey of approximately 1,150 km. In the laboratory, NaOH digestion (10 M NaOH at 60°C for 24 h) was used to isolate plastic litter from the organic tissue. The NaOH method has been shown to digest organic matter with an efficacy of >90% [51], and the tests of this protocol prior to its use confirmed such high efficiencies ($96.6 \pm 0.9\%$, $n = 5$, data not shown). Importantly, NaOH digestion has a minimal impact on the chemical and physical states of plastics, especially when compared to strong acid digestion which, while also being an effective digestant of organic matter, can discolor or degrade plastics. Post-digestion, plastics, and a minimal amount of partially digested tissue were rinsed from the NaOH through 250 μm mesh stainless steel sieves under running water and placed on filter paper to dry. Samples were then brought to the laboratory (Roskilde University, Denmark), and suspected plastic pieces were separated from tissue residue under light dissection microscope. The chemical composition of all suspected plastics was identified nondestructively by attenuated total reflectance Fourier transform infrared (ATR-FTIR) spectroscopy, a standard analytical technique for identifying the chemical composition of samples larger than 0.5 mm. Scans were run at a resolution of 2 cm^{-1} between 4,000 and 650 cm^{-1} on a Bruker Alpha FT-IR instrument (Bruker, Billerica, MA, USA) fitted with a diamond internal reflectance element. Spectra were compared with standard references on the same instrument and processed using Opus software supplied by Bruker.

In total, suspected plastics were recovered from the gastrointestinal tracts of 11 perch (55%) and 7 tilapia (35%). However, some plastics were too small (i.e., <0.5 mm) to have their chemical structure confirmed by ATR-FTIR. In addition, spectroscopy of some suspected plastic samples showed that their compositions most closely resembled cellulose, suggesting these samples were likely plant material or paper originating from perhaps newspaper, tissues, or cigarette filters. Thus 20% of each fish species (i.e., four individuals) contained confirmed MPs within their gastrointestinal tracts. The polymers recovered from the fish were polyethylene, polyurethane, polyester, copolymer (consisting of polyethylene and polypropylene), and silicone rubber (Fig. 3). The common use of such materials includes packaging, clothing, food and drink containers, insulation, and industrial applications (Table 2). Given the dimensions of the recovered plastics (0.5–5 mm, Fig. 3), it is likely that the MPs ingested by the fish are secondary MPs which have resulted from the degradation and breakdown of larger plastic pieces [38]. A likely source of the input of such materials into the Mwanza Gulf area is from the drainage ditches that are filled with urban waste, including plastic products (Fig. 2). This may be a particular problem during heavy rain when input into the lake is increased. In common with other studies conducted at freshwater sites [10, 25], it appears that the nature of the plastic pollution is related to the usage and waste by the local human population.

This work provided the first evidence that MPs are present in the African Great Lakes and that they are ingested by economically important fish species. In addition to confirming the ingestion of MPs by freshwater fish species [48], the chemical composition of the MPs was determined. However, this is only a preliminary study and only limited conclusions can be drawn. With plastics confirmed in only 20% of both species, the study likely underestimates the true extent of plastic ingestion by Nile perch and Nile tilapia, especially when considering the constraints of ATR-FTIR analysis and the inability to confirm the identity of the smaller-sized suspected "MPs." Similarly, it is not possible to determine whether the feeding preferences of the two species effected their ingestion of plastics. Thus, while this study provides evidence for the ingestion of secondary MPs by fish populations, it is clear that further research needs to be undertaken in Lake Victoria to fully characterize the extent of MP pollution.

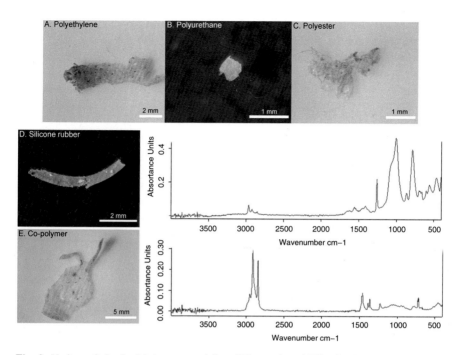

Fig. 3 Variety of plastic debris recovered from Nile perch and Nile tilapia. Images (**a–e**) are examples of the range of polymers isolated after NaOH digestion of the gastrointestinal tissue. In each case the identity of the polymer was confirmed by ATR-FTIR spectroscopy. Spectra attributed to silicone rubber (**d**) and polyethylene/polypropylene copolymer (**e**) debris are shown next to their respective plastic samples

Table 2 Polymers recovered from the gastrointestinal tracts of sampled fish and their common uses and potential source of plastic pollution in Lake Victoria

Polymer	Common uses and potential sources
PE/PP copolymer[a]	Packaging, carrier bags
Polyethylene	Carrier bags, food wrappers, beverage bottles
Polyester	Beverage bottles, textile (clothing, carpets, curtains)
Polyurethane	Insulation, sealants, packaging
Silicone rubber	Industrial sealants. O-rings, molds, food storage

[a]PE (polyethylene)/PP (polypropylene)

3.3 Plastics and MPs in Lake Victoria

Together these two cases provide compelling evidence that plastic debris in the Tanzanian waters of Lake Victoria is subject to degradation and the products of that breakdown are available for ingestion by resident piscine populations. While Ngupula et al. [16] found fishing to be the predominant source of debris, the identification of different polymers from the fish intestinal tracts suggests a wider range of inputs related to urban waste [17].

The consequence of plastic debris and MPs in the lake ecosystem requires further research. Other types of solid waste, such as those originating from paper production and agriculture, were found to interfere with the distribution of macroinvertebrate communities in the Kenyan waters of Lake Victoria [52]. Future investigations could also consider the trophic transfer of MPs through the freshwater food chain, particularly in the case of Nile perch which are known to feed on smaller fish (*haplochromine* cichlids) and gastropods, as well as any potential "vector effect" that facilitates the movement of adhered contaminants through the food chain [53]. These are important aspects to study primarily because the top predators in these food chains are the local residents that ultimately consume the fish.

Given the existing population density surrounding Lake Victoria and its estimated growth, the prevalence of plastic debris and subsequently MPs is also likely to increase. The reliance on the lake as a resource means that any potential impacts of MPs on the ecosystem and biota need to be researched, assessed, and, if possible, mitigated. However, research activities should not be confined only to Lake Victoria. A number of African freshwater bodies are just as likely or even more likely to be impacted by MP pollution. Potentially, the main message to be taken from these case studies is the relative simplicity by which they were accomplished. In particular, the purchase of fish from market and subsequent dissections required little specialist scientific equipment and could be replicated in other locations. In the following section, we consider the current challenges to MP research and mitigation in Africa and discuss future research needs.

4   Current Challenges and Future Research Needs

With only two case studies available to highlight plastic and MP pollution in African waters, the most obvious challenge and research need is the lack of data. More studies are urgently required to assess the extent of MP pollution in African freshwaters, as well as their sources and their fate. However, let us assume that this lack of data does not indicate that MPs are not present in the environment and that further research would describe their presence. In this case, the more immediate challenges may be how to mitigate MP pollution rather than just report it. We argue that effective waste management, increased public awareness, and political will are all necessary to avoid deleterious impacts. However, it is the combination of these factors rather than each one in isolation that is likely to affect change.

4.1 Current Challenges

4.1.1 Waste Management

Unlike most developed nations where plastic waste is often separated from other wastes prior to disposal [54], the management of solid wastes in many developing countries can be considered as problematic often due to inappropriate technology and infrastructure [55]. Thus while a significant proportion of plastics in developed countries are collected and recycled [6], in most African countries, even in the presence of reuse and/or recycling practices, effective plastic waste management often lacks a legal foundation [24]. This results in urban and industrial wastes in developing countries being sent to disposal sites or dumped as mixed bulks [56]. This type of dumping of refuse has been documented as a major cause of pollution in African waters and is a recognized source for MP pollution (e.g., Fig. 2b).

In order to improve waste management practices, sustainable approaches should be a priority. Examples of these approaches could include establishing permanent recycling stations or working with communities to promote recycling and change their perception of plastic from disposable single-use items. However, such approaches require time and effort, and moreover do necessarily have an impact on the current level of plastic waste in the inland water bodies. Following the characterization of plastic litter in Mongolia's Lake Hovsgol, local plans to regulate waste management and reduce waste production were suggested [25]. Based on the analyses and observations made in the two case studies presented in this chapter, similar proposals could certainly be made for this affected area and potentially implemented in other areas, following appropriate initial data collection and analysis.

One methodology that has been proposed for quickly assessing the impact of waste in the environment is the rapid environmental assessment (REA). The method involves scoring the abundances of key indicator species and the magnitude of environmental pressures concurrently on the same logarithmic assessment scale [57]. High pressure scores coupled with decreases in biological abundances indicate that urgent action is mediated. REAs were used to assess potential impacts and threats in the coastal region of Kerkennah, Tunisia. Solid waste densities, including plastics, were ranked with high scores, indicating the need for action, but scores for other pressures and biological abundance decreases were not determined to be high enough for remediation actions to take place. In this example the authors suggested that beach rubbish and coastal debris should be cleaned up, but further action was not needed at the present time [57]. While the REA approach demands a certain level of taxonomic knowledge, this is not prohibitory for the involvement of "non-experts" as the focus is taxonomic breadth rather than depth (i.e., broadscale). In Kerkennah, the training of team members without specific taxonomic or technical expertise was achieved via a 1–2 h PowerPoint presentation followed by trial REAs. Following training, assessment at each site was typically conducted in approximately

1 h [57]. While REAs capture low-resolution data, they do provide a means of grading levels of management urgency and response. Moreover, the surveyors need not be experts and could be sourced from the local community within a program set by the municipality or regional government – although such action requires political will and public awareness (as discussed in the following sections). Thus REAs with criteria (pressures and species indicators) tailored for site-specific thresholds could become a valuable tool in determining which African freshwater locations require remediation from MP pollution.

4.1.2 Political Will and Governance

Most of African freshwater bodies are transboundary (see Table 1), and therefore their management requires cooperation and effective, coherent regional environmental policies [58]. However, the management of most African transboundary lakes and rivers ecosystems is largely compromised by conflicting political standings among the riparian countries [59]. A good example of this is Lake Victoria which is shared by Tanzania, Kenya, and Uganda. Its management has been challenging due to a lack of good cooperation and harmonized policies mainly following the collapse of East African Community of 1977. Despite its reformation, there are still country-specific political issues hindering the management of the lake. This is also the case for other African Great Lakes like Lake Tanganyika and Lake Malawi. However, when policies, conventions, and cooperations do occur, the major focus is often on how natural resources can be shared [60], rather than the control of pollutants. Thus, at an international level, the political will to combat issues like MP pollution is not strong and is equally problematic at the local level. In most African countries, MP pollution is not recognized as emergent issue of concern, although the efforts to levy, reduce, and ban the use of plastic bags [26, 27] would suggest that the plastic issue is not entirely ignored.

It is perhaps stereotypical to consider, but in many African nations, the challenges faced are greater than MP pollution – war, famine, literacy rate, infrastructure, clean drinking water, poverty, and corruption [61]. Moreover, most African countries have insufficient budgets from which to plan and execute governmental projects including research activities. A number of countries receive financial aid, and under these circumstances, and understandably, the study of MP pollution is not of the highest priority. Based on this, the current financial challenges of working with MPs in African waters may not be solved by local budgets but rather by bringing together different stakeholders (i.e., local community, local and national governments, NGOs, researchers), in order to first collect data, evaluate steps forward, and implement effective measure to halt MP pollution.

4.1.3 Public Awareness

The role of the general public through awareness and active involvement (i.e., citizen science) is discussed in detail elsewhere in this book, both with an historical overview and specifics related to MP pollution (see Syberg et al. this volume [62]). Briefly it could be suggested that in comparison to other environmental issues, the public has been invaluable in assessing the magnitude of plastics and MP pollution through volunteer beach cleanups and surveys that provide data for monitoring programmers, as well as carrying out the practical task of removing beach litter. In the USA most information regarding the abundance and distribution of beach debris has been derived from volunteer beach cleaning efforts [63], and such public involvement is also occurring elsewhere. Public collaboration with scientific research has taken place in a number of locations worldwide, for instance, the collection of marine litter in the Firth of Forth, Scotland [64], collection of beach debris along the coast of southeast Chile [65], and many volunteers mobilized for beach surveys in South Africa [66]. However, to the best of our knowledge, such public-involving initiatives have not been attempted in areas surrounding African freshwaters. Part of this problem may be, as has been discussed, a scarcity of information regarding the scale of potential MP pollution, which results in a lack of funding and a lack of awareness.

As discussed, funding for environmental issues may not be the highest priority in most African countries, but NGOs which could collaboratively work with various public sectors have paid little or no attention in raising public awareness in the issue of plastic waste management [67]. Similarly, the opportunities for 3Rs (reduce, reuse, and recycle) are not well explored and advocated in developing countries [68]. It has been suggested that improved education on the issues of waste management in developing countries, and the preparation and training of environmental professionals and technicians, could be the way forward. Some developing countries have reported positive effects from investing in education, such as citizens assuming responsibility and higher status of waste workers, which have resulted in cleaner cities [68]. Such programs would potentially have similar results in urbanized regions around African freshwaters, and the downstream effect of cleaner cities would be less urban waste from which to produce MPs. But as mentioned earlier in this section, the increase of awareness and education of the population must be coupled with an increase in effective waste management and ultimately coherent regional political action.

4.2 Future Research Needs

To discuss the future research needs, we revisit the themes of this chapter – presence, sources, and fate of MPs in African inland waters. As mentioned several times, there is a dearth of information regarding the prevalence of MPs within

Africa's freshwaters. Filling this knowledge gap must therefore be the highest priority and an absolute necessity to further understandings of sources and fate. The two studies described in detail in this chapter have been conducted in the same region and found that the sources of plastic (and MP) pollution were linked to urban refuse and fishing activities. This echoes the findings of studies in other freshwater areas, where type of plastic and MPs reflect the usages and anthropogenic inputs of the local populations [10, 13, 25]. The population of Mwanza is estimated to be 1.12 million people by 2020 (Table 1), and while not an insignificant number, this is by no means the largest urban center close to a freshwater body. We, therefore, suggest likely candidates for future research are locations with high population densities.

The River Nile flows through a number of heavily populated cities, most notably, Khartoum in Sudan (almost six million inhabitants estimated by 2020), Alexandria (5.23 million), and, of course, Cairo (20.57 million) (Fig. 1, Table 1). While MPs have not been described in the Nile, other pollutants (i.e., trace metals Cd, Cr, Cu, Fe, Hg, Mn, Pb, and Zn) were found in the abiotic compartments and the tissues of resident fish populations [18]. It is worth noting that MPs have been shown to adsorb trace metals in the environment [69, 70], and within the laboratory, polyethylene MPs were shown to alter the bioavailability and uptake of Ag to freshwater zebra fish [71]. The River Congo similarly flows through densely populated cities, notably Kinshasa (14.12 million inhabitants) and Brazzaville (2.21 million), and these waters would also be suspected of having MPs present. Elevated trace metal concentrations in Congo sediments were found in the vicinity of urban runoff and domestic and industrial wastewater discharge into the river basin [72]. It would seem obvious to expect MPs to be present alongside other pollutants of urban origin in both these rivers.

How to determine the prevalence of MPs requires thought, and there are various sampling techniques to assess MP abundances to consider: (1) shoreline combing, (2) sediment sampling, (3) water trawls, (4) observational surveys, and (5) biological sampling. In different locations, some may be more or less relevant based on practical (the availability of personnel and equipment) and economic factors (i.e., funding). In our study (Case Study II [17]), reporting the presence of MPs in Lake Victoria, biological sampling was considered to be the most suitable technique as it required little specialist field equipment (i.e., mantra trawls or trawl nets), and the laboratory apparatus required to digest gastrointestinal tracts is relatively common. Additionally, the study was inexpensive as fish were purchased from the local market and the research could be conducted within a short space of time. However, it is necessary to select suitable biological indictors. Nonselective feeders provide a better reflection of MPs in the environment [32, 36]. For instance, the omnivorous fish, Nile tilapia, was used in Lake Victoria, and water-filtering mussels (*Mytilus edulis*) and sediment-dwelling lugworms (*Arenicola marina*) have been shown to take up MPs from their respective environments [73]. Studies such as the one we conducted in Lake Victoria only present a "snapshot" of MP pollution, and longitudinal studies are required to describe temporal and spatial differences. Where possible, a combination of techniques may be more advisable particularly to present a complete picture of MPs in the environment. However, with the current lack of information, reporting the presence of MPs from any compartment of African freshwater systems would be a welcome addition to the literature.

As described by Wagner et al. [8], information on the fate of MPs in freshwaters is scarce, if not absent. Some common questions that need to be addressed in all freshwaters and are still outstanding in marine waters include (1) the behavior of MPs in environment – how they distribute and where they settle; (2) interactions with biota, such as rates of excretion, accumulation, and infiltration in tissue; (3) effects of MP exposure in order to determine environmental hazard; and (4) interaction between MPs and other pollutants, the so-called vector effect. Such considerations are as important in African freshwaters as elsewhere, but as in most locations, regional concerns are also noted. As degradation rates of MPs are influenced by the amount and strength of UV radiation [74], MPs in African freshwaters, largely located in the tropics, are likely to be degraded faster than in more temperate conditions as reactions, such as photolysis, thermo-oxidation, and photooxidation, are accelerated in strong UV light [74, 75]. Degradation rates for MPs under these conditions and how this affects the aforementioned questions of distribution, biotic interactions, interactions with waterborne chemicals, and vector interactions should be determined.

In order to prevent and mitigate deleterious effects, the challenges of MP pollution cannot be dealt with by solely focusing on their presence and impacts in the environment, but rather investigation of the entire chain from production to disposal is mandatory [76]. Thus questions of fate must be integrated into the requirement to report the presence and understand the sources. We recommend the following focus areas to assess the current state of MPs in African inland waters:

1. Establishing a more complete picture of MP pollution in African freshwaters with the prioritization of locations with dense urban populations
2. Environmental monitoring programs that encompass water, sediment, and biota sampling and that consider spatial and temporal distributions
3. Life cycle assessments of plastics that consider production through disposal and fate in the environment
4. Interactions between MPs and (a) environmental factors, (b) other pollutants, and (c) resident biota

5 Conclusions

Knowledge regarding the presence, sources, and fate of MPs in freshwaters is being gathered apace in different parts of the world, but this information is currently lacking in Africa. Owing to the human pressures that increased urbanization has placed on many inland rivers and lakes, in combination with ineffective waste management and a general lack of awareness (although there are some notable exceptions, e.g., plastic bag bans), the potential for MP pollution is great.

The question then becomes, not if MPs are present, but where and how to sample them. The marine and estuarine research conducted in South Africa provides a potential guide via beach combing, water and sediment collection, and biological sampling. However, such efforts may be difficult in the absence of personnel, apparatus, and, of course, funding. Thus, the study of Biginagwa et al. [16], exemplified in Case Study II, in which MPs were extracted and identified from suitable biological indicators that inhabit the urbanized catchment area is offered as a model for research in other areas that can be conducted in a cost- and time-effective manner.

The confirmation of MPs is only the first step, albeit necessary for further understanding of sources and fate. Mitigating the effects of MPs requires the coming together of numerous interested stakeholders, not least the local populations. In the place where our species first evolved, it now falls on the current generation to preserve Africa's freshwaters for the future.

References

1. Oppenheimer S (2012) Out-of-Africa, the peopling of continents and islands: tracing uniparental gene trees across the map. Philos Trans R Soc B Biol Sci 367:770–784. doi:10.1098/rstb.2011.0306
2. Oppenheimer S (2009) The great arc of dispersal of modern humans: Africa to Australia. Quat Int 202:2–13. doi:10.1016/j.quaint.2008.05.015
3. Crutzen PJ (2002) Geology of mankind. Nature 415:23. doi:10.1038/415023a
4. Steffen W, Crutzen J, McNeill JR (2007) The Anthropocene: are humans now overwhelming the great forces of nature? Ambio 36:614–621. doi:10.1579/0044-7447(2007)36[614:TAAHNO]2.0.CO;2
5. Zalasiewicz J, Williams M, Haywood A, Ellis M (2011) The anthropocene: a new epoch of geological time? Philos Trans A Math Phys Eng Sci 369:835–841. doi:10.1098/rsta.2010.0339
6. Hopewell J, Dvorak R, Kosior E (2009) Plastics recycling: challenges and opportunities. Philos Trans R Soc B Biol Sci 364:2115–2126. doi:10.1098/rstb.2008.0311
7. Wright SL, Thompson RC, Galloway TS (2013) The physical impacts of microplastics on marine organisms: a review. Environ Pollut 178:483–492. doi:10.1016/j.envpol.2013.02.031
8. Wagner M, Scherer C, Alvarez-Muñoz D, Brennholt N, Bourrain X, Buchinger S, Fries E, Grosbois C, Klasmeier J, Marti T, Rodriguez-Mozaz S, Urbatzka R, Vethaak A, Winther-Nielsen M, Reifferscheid G (2014) Microplastics in freshwater ecosystems: what we know and what we need to know. Environ Sci Eur 26:12. doi:10.1186/s12302-014-0012-7
9. Eerkes-Medrano D, Thompson RC, Aldridge DC (2015) Microplastics in freshwater systems: a review of the emerging threats, identification of knowledge gaps and prioritisation of research needs. Water Res 75:63–82. doi:10.1016/j.watres.2015.02.012
10. Eriksen M, Mason S, Wilson S, Box C, Zellers A, Edwards W, Farley H, Amato S (2013) Microplastic pollution in the surface waters of the Laurentian Great Lakes. Mar Pollut Bull 77:177–182. doi:10.1016/j.marpolbul.2013.10.007
11. Faure F, Corbaz M, Baecher H, de Alencastro L (2012) Pollution due to plastics and microplastics in Lake Geneva and in the Mediterranean Sea. Arch Sci 65:157–164
12. Imhof HK, Ivleva NP, Schmid J, Niessner R, Laforsch C (2013) Contamination of beach sediments of a subalpine lake with microplastic particles. Curr Biol 23:R867–R868. doi:10.1016/j.cub.2013.09.001

13. Morritt D, Stefanoudis PV, Pearce D, Crimmen OA, Clark PF (2014) Plastic in the Thames: a river runs through it. Mar Pollut Bull 78:196–200. doi:10.1016/j.marpolbul.2013.10.035
14. Dris R, Gasperi J, Rocher V, Saad M, Renault N, Tassin B (2015) Microplastic contamination in an urban area: a case study in Greater Paris. Environ Chem 12:592–599. doi:10.1071/EN14167
15. Lechner A, Ramler D (2015) The discharge of certain amounts of industrial microplastic from a production plant into the River Danube is permitted by the Austrian legislation. Environ Pollut 200:159–160. doi:10.1016/j.envpol.2015.02.019
16. Ngupula GW, Kayanda RJ, Mashafi CA (2014) Abundance, composition and distribution of solid wastes in the Tanzanian waters of Lake Victoria. Afr J Aquat Sci 39:229–232. doi:10.2989/16085914.2014.924898
17. Biginagwa FJ, Mayoma BS, Shashoua Y, Syberg K, Khan FR (2016) First evidence of microplastics in the African Great Lakes: recovery from Lake Victoria Nile perch and Nile tilapia. J Great Lakes Res 42:146–149. doi:10.1016/j.jglr.2015.10.012
18. Osman AGM, Kloas W (2010) Water quality and heavy metal monitoring in water, sediments, and tissues of the African Catfish *Clarias gariepinus* (Burchell, 1822) from the River Nile, Egypt. J Environ Prot (Irvine, Calif) 1:389–400. doi:10.4236/jep.2010.14045
19. United Nations, Department of Economic and Social Affairs PD (2014) No Title. In: World Urban. Prospect. 2014 Revis. Cust. data Acquir. via website. https://esa.un.org/unpd/wup/
20. Okot-Okumu J (2012) Solid Waste Management in African cities – East Africa. INTECH Open Access Publisher, pp 3–20
21. Barrios S, Bertinelli L, Strobl E (2006) Climatic change and rural-urban migration: the case of sub-Saharan Africa. J Urban Econ 60:357–371. doi:10.1016/j.jue.2006.04.005
22. Henry RK, Yongsheng Z, Jun D (2006) Municipal solid waste management challenges in developing countries – Kenyan case study. Waste Manag 26:92–100. doi:10.1016/j.wasman.2005.03.007
23. Parrot L, Sotamenou J, Dia BK (2009) Municipal solid waste management in Africa: strategies and livelihoods in Yaoundé, Cameroon. Waste Manag 29:986–995. doi:10.1016/j.wasman.2008.05.005
24. Lederer J, Ongatai A, Odeda D, Rashid H, Otim S, Nabaasa M (2015) The generation of stakeholder's knowledge for solid waste management planning through action research: a case study from Busia, Uganda. Habitat Int 50:99–109. doi:10.1016/j.habitatint.2015.08.015
25. Free CM, Jensen OP, Mason SA, Eriksen M, Williamson NJ, Boldgiv B (2014) High-levels of microplastic pollution in a large, remote, mountain lake. Mar Pollut Bull 85:156–163. doi:10.1016/j.marpolbul.2014.06.001
26. Njeru J (2006) The urban political ecology of plastic bag waste problem in Nairobi, Kenya. Geoforum 37:1046–1058. doi:10.1016/j.geoforum.2006.03.003
27. Clapp J, Swanston L (2009) Doing away with plastic shopping bags: international patterns of norm emergence and policy implementation. Environ Polit 18:315–332. doi:10.1080/09644010902823717
28. Dikgang J, Leiman A, Visser M (2012) Resources, conservation and recycling analysis of the plastic-bag levy in South Africa. Resour Conserv Recycl 66:59–65. doi:10.1016/j.resconrec.2012.06.009
29. Ryan PG (1988) The characteristics and distribution of plastic particles at the sea-surface off the southwestern Cape Province, South Africa. Mar Environ Res 25:249–273. doi:10.1016/0141-1136(88)90015-3
30. Ryan PG, Moloney CL (1990) Plastic and other artefacts on South African beaches: temporal trends in abundance and composition. South African J Sci 86:450–452
31. Nel HA, Froneman PW (2015) A quantitative analysis of microplastic pollution along the south-eastern coastline of South Africa. Mar Pollut Bull 101:274–279. doi:10.1016/j.marpolbul.2015.09.043
32. Ryan PG (1987) The incidence and characteristics of plastic particles ingested by seabirds. Mar Environ Res 23(3):175–206

33. Ryan PG (2008) Seabirds indicate changes in the composition of plastic litter in the Atlantic and south-western Indian Oceans. Mar Pollut Bull 56:1406–1409. doi:10.1016/j.marpolbul. 2008.05.004

34. Dekiff JH, Remy D, Klasmeier J, Fries E (2014) Occurrence and spatial distribution of microplastics in sediments from Norderney. Environ Pollut 186:248–256. doi:10.1016/j. envpol.2013.11.019

35. Naidoo T, Glassom D, Smit AJ (2015) Plastic pollution in five urban estuaries of KwaZulu-Natal, South Africa. Mar Pollut Bull 101:473–480. doi:10.1016/j.marpolbul.2015.09.044

36. Naidoo T, Smit A, Glassom D (2016) Plastic ingestion by estuarine mullet *Mugil cephalus* (Mugilidae) in an urban harbour, KwaZulu-Natal, South Africa. Afr J Mar Sci 2338:1–5. doi:10.2989/1814232X.2016.1159616

37. Jambeck JR, Geyer R, Wilcox C, Siegler TR, Perryman M, Andrady A, Narayan R, Law KL (2015) Plastic waste inputs from land into the ocean. Science (80) 347:768–771

38. Derraik JG (2002) The pollution of the marine environment by plastic debris: a review. Mar Pollut Bull 44:842–852. doi:10.1016/S0025-326X(02)00220-5

39. Cole M, Lindeque P, Halsband C, Galloway TS (2011) Microplastics as contaminants in the marine environment: a review. Mar Pollut Bull 62:2588–2597. doi:10.1016/j.marpolbul.2011. 09.025

40. Ivar Do Sul JA, Costa MF (2014) The present and future of microplastic pollution in the marine environment. Environ Pollut 185:352–364. doi:10.1016/j.envpol.2013.10.036

41. Gregory MR (1996) Plastic scrubbers' in hand cleansers: a further (and minor) source for marine pollution identified. Mar Pollut Bull 32:867–871. doi:10.1016/S0025-326X(96)00047-1

42. Fendall LS, Sewell MA (2009) Contributing to marine pollution by washing your face: microplastics in facial cleansers. Mar Pollut Bull 58:1225–1228. doi:10.1016/j.marpolbul. 2009.04.025

43. Chang M (2015) Reducing microplastics from facial exfoliating cleansers in wastewater through treatment versus consumer product decisions. Mar Pollut Bull 101:330–333. doi:10. 1016/j.marpolbul.2015.10.074

44. Sitoki L, Gichuki J, Ezekiel C, Wanda F, Mkumbo OC, Marshall BE (2010) The environment of Lake Victoria (East Africa): current status and historical changes. Int Rev Hydrobiol 95: 209–223. doi:10.1002/iroh.201011226

45. Canter MJ, Ndegwa SN (2002) Environmental scarcity and conflict: a contrary case from Lake Victoria. Glob Environ Polit 2:40–62

46. Galgani F, Souplet A, Cadiou Y (1996) Accumulation of debris on the deep sea floor off the French Mediterranean coast. Mar Ecol Ser 142:225–234. doi:10.3354/meps142225

47. Lusher AL, McHugh M, Thompson RC (2013) Occurrence of microplastics in the gastrointestinal tract of pelagic and demersal fish from the English Channel. Mar Pollut Bull 67:94–99. doi:10.1016/j.marpolbul.2012.11.028

48. Sanchez W, Bender C, Porcher JM (2014) Wild gudgeons (*Gobio gobio*) from French rivers are contaminated by microplastics: preliminary study and first evidence. Environ Res 128: 98–100. doi:10.1016/j.envres.2013.11.004

49. Ogutu-Ohwayo R (1990) The decline of the native fishes of lakes Victoria and Kyoga (East Africa) and the impact of introduced species, especially the Nile perch, Lates niloticus, and the Nile tilapia, *Oreochromis niloticus*. Environ Biol Fish 27:81–96. doi:10.1007/BF00001938

50. Njiru M, Muchiri M, Cowx IG (2004) Shifts in the food of Nile tilapia, *Oreochromis niloticus* (L.) in Lake Victoria, Kenya. Afr J Ecol 42(3):163–170

51. Cole M, Webb H, Lindeque PK, Fileman ES, Halsband C, Galloway TS (2014) Isolation of microplastics in biota-rich seawater samples and marine organisms. Sci Rep 4:4528. doi:10. 1038/srep04528

52. Muli JR, Mavuti KM (2001) The benthic macrofauna community of Kenyan waters of Lake Victoria. Hydrobiologia 458(1):83–90

53. Syberg K, Khan FR, Selck H, Palmqvist A, Banta GT, Daley J, Sano L, Duhaime MB (2015) Microplastics: addressing ecological risk through lessons learned. Environ Toxicol Chem 34: 945–953. doi:10.1002/etc.2914
54. González-Torre PL, Adenso-Díaz B (2005) Influence of distance on the motivation and frequency of household recycling. Waste Manag 25:15–23. doi:10.1016/j.wasman.2004.08. 007
55. Matete N, Trois C (2008) Towards zero waste in emerging countries – a South African experience. Waste Manag 28:1480–1492. doi:10.1016/j.wasman.2007.06.006
56. Sharholy M, Ahmad K, Mahmood G, Trivedi RC (2008) Municipal solid waste management in Indian cities – a review. Waste Manag 28:459–467. doi:10.1016/j.wasman.2007.02.008
57. Price ARG, Jaoui K, Pearson MP, Jeudy de Grissac A (2014) An alert system for triggering different levels of coastal management urgency: Tunisia case study using rapid environmental assessment data. Mar Pollut Bull 80:88–96. doi:10.1016/j.marpolbul.2014.01.037
58. Botts L, Muldoon P, Botts P, von Moltke K (2001) The great lakes water quality agreement: its past successes and uncertain future. Knowl Power Particip Environ Policy Anal Policy Stud Rev Annu 12:121–143
59. UNU-INWEH (2011) Transboundary Lake Basin Management: Laurentian and African Great Lakes. UNU-INWEH, Hamilton, Ontario, Canada, 46 p. ISBN 92-808-6015-1
60. Abila R, Onyango P, Odongkara K (2006) Socio-economic viability and sustainability of BMUs: case study of the cross-border BMUs on Lake Victoria. Great Lakes of the World IV. Aquatic Ecosystem Health and Management Society, Bagamoyo, Tanzania. Burlington (ON)
61. Bashir NHH (2013) Plastic problem in Africa. Jpn J Vet Res 61(Supple: S1–S11). doi:10. 14943/jjvr.61.suppl.s1
62. Syberg K, Hansen SF, Christensen TB, Khan FR (2017) Risk perception of plastic pollution: importance of stakeholder involvement and citizen science. In: Wagner M, Lambert S (eds) Freshwater microplastics: emerging environmental contaminants? Handbook of environmental chemistry. Springer, Heidelberg. doi:10.1007/978-3-319-61615-5_10
63. Moore SL, Gregorio D, Carreon M, Weisberg SB, Leecaster MK (2001) Composition and distribution of beach debris in Orange County, California. Mar Pollut Bull 42:241–245. doi:10. 1016/S0025-326X(00)00148-X
64. Storrier KL, McGlashan DJ (2006) Development and management of a coastal litter campaign: the voluntary coastal partnership approach. Mar Policy 30:189–196. doi:10.1016/j.marpol. 2005.01.002
65. Bravo M, de los Ángeles Gallardo M, Luna-Jorquera G, Núñez P, Vásquez N, Thiel M (2009) Anthropogenic debris on beaches in the SE Pacific (Chile): results from a national survey supported by volunteers. Mar Pollut Bull 58:1718–1726. doi:10.1016/j.marpolbul.2009.06.017
66. Ryan PG, Moore CJ, van Franeker JA, Moloney CL (2009) Monitoring the abundance of plastic debris in the marine environment. Philos Trans R Soc Lond Ser B Biol Sci 364: 1999–2012. doi:10.1098/rstb.2008.0207
67. Scheinberg A, Wilson DC, Rodic L (2010) Solid Waste Management in the World's Cities. UN-Habitat's State Water Sanit World's Cities Ser Earthscan UN-Habitat, London, pp 1–228
68. Guerrero LA, Maas G, Hogland W (2013) Solid waste management challenges for cities in developing countries. Waste Manag 33:220–232. doi:10.1016/j.wasman.2012.09.008
69. Ashton K, Holmes L, Turner A (2010) Association of metals with plastic production pellets in the marine environment. Mar Pollut Bull 60:2050–2055. doi:10.1016/j.marpolbul.2010.07.014
70. Holmes LA, Turner A, Thompson RC (2012) Adsorption of trace metals to plastic resin pellets in the marine environment. Environ Pollut 160:42–48. doi:10.1016/j.envpol.2011.08.052
71. Khan FR, Syberg K, Shashoua Y, Bury NR (2015) Influence of polyethylene microplastic beads on the uptake and localization of silver in zebrafish (*Danio rerio*). Environ Pollut 206: 73–79. doi:10.1016/j.envpol.2015.06.009
72. Mwanamoki PM, Devarajan N, Thevenon F, Birane N, de Alencastro LF, Grandjean D, Mpiana PT, Prabakar K, Mubedi JI, Kabele CG, Wildi W, Poté J (2014) Trace metals and persistent organic pollutants in sediments from river-reservoir systems in Democratic Republic

of Congo (DRC): spatial distribution and potential ecotoxicological effects. Chemosphere 111: 485–492. doi:10.1016/j.chemosphere.2014.04.083

73. Van Cauwenberghe L, Claessens M, Vandegehuchte MB, Janssen CR (2015) Microplastics are taken up by mussels (*Mytilus edulis*) and lugworms (*Arenicola marina*) living in natural habitats. Environ Pollut 199:10–17. doi:10.1016/j.envpol.2015.01.008

74. Andrady AL (2011) Microplastics in the marine environment. Mar Pollut Bull 62:1596–1605. doi:10.1016/j.marpolbul.2011.05.030

75. Bonhomme S, Cuer A, Delort AM, Lemaire J, Sancelme M, Scott G (2003) Environmental biodegradation of polyethylene. Polym Degrad Stab 81:441–452. doi:10.1016/S0141-3910 (03)00129-0

76. Vidanaarachchi CK, Yuen STS, Pilapitiya S (2006) Municipal solid waste management in the Southern Province of Sri Lanka: problems, issues and challenges. Waste Manag 26:920–930. doi:10.1016/j.wasman.2005.09.013

Modeling the Fate and Transport of Plastic Debris in Freshwaters: Review and Guidance

Merel Kooi, Ellen Besseling, Carolien Kroeze, Annemarie P. van Wenzel, and Albert A. Koelmans

Abstract Contamination with plastic debris has been recognized as one of today's major environmental quality problems. Because most of the sources are land based, concerns are increasingly focused on the freshwater and terrestrial environment. Fate and transport models for plastic debris can complement information from measurements and will play an important role in the prospective risk assessment of plastic debris. We review the present knowledge with respect to fate and transport modeling of plastic debris in freshwater catchment areas, focusing especially on nano- and microplastics. Starting with a brief overview of theory and models for nonplastic particles, we discuss plastic-specific properties, processes, and existing mass-balance-, multimedia-, and spatiotemporally explicit fate models. We find that generally many theoretical and conceptual approaches from models developed earlier for other types of (low density) particles apply also to

---

This chapter has been externally peer reviewed.

M. Kooi (✉)
Aquatic Ecology and Water Quality Management Group, Wageningen University & Research, 6700 AA, Wageningen, Netherlands
e-mail: merel.kooi@wur.nl

E. Besseling and A.A. Koelmans
Aquatic Ecology and Water Quality Management Group, Wageningen University & Research, 6700 AA, Wageningen, Netherlands

Wageningen Marine Research, 1970 AB IJmuiden, Netherlands

C. Kroeze
Water Systems and Global Change Group, Wageningen University & Research, 6700 AA, Wageningen, Netherlands

A.P. van Wenzel
KWR Watercycle Research Institute, Nieuwegein, Netherlands

Copernicus Institute, Utrecht University, Utrecht, Netherlands

M. Wagner, S. Lambert (eds.), *Freshwater Microplastics*,
Hdb Env Chem 58, DOI 10.1007/978-3-319-61615-5_7,
© The Author(s) 2018

plastic debris. A unique feature of plastic debris, however, is its combination of high persistence, low density, and extremely wide size distribution, ranging from the nanometer to the >cm scale. This causes the system behavior of plastic debris to show a far wider variety than most other materials or chemicals. We provide recommendations for further development of these models and implications and guidance for how fate and transport models can be used in a framework for the tiered risk assessment of plastic debris.

Keywords Fate, Freshwater, Microplastics, Modeling, Nanoplastics

1 Introduction

Contamination of the environment with plastic debris has received increasing attention from the public, environmentalists, scientists, and policy makers since the 1970s [1, 2]. Model predictions suggest that currently over 5 trillion plastic particles float on the ocean surface [3] and that in 2010 alone between 4.8 and 12.7 million metric tons of plastic entered the ocean [4]. Plastics occur in a wide range of sizes, and particles can therefore be ingested by a variety of terrestrial [5] and aquatic species [6]. Ingestion of microplastics, particles <5 mm in length [1], can negatively affect hatching, growth rates, and food ingestion [7, 8]. Besides the potential effect of ingestion, plastic particles can act as vectors for organic pollutants [9] or function as floaters for (invasive) rafting species [10]. The occurrence and distribution of plastic debris in the marine environment has been studied even in the most remote areas, such as the arctic [11] and the ocean floor [12]. However, even though rivers are recognized as a major source of marine litter [13–15], the occurrence of plastic debris in freshwater systems just started to receive attention [16, 17].

Microplastics have been found in freshwater systems around the world, as summarized in a recent review by Eerkes-Medrano et al. [17]. Occurrence of microplastics in freshwater systems ranges from remote lakes [18] to industrial rivers such as the Rhine [15, 19] or St. Lawrence River [20]. Sources of plastic debris in freshwater systems have not been studied extensively but likely include effluents from wastewater treatment plants (WWTP), sewage sludge, shipping activities, atmospheric fallout, direct disposal from the public, beach littering, and runoff from agricultural, recreational, industrial, and urban areas [16, 21]. High loads are estimated to enter the marine environment: for example, an average of 1,533 t plastic per year was estimated to enter the Black Sea from the Danube [13], and an average of 208 t plastic per year was estimated to enter the Mediterranean from the Rhone [22]. However, river loads exhibit a high degree of variation. For example, rain events were shown to increase the plastic concentration up to 150 times in an urban part of the Rhone catchment [22]. Also, total loads in the Danube varied between 10.9 ± 43.6 and 2.2 ± 3.0 g (mean ± SD) per 1,000 m^3 from 2010 to 2012 [13], indicating both the uncertainty in the load estimates and the temporal change of plastic loads. Transport of plastic near the bottom of the river

[23], plastic deposited in river sediments [15], and fragmentation increase the uncertainty with respect to loads even further.

Besides microplastics, nanoplastics are likely to be present in the freshwater environment [24]. No formal size definition has been set for nanoplastics, resulting in different classifications such as <100 nm [24, 25], <1 µm [26], and <20 µm [16]. Hereafter, we will use <100 nm as a size cutoff for nanoplastics, to comply with the definition of engineered nanoparticles [24]. Nanoplastics can be either directly released into to the environment (e.g., as a by-product of thermal cutting, 3D printing) or indirectly via the degradation of larger plastics [24, 27–29]. Several studies have shown that nanoplastics can be ingested by a variety of organisms, although systematic effects remain unknown (summarized in [24, 30]). Despite the attention to plastic pollution and the potential harm it causes in the environment, to date no proper environmental risk assessment (ERA) framework is available for this anthropogenic pollutant. So far, microplastics have been found to be ingested by freshwater organisms such as fish [31–33] and mud snails [34] (see [8] for further detail). However, effect assessments are scarcely done for freshwater species [16, 17]. Retrospective exposure assessments have also not been done yet for plastic debris, because of the difficult, time-consuming, and costly detection methods currently available. However, exposure assessments can also be based on quantitative model estimates of plastic debris loads and distributions. To our knowledge, only one transport-fate model has been developed for plastic debris from nano- to 1-cm-sized particles [35, 36], one for microplastics [37] in rivers, and none for lakes. However, other types of models simulating particle transport in rivers do exist, and they can be used as inspiration for new plastic debris transport models for the freshwater environment.

The aims of this review are (a) to identify how existing particle transport models can serve as examples for new plastic transport models, (b) to identify the properties and processes that are relevant for the modeling of plastic debris in freshwater systems, (c) to review the existing models that (to some extent) already take into account these properties and processes, and (d) to provide recommendations for the further development of these models and guidance of how these models can be used in the framework of an ERA. We first briefly discuss existing particle transport and fate models for different particle types such as sediment or organic matter (Sect. 2). We identify what characterizes plastic debris from a transport modeling perspective and how this differs from other (traditional, natural) particles (Sect. 3), followed by a critical review of the fate models for freshwater systems published in the peer-reviewed literature (Sect. 4). In Sect. 5, we include a short review on data and knowledge gaps in relation to plastic modeling and discuss what kind of model categories are highly relevant for plastic debris. We also discuss the possible role of fate modeling in a future risk assessment framework for plastic debris in freshwater systems. The terms "plastic debris," "plastics," and "plastic particles" are used interchangeably in this review and do not refer to a specific size class. Macroplastics, microplastics, and nanoplastics refer to particles >5 mm in size, particles between 5 and 100 nm in size, and particles <100 nm, respectively.

2   Modeling the Transport of Particles in Aquatic Systems

Few models exist that simulate the transport and fate of plastic debris in freshwater systems. Plastic debris includes buoyant macroplastic items like bottles, food wrappers and containers, plastic cutlery, and expanded polystyrene (PS), larger polyethylene (PE), or polypropylene (PP) items that float at the surface [13, 38] and will be transported under the influence of water flow and wind (discussed in Sect. 4 and Fig. 1). Non-buoyant plastics or buoyant plastics that become more susceptible to vertical mixing due to their small size (i.e., microplastic and nanoplastic) will become submerged and may be subject to settling in a fashion similar to that of natural colloids and suspended solids (Fig. 2). Hence, such natural particles may serve as a proxy for some classes of plastic debris, and models simulating the transport of such natural particles can form the basis for the development of transport models for plastic debris. In this section, we summarize modeling methods for (submerged) particles in freshwater systems in general. This includes how different materials, aquatic systems, processes, and scales can be modeled. It is beyond the scope of this review to strive for completeness with respect to the large number of specific particle transport models that have been presented before, especially since excellent reviews on transport models already exist for sediment [39–42], algae [43], microorganisms [44], and nanomaterials [45, 46]. These reviews describe the present top models such as SWAT, WASP, HSPF, ANSWERS, and WEPP, all of which include suspended solids [39, 42].

Key Processes Affecting Particle Transport in Freshwater Systems Particles can enter an aquatic system via external inputs including, for example, WWTP

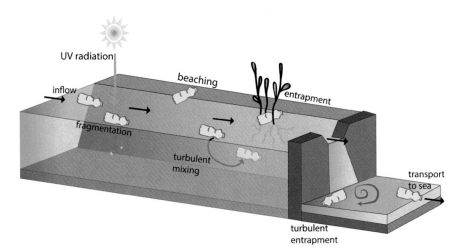

Fig. 1 Schematic representation of the different processes playing a role in the transport of macroplastic in a river and lake. Turbulent water movement below a weir can "capture" plastic debris for a certain period of time. The scaling of the different components is not representative, and not all processes happen to each plastic piece or in a fixed order

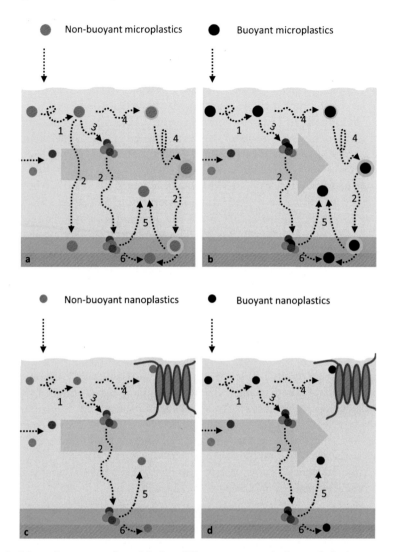

Fig. 2 Schematic representation of the key different processes playing a role in the transport of non-buoyant microplastics (**a**), buoyant microplastics (**b**), non-buoyant nanoplastics (**c**), and buoyant nanoplastics (**d**) in a river or lake. Processes include (*1*) turbulent transport, (*2*) settling, (*3*) aggregation, (*4*) biofouling, (*5*) resuspension, and (*6*) burial. Aggregates can be formed with, for example, sediment, algae, organic matter, or dissolved substances. The scaling of the different components is not representative, and not all processes happen to each particle or in a certain order. Other processes, such as removal by ingestion, relocation, and hydrodynamic alteration by ingestion and excretion (e.g., zooplankton, mussels), can also affect particle fate but are not depicted here

effluents, atmospheric deposition, groundwater, or surface runoff [16, 21]. Several fundamental processes drive the subsequent transport of particles in streams. In fluid mechanics, the collective motion of particles in a fluid is the result of advective, dispersive, and diffusive mass transfer. Advection refers to the longitudinal transport based on the average flow velocity. Dispersive mass transfer is the turbulent spreading of mass from highly concentrated areas to less concentrated areas [47]. This results from nonideal flow patterns (i.e., deviations from plug flow) and is a macroscopic phenomenon. Dispersion is multidirectional, as it covers the distribution of all flow directions and velocities [48]. Diffusion is the transport of particles from a high to a low concentration caused by random molecular (Brownian) motion, which is a microscopic phenomenon. The combined transport of substances in rivers due to advection and dispersion is commonly described by the one-dimensional advection-dispersion equation [49, 50]. Besides being transported in the water, suspended solids can be removed from the fluid by settling, which can be modeled with Stokes law or a modification of that law [51, 52]. Particles and substances can reenter the water column by erosion/resuspension of the riverbed. Advection, dispersion, diffusion, settling, and resuspension depend on particle properties such as size, density, shape, fractal dimension, and porosity [53–55]. During transport, the aforementioned particle properties can change due to aggregation or biofouling, which will further influence their fate [56–58]. Aggregation is usually modeled using a von Smoluchowski particle interaction model where the formation of aggregates is described kinetically as a function of the colliding particle concentrations, their sizes and densities, their collision frequencies, and attachment efficiencies [56, 59, 60]. Many studies assume spherical particles or aggregates, although it has been suggested that fractal dimensions should be included in these models [52, 61]. For aggregates, this would result in more realistic collision radii and hydrodynamics, therewith providing better collision frequency and sedimentation estimates [61]. The relative importance of the different transport processes is dependent on the particle properties. For example, Brownian motion is important for nanoparticle aggregation [36, 56], whereas it will be negligible when studying the transport of larger particles [47]. Likewise, biofouling has a faster effect on the buoyancy of microscopic particles, which have a large surface-to-volume ratio, compared to macroscopic particles [62].

Type of Aquatic System Several system properties affect the occurrence of the abovementioned processes. Transport of particles in rivers and lakes differs in many aspects. Rivers have a downstream discharge driven by an elevation gradient. Although natural rivers are turbulent, the time-averaged motion of the water is in the longitudinal direction. In rivers, the advection flow component is usually higher than the dispersion component [47]. Due to sedimentation and burial of suspended solids and associated contaminants, rivers often act as a sink for these contaminants. Compared to rivers, lakes have a very low, if any, flow velocity, enhancing sedimentation processes. Water residence times can be days to $>10^3$ years [63] compared to days or weeks in rivers. Therefore, biological and chemical processes are usually more important for the fate of particles in lakes. Mixing processes in lakes, and therefore the importance of sedimentation versus resuspension, can be complex to model because of vertical stratification, the effect of wind, and the lake geometry [47]. Dams in rivers or lakes can increase the water retention time and lower the flow velocity, enhancing sedimentation of suspended solids [64]. Also, water withdrawal for various human uses [65, 66], which is highly different for different regions [67], influences the fate of plastics as these abstracted particles are no longer carried to the ocean.

Particle Type With respect to modeling natural particles in freshwater systems, different particle types can be distinguished, such as sediment [39, 68, 69], algae [43], microorganisms [44], particulate organic matter [70, 71], nanoparticles [59, 63, 72, 73], and seeds [74]. The properties of these materials, such as size, shape, density, porosity, fractal dimension, and attachment efficiency, influence their hydrodynamic behavior and thereby their fate [75]. Some of them approach the properties of categories of plastic particles, which may cause them to have similar hydrodynamic behavior and a mutual applicability of modeling approaches and results. For instance, some plastic particles may become captured in low-density aggregates or flocs, as has been shown for the marine environment [57, 76], which affects the hydrodynamics of the resulting new composite particles [77, 78]. This implies that the transport of the plastic-inclusive floc or aggregate may become indistinguishable from that of a fully natural floc or aggregate. The implications of similarities and differences of plastic compared to natural solid materials for fate and transport will be further discussed in the next section.

3 Plastic Debris: Properties and Processes Relevant for Fate Modeling

Key Properties Relevant for Fate Modeling Plastic debris comprises a highly diverse mixture of particle sizes and shapes, made out of different polymers. The size ranges from >10 cm for fishing nets, bottles, and plastic bags to nanosized particles <100 nm. Nanoplastics have so far not been detected in natural waters but are likely to be present [24, 36]. The density of plastics ranges from 50 kg m^{-3} for extruded polystyrene foam to 1,400 kg m^{-3} for PVC. It can be expected that the composition of plastic in rivers is related to the production volumes of the different polymers, of which polyethylene (38%), polypropylene (24%), PVC (19%), and polystyrene (6%) are produced most [1]. Recent data partly confirmed these relative proportions of polymers in river sediments of the river Rhine [15], in the reservoir of the Three Gorges Dam [79], and floating on the river Seine [38]. Besides the size and density, the shape of plastics is also highly variable, ranging from small lines and fibers to irregular fragments to granules [80]. Microplastics have often been classified as fragments, fibers, spheres, pellets, lines, sheets, flakes, and foam [13, 15, 22, 79, 81], of which fragments are most abundant [15, 22, 79]. The size, shape, and density of particles will influence their transport behavior and fate in the aquatic environment.

The unique nature of plastic debris can be illustrated by comparison with properties of other types of particles present in water systems. Plastic can be considered to be unique with respect to fate processes because:

- Other particles can be similar sized but then have higher density (metal-based nanoparticles and colloids, suspended sediments, clays, minerals).
- Other particles can have similar density but are far less persistent (wood, algae, detritus, exopolymers, organic matter flocs, or organic colloids).
- Other particles do not exist in a nm to > cm size range with all other properties being similar to those of plastics.

We argue that the combination of low density (often near that of water), persistence, wide size range, and variable shape is what makes plastic particles and thus fate model simulation results different from those for other particles. At the same time, low-density nanomaterials (fullerenes, carbon nanotubes) or natural organic particles like cellulose can have a hydrodynamic behavior similar to that of some specific plastic particles.

Processes Specifically Relevant for the Modeling of Plastic Debris Once in the aquatic environment, plastics will be transported downstream. Floating macroplastic can be assumed to be transported with the flow (Fig. 1), i.e., to estuaries, to sea, or to lake reservoirs, where reduced flow conditions, fouling, embrittlement, and fragmentation may trigger sedimentation and further dispersion. Larger items will also accumulate on riverbanks due to wind or reduced flow or dispersive flow patterns in river bends. Vegetation or trees near the shores may serve as a temporary sink for large plastic debris [82], which later on may be released again to the main stream. Non-buoyant plastic debris is subject to the advective, dispersive, and sedimentation processes as described in the previous section. A unique feature here is that a high proportion of the plastic will have a density not that different from that of water, in contrast to natural suspended (mineral) solid particles of the same size. The variety of plastic sizes and densities, however, still varies enormously, leading to a wide variety of transport patterns for individual particles in the mixture.

Biofouling of plastics has been reported for freshwater samples [83, 84] and also is a well-researched phenomenon in marine waters [57, 58, 62, 84]. Plastic debris of all sizes and densities will be fouled and colonized by microbes, forming biofilms, which can lead to significant changes in particle buoyancy. For instance, increased settling as a result of biofouling has recently been shown for marine particles [57, 58, 62], and it is plausible that the same holds for plastics in the freshwater environment (Fig. 2). The recent detection of microplastics in rivers and lake sediment [15, 20, 85] confirms that particles with a density higher as well as lower than water can settle and be buried in the sediment. Recent model analysis showed that this also can be explained on a theoretical basis [35, 36]. Buoyant plastics will only settle when they are incorporated in aggregates with a density larger than the water density. This is an important phenomenon, which is mechanistically explained by biofouling causing an overall increase in density and attachment efficiency with other particles. Heteroaggregation with natural colloids, clays, and other high-density suspended particles will lead to faster sedimentation of the plastic particles that are captured in the aggregate [35, 36] (Fig. 2).

Another unique feature is the high persistence of plastic. Other particles with similar density and size, e.g., wood, algae, detritus, or other natural organic matter solids, disappear through degradation and mineralization within rather short time scales. Plastic debris however, once buried in the sediment, will only be mineralized on very long time scales, rendering them highly accumulative, bioavailable, and also subject to further transport. As long as plastic particles are close to the sediment surface, they can be resuspended if the flow velocity is high enough to exceed the critical shear stress [86]. However, after prolonged sedimentation, the particles could become "buried." Buried plastic debris would not resuspend anymore, unless turbulence would increase sharply due to storm events or flash floods, for example.

In the laboratory under accelerated weathering conditions, plastics have been shown to become brittle and fragment [28], and it is likely that this also occurs in freshwater systems. This process however is very slow in nature [87] and probably much slower than the typical residence times of plastic in rivers. In lakes with a large retention time, weathering is potentially important though. Fragmentation is caused by photodegradation, thermo-oxidation, hydrolysis, physical abrasion, and/or biodegradation [1, 88]. Most of these processes require either light, friction, or oxygen to act on the surfaces of the particles, which implies that once buried in anoxic sediment layers, plastic will be preserved for at least decades [89]. Several model categories exist that can use the above mechanistic evidence to simulate the fate of plastic debris in rivers, some of which already have been published in the literature [36, 37, 72].

4 Models for Fate and Transport of Microplastics in Freshwater Systems

In this section, four categories of models will be discussed: emission-based mass balance modeling, global modeling, multimedia modeling, and spatiotemporally explicit modeling. The models differ in their aim, design, scale, level of detail, and state of validation (Table 1). We classified the models based on their major characteristics, but some overlap in these classifications can be found. For example, a global model can also be referred to as spatiotemporally explicit yet on a much larger scale, and a small-scale spatiotemporally explicit model can cover plastic transport in water and sediment, rendering it "multimedia."

Table 1 Summary of the main features of the currently existing plastic debris models for freshwater systems

	Mass flow models	Global modeling	Multimedia models	Spatiotemporally explicit models	
Reference	Van Wezel et al. [90]	Siegfried et al. [91, 92]	This study, based on Meesters et al. [72]	Nizzetto et al. [37]	Besseling et al. [35, 36]
Plastic size range[a]	Microplastics	Microplastics	Nanoplastic (<100 μm), microplastic	Microplastic, 0.005–0.5, separated in five size classes	Nano- and microplastics. Ten sizes modeled; from 100 nm to 10 mm
Plastic density	All	All	All	Non-buoyant	All
Media included	Effluents	Water	Air, water, soil, sediment	Soil, effluents, water, sediment	Water, sediment
Processes included	Emissions (personal care products), plastic removal in WWTP	Emissions (personal care products, care tires), plastic removal in WWTP, during river transport and by water abstraction	Assumed emissions (1,000 t)	Emissions from sewage sludge, surface runoff, WWTP effluents, advection, settling, resuspension, store depletion	Assumed emissions upstream, advection, dispersion, biofouling, aggregation, degradation, settling, resuspension, burial
Spatial resolution	zero-D	1°latitude by 1°longitude (input) and basis totals (output)	zero-D	10,000 km^2 divided in eight segments	40 km river stretch divided in 477 segments of on average 87.7 m
Temporal resolution	Steady state	Annual totals	Steady state	Daily, simulation for 2008–2014	0.01 day, modeled until steady state was reached
Validation type[b]	c,d	c,d	c,d	b,c,d; model was validated for sediment particles and hydrology	b,c,d; model was validated for CeO$_2$ submicron particles and hydrology
Key assumptions	Generic, all water ends in WWTP, all used cosmetics end in WWTP, no secondary plastics, no other sources	Homogeneous distribution of parameters per catchment	All processes can be captured by first-order relations	Homogeneous distribution of MP in segment, lumped rainfall and temperature for catchment, pristine particles	Constant concentration upstream, (near-) spherical particles. Dominance of hetero- over homoaggregation

Model is based on	Mass balance point model	Global NEWS – Nutrient Export from WaterSheds	SimpleBox and SimpleBox4Nano (SB4N)	Hydrobiogeochemical spatiotemporally multi-media model, INCA-contaminants [93], with surface runoff [94] and sediment transport [95] modules	DUFLOW water quality modeling suite, NanoDUFLOW [96]
Review	Simple model which is easily applicable and adjustable. Drawback is the large uncertainty in estimates and the lack of spatial or temporal resolution. Validation with data still uncertain	This model is applicable to >6,000 rivers worldwide. For each river basin, the model calculates plastic in wastewater, removal during treatment, and removal during river transport	Simple model which is easily applicable and adjustable. It includes soil, water, and atmosphere compartments. Drawback is the large uncertainty in estimates and the lack of spatial or temporal resolution	Spatiotemporally explicit, with a high quality on the hydrodynamic processes, including surface runoff. Present version assumes pristine particles, which can result in underestimation of settling because biofouling and aggregation are not included	Spatiotemporally explicit, with a high quality on the hydrodynamic processes, including aggregation and nanoplastic behavior. Promising validation for submicron particles has been performed. Present version assumes near sphericity of the particles, making it less suitable for microplastic fibers

Review is based on the current available versions of the model

aSize range not indicated when not specified in this table

bValidation types a, b, and c refer to (a) agreement with empirical data, (b) agreement with hydrology and other particles, (c) conform design criteria, and (d) in agreement with state-of-the art knowledge

4.1 Emission-Based Mass Flow Modeling

Emission-based mass flow or mass balance models have been used for chemicals [97] and have recently been implemented for engineered nanoparticles as well [45, 98, 99]. The latter category of models is of particular interest for this review, because mass flow models for plastic particles can relatively easily be developed along the same lines. Based on estimates of nanoparticle emissions from products, environmental fluxes are calculated to the major compartments like air, soil, water, sediment, and several technical compartments [45]. The compartments typically are considered homogeneous and well mixed [45, 98]. Deposition and removal of particles within compartments are modeled as constant annual flows into a sub-compartment of each box considered. Similar mass flow model applications that calculate environmental concentration for plastic debris in all media (air, soil, water, and sediment) have not been published yet. However, the essence of the approach has been used to estimate concentrations of microplastics from cosmetics in WWTP effluents in the Netherlands [90] and mass emissions of microplastics from cosmetics from Europe to the North Sea [100]. The first study is discussed in detail below.

Mass Flow Modeling of Microplastic Concentrations in WWTP Effluents With the use of a mass flow modeling approach, Van Wezel et al. [90] estimated the emission of microplastics from consumer products to the surface water via WWTP effluents (Table 1). Based on the known use of microplastics in cosmetics and personal care products, cleaning agents, and paints and coatings, emissions were estimated. Per product category, data on the use of the product, the market penetration, and concentration of microplastics in the product were collected. It was estimated that during the wastewater treatment, between 40 and 96% of the microplastics would be retained by the WWTP. The model calculated the predicted concentration of microplastics in a WWTP effluent as the product of the concentration of microplastics in a product, the daily usage of that product, the fraction of microplastics removed during the wastewater treatment, and the market penetration of the products, divided by the volume of wastewater produced. The estimated effluent concentration of microplastic ranged from 0.2 µg L^{-1} for the conservative estimate to 66 µg L^{-1} for the maximum scenario.

Measured concentrations of microplastics in WWTP effluents range from 20 to 150 particles L^{-1}, as reported after a Dutch monitoring campaign [90]. These particle numbers were converted to mass, based on the size range, the volume assuming cubic shapes, and an average density. To validate the model, the model outcomes were compared with the observations of the monitoring [90]. Three different particle number-to-mass conversion categories were used, classified at "little and light," "intermediate," and "big and heavy" particles, the names relating to the assumed particle size, volume, and density. The model coincided best with observations when "big and heavy" particles were assumed to be measured, that is, particles with a relatively high density, large size, and large volume. However, the measured concentrations include both primary and secondary plastics (i.e., produced and fragmented/weathered particles, respectively), whereas the model only included primary plastics, which may have interfered with the reported validation.

Current knowledge on the use of nano- and microplastics in consumer products is limited, so a generic approach with many assumptions was used in this mass flow modeling study, contrasting with the approaches used with more advanced multi-media mass balance models, life cycle perspective models, or probabilistic material flow models. More reliable data to feed the models are needed to improve the emission estimates [90].

4.2 Global River Models

River pollution is a worldwide problem. Human activities on the land pollute rivers in all continents. A number of global river pollution models exist. One of these is the Global NEWS (Nutrient Export from WaterSheds) model [101, 102]. Global NEWS is a model that calculates river export of nutrients from land to sea as a function of human activities on the land. Global NEWS includes more than 6,000 river basins using hydrology from the water balance model [103]. It calculates river export at the river mouth. The model input is mostly on a grid of 1 degree longitude by 1 degree latitude. It has been used to simulate trends in river pollution for the period 1970–2050, taking into account change in land use, food production, urbanization, and hydrology [103–105]. Results indicate that over time, most rivers worldwide become more polluted.

Global river export models for nutrients, like Global NEWS, have been under development for more than 20 years. For other pollutants global river export models do not have such a long history. As a result, the Global NEWS approach has been taken as an example and inspiration for other pollutants [44]. Nutrients in rivers can have point sources (e.g., pipes draining into the river) or diffuse sources (e.g., runoff from soils or atmospheric deposition [65]). This is the case for nutrients, but also for other pollutants, like plastic debris. Model structures for point sources of one pollutant can easily serve as an example for other pollutants. The same holds for diffuse sources.

A river export model for microplastics, inspired by the Global NEWS model, is currently under development (Table 1). Preliminary results for point source inputs of microplastics to European seas have been presented [91]. This plastic model calculates point source inputs of microplastics from sewage to rivers. In addition, it simulates river transport of microplastics as a function of population, sewage connection, wastewater treatment, and river retention. River retention is derived from [36]. First results indicate that car tires are important point source inputs of microplastics in European rivers.

4.3   Multimedia Modeling

Multimedia models for chemicals are built by setting up a mass balance equation for each compartment that calculates the fluxes of transport via all exchange processes among compartments that are considered relevant. The fluxes are calculated based on first-order kinetic process rate parameters and concentration or fugacity gradients. The model equations are commonly solved by simple matrix algebra assuming steady state, but they also can be temporally resolved. Common multimedia models for nanosized particles are MendNano [106] and SimpleBox4Nano (SB4N) [72, 73]. These models calculate steady-state concentrations in the compartments atmosphere, surface water, soil, and sediment. In this review we discuss SB4N in more detail, as a first plastic implementation has already been made for this model (Fig. 3). SB4N models the partitioning between dissolved and particulate forms of the chemical as nonequilibrium colloidal behavior, instead of equilibrium speciation. Within each compartment, particles can occur in different physical−chemical forms (species): (a) freely dispersed, (b) heteroaggregated with natural colloidal particles, smaller than 450 nm, or (c) attached to natural particles larger than 450 nm. All these particle forms are subject to gravitational forces in aqueous media. Because SB4N is a spreadsheet model, it can easily be implemented for plastic debris of all sizes, as long as the parameter values are known. One of the advantages is that the model stems from SimpleBox, which is an established model already used in the risk assessment of chemicals [108]. A limitation is that the model only calculates average background concentrations.

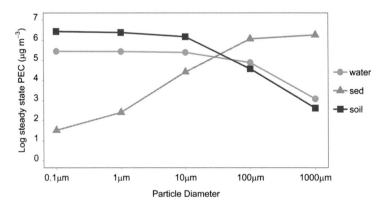

Fig. 3 Multimedia distribution of plastic debris of size 100 nm to 1 mm, between atmosphere, soil, water, and sediment on a regional scale, for the river Rhine catchment, simulated with SB4N [72, 73]. Concentrations are given on the log scale. Predicted environmental concentrations (PEC) assume a yearly emission in the catchment of 20 kt (based on data provided in [107]) in total, a (fouled) plastic density of 1,100 kg/m^3, negligible degradation and fragmentation due to short particle residence time in the system, and an attachment efficiency for heteroaggregation of 0.01 [35, 36]

We provide a first implementation for plastic in SB4N in this review (Fig. 3), which has not been published before. SB4N was parameterized for the river Rhine catchment, assuming initial emissions to the compartments soil and surface water of 50% of total emission, each. We assumed that no direct emission to sediment occurs, implying that plastic particles reach the sediment only through settling from the water column. Further assumptions are provided in the caption of Fig. 3. With all parameters at the same value, an increase in particle diameter results in more removal from water and soil and increased concentrations in sediment (more settling, Fig. 3).

4.4   Spatiotemporally Explicit Models

To date, two models have been presented that are able to simulate the transport of plastic debris in freshwater rivers with high spatial and temporal resolution [35–37]. Both models are framed by the authors as theoretical models, that is, they are supposed to be valid with respect to the design criteria and in agreement with existing theory, but they are not yet validated against measured data for plastic debris (Table 1).

Modeling the Transport of Plastic Debris in the Dommel River (The Netherlands)
The model by Besseling et al. [35, 36] is the first model that simulated the fate of nano- up to centimeter (i.e., macroplastic)-sized plastic particles in a river (see [24] for review). The model is based on the NanoDUFLOW hydrological model [96, 109] and includes advective transport of particles, their homo- and hetero-aggregation, biofouling, sedimentation/resuspension, degradation of plastic, and burial in the sediment. This implies that all processes mentioned in Sect. 3 were accounted for. Although not yet formally validated for plastic particles because of lacking monitoring data, earlier model simulations for nano-CeO_2 showed good agreement with measured nano-CeO_2 submicron particles in the same river [96]. The model can be implemented for other catchments using DUFLOW Modeling Studio [110] and allows for the inclusion of tributaries and diffuse as well as point sources (e.g., WWTPs) [96].

To simulate the transport of plastic debris, parameter values were set based on literature data. Data for the attachment efficiency for heteroaggregation are scarce and therefore were also determined experimentally. A 40 km stretch of the river Dommel (the Netherlands) was modeled with a spatial resolution of 477 sections of an average 87.7 m length and with section widths ranging from 8 to 228 m. The effect of all processes was calculated per section and the result was passed on to the next. An upstream point source with known mass concentration was used as a boundary condition at time zero, based on an average order of magnitude of published concentrations of microplastics in freshwaters. Scenario studies aimed at identifying how plastic debris of all sizes and densities would be distributed along the river. Realistic flow data were used. Impacts of long-term variability in weather conditions were not accounted for given the short simulation times required to reach steady state in the water column.

The scenario studies showed that the attachment efficiency to suspended solids or other plastic particles, biofilm formation on the plastic particles, and polymer type of the plastic particles had only a small effect on the modeled fate and retention [36]. Particle size, however, had a much stronger effect. Both the occurrence of accumulation "hot spots" in river sediment and overall retention in the 40 km river stretch were found to be highly affected by particle size. The larger micro- and millimeter-sized plastic particles showed high up to complete retention in the river due to direct Stokes settling [36]. Nanoplastic appeared to be retained and transported to an equal extent, due to the predicted fast heteroaggregation with natural solids. These solids have a higher density than the plastic and the water, causing subsequent sedimentation of these aggregates that captured the plastic particles. Retention, however, was lowest for the intermediate size class of plastic particles around 5 (1–10) μm due to the trade-off between these "direct" and "indirect heteroaggregate" Stokes settling mechanisms. The authors emphasized the importance of this size selection mechanism in rivers. The model was also applied to particles with a density equal to water, which resulted in no particle settling. This scenario was taken as a proxy for buoyant particles.

Freshwater organisms might be exposed relatively more to such intermediate size classes, whereas they would be depleted in the mixture of particles that is exported to sea under discontinuous input regimes [36]. The fact that the model accounted for all known processes that are relevant and that it was in agreement with data for another particle type (nano-CeO_2 submicron particles [96]) contributes to the credibility of the results. The model however was set up for (near-) spherical particles. This means that it is already applicable for spherules, microbeads, or secondary plastics (e.g., car tire dust) that can be assumed to approach sphericity, but it may not yet simulate particles with diverging shapes like fibers or thin films with the same level of accuracy. Furthermore, parameters for heteroaggregation are still poorly known, which also calls for further refinement.

Modeling the Transport of Microplastic Debris in the Thames River Catchment (UK) Nizzetto et al. presented a spatiotemporally explicit model that was applied to the Thames River catchment [37]. The study is framed as purely theoretical as empirical data on microplastic emissions and concentrations were not available. The model is based on an existing hydrobiogeochemical multimedia model, INCA-contaminants [93], with a sediment transport module [95], a rainfall-runoff module [94], and the possibility to add direct effluent inputs from, for instance, WWTPs. It is a lumped model as it assumes homogeneous rainfall and temperature distributions.

The model accounted for surface runoff and effluent inputs and reentrance to the system by resuspension. Whether particles are transported by surface runoff depended on the microplastic pool available for mobilization, the transport capacity of the overland flow both for microplastics and sediment, and the detachment of plastics through splash erosion and flow erosion. In the stream, the particles are assumed to be uniformly mixed within each section, and the transport processes advection, settling, and size-dependent resuspension from the sediment bed were taken into account.

Microplastic properties were defined by dimension and density. The study assumed plastic particles were pristine, that is, effects of biofouling were not taken into account. The model showed that the transport of microplastics is strongly related to flow regimes, especially for the larger (> 0.2 mm) particles. The transport dynamics were more influenced by size than by density, which confirms the findings by Besseling et al. [35, 36]. Average retention of particles was size dependent, decreasing with decreasing particle size and starting with 90–100% retention for particles >0.2 mm. Particles <0.2 mm were less well retained, and a large portion was expected to end up in the marine environment. The particle size range of the simulated particles was 0.05–0.7 mm; densities ranged from 1,000 to 1,300 kg m^{-3}. The model did not include biofouling, aggregation, or fragmentation. These processes influence the hydrodynamic behavior and size distribution of the particles but according to the authors should be better understood before they can be included in the model. Nanoparticles are also not included in the model yet [37].

Comparison of the Besseling (DUFLOW) and Nizzetto (INCA-Plastic) Models Both the DUFLOW and INCA-Plastic models were in accordance with their design criteria and study aim. The NanoDUFLOW model seems more complete as it includes aggregation, which has been shown to be a crucial process, especially for submicron particles [59, 60]. The model by Besseling et al. [35, 36] also accounted for biofouling, which also has been shown to affect the settling behavior of plastic particles. Given the study aim, Besseling et al. did not provide long-term simulations that accounted for the impacts of weather conditions. However, in principle DUFLOW can accommodate point and diffuse sources like WWTPs, tributaries [96], or runoff [111]. The latter processes were already accounted for in the INCA-Plastic implementation by Nizzetto et al., which is a relevant merit of that study. Both model outcomes agree on the important effect of particle size on retention and on a high retention for particles >0.2 mm. A contrasting conclusion, however, is that the INCA-Plastic model predicted that smaller particles would be less well retained in the river and thus exported to sea, whereas the NanoDUFLOW model reported an increased retention again for particles smaller than 5 μm. This difference can be explained from the fact that NanoDUFLOW accounted for aggregation of these small plastic particles, which allowed for the simulation of the increased sedimentation of these small plastic particles captured in heteroaggregates. This emphasizes the need to include this process. It has been shown that because heteroaggregation captures virtually all free nanosized particles, uncertainty with respect to the exact parameterization of heteroaggregation is of minor importance [73, 96, 112]. The conclusions of both studies depend on the modeled scenario's and parameters' variability. Also laboratory experiments have shown that processes like biofouling and aggregation [57, 62, 84] and particle properties like density, size, and shape [52, 55] significantly influence particle fate.

5 Recommendations and Guidance for the Development of Fate Models for Plastic Debris from a Risk Assessment Perspective

5.1 Data and Knowledge Gaps with Respect to Further Model Development

Quality Criteria for Analysis and Detection To date, few studies have measured concentrations and characteristics of plastic debris in the freshwater environment, which implies that more and also better data are of utmost importance. Quality assurance criteria are common in analytical chemistry or ecotoxicology [113, 114] but are less self-evident for monitoring of plastic debris which is a relatively young field of science [115].

There also is an urgent need to standardize the units used to quantify abundance of plastic debris [81]. For instance, for freshwater systems, concentrations of plastics in water and sediment have until now been reported in mass per unit of volume of water [13], mass per mass of sediment [15], particles per volume of water [13, 19], particles per surface area of water [18, 19, 83, 116], and particles per mass of sediment [15]. Utility of data for modeling would improve enormously if studies would at least mention both mass and particle count data and, when taking water samples, mention the sampling depth and sampling net dimensions, which would enable a surface-to-volume conversion or vice versa. This conversion only holds under the assumption that particles are evenly distributed over the sampled depth, which is also often assumed in models.

Depending on the aims of the modeling, measured plastic abundances should meet specific requirements. To validate mass flow analysis, an estimate of the total mass of plastic per unit of volume of the modeled media would be required. Multimedia models like SB4N [72] model the free, <0.45 μm aggregated, and >0.45 μm aggregated species, and validation ideally would require mass concentrations for these size classes. Because the latter models start with emission data, i.e., from production figures, the modeling will usually relate to a specific polymer type. For deterministic spatiotemporally explicit modeling, sufficient detail with respect to actual size and polymer density distributions is required because such approaches aim to simulate the reality as closely as possible. This implies that analysis and characterization of plastic in environmental samples would need to include (a) sufficient detail in the particle size and density distributions and (b) sufficient detail in the classification of shape, i.e., like fibers, fragments, and spherules [81]. What is to be considered as "sufficient" in this respect depends on the more specific aim of the modeling and is beyond the scope of this review. Given that particle interactions as well as potential ecological effects across different species traits are size dependent, standardization of methods, including those for nano- and micrometer-sized plastic particles, is considered very important.

Recommendation for Model Validation Validation would require sufficient data to verify the credibility of the model with statistical rigor. For mass flow or multimedia models, limited data per system yet for a high number of aquatic systems would be preferred. For spatiotemporally resolved models, however, it would be preferred to sample one catchment in detail. Such a case study catchment could then be used to calibrate and validate models, which could later be applied to other catchments. As for process parameters, little is known yet especially about the time scales of aggregation, fragmentation, and biofouling. This means that experimental work is needed, after which the parameter values obtained from these experiments can be applied in models. The development of fate models for freshwater may also benefit from experimental and model studies on marine plastic aggregation, fragmentation, and biofouling.

5.2 Comparing the Models: What Model for Which Question?

In the previous sections, we described different categories of models in detail. Here, we briefly discuss what category of model is needed for which type of question or application. In essence, this categorization does not differ from that for soluble chemicals or engineered nanomaterials.

For emission-based regional estimates of environmental concentrations of plastic debris, mass balance, mass flow, and especially mechanistic multimedia models are adequate. Recently, for nanoparticles such models have been developed, like the SB4N model [63, 72, 73] and the MendNano model [106]. It is highly recommended that such mechanistic multimedia models are adapted for plastic as well. Being neither temporal nor spatially explicit, such models are screening level models that can be used to assess relative concentrations among classes of nano- and microplastics or among plastic emission scenarios. Hence, such models are useful to calculate regional average or background concentrations (PECs, predicted environmental concentrations) for different plastic types, for different regions, or for different future emission scenarios. Multimedia particle models can also be used to detect the parameters to which the model output is most sensitive or to quantify uncertainty in PECs, which than can be applied in probabilistic risk assessments (discussed below).

Compared to the output provided by multimedia models, more realistic estimates of local environmental fate and concentrations can be obtained with spatiotemporally explicit models. However, fate models that are spatially explicit only yield better estimates if data on spatial variability in emission intensities are available. At present, there is only limited information on such spatial variation. Furthermore, estimating regional average concentrations still requires definition of what is defined as "a region." Different models use different scales, which means that the research question defines what model is most adequate. Global river models like

Global NEWS link mass flow models for river catchments, which thus accounts for spatial variation among catchments on a global scale, but not within catchments [91, 117]. As for applications, such models can rank catchments, regions, countries, or continents with respect to emission intensity to the marine environment [92]. The multimedia model SB4N can also accommodate various spatial scales, like regional, continental, and global, but always calculates one average concentration for soil, sediment, air, lake, river, and seawater. It is possible, however, to run models like SB4N for a certain grid, within an overarching model that provides input on a scale of, for instance, 200 × 200 km [118].

For more accurate local estimates of concentrations of plastic debris, system-specific zero-D mass balance approaches can be used for smaller systems, like lakes [63]. However, to better account for variability, spatiotemporally explicit models in 1, 2, or 3 dimensions can be used. As far as we know, the 1-D NanoDUFLOW model discussed above is the most elaborated model available. By defining small segments in a river, full hydrology can be taken into account. This is important for answering questions with respect to "hot spot" locations, quantifying which plastic types and sizes can be expected where (including nanosized plastic), calculation of retention versus flow-through to sea, and prospective assessments of fate and exposure on a detailed local scale. It has been argued recently that such models may be able to predict biologically relevant nanoparticle aggregate species as a function of time and space, which in turn can be linked to exposure by biota inhabiting the water system in question [24]. We propose that a similar approach also is possible for plastic debris, although further validation of fate models as well as further assessment of what has to be considered bioavailable and ecologically relevant is required. These last steps are particularly important when models are used in the framework of a formal risk assessment.

5.3   Fate and Exposure Models in the Context of ERA for Plastic in Freshwater Systems

To date, no ERA framework has been defined or applied to plastic debris. Here, we postulate that for plastic debris the same basic components of ERA can be used as for traditional chemicals and engineered nanomaterials: problem definition stage, an exposure assessment, an effect assessment, and a risk characterization step [119, 120]. For plastic debris, exposure presently is difficult to measure, so there is a relatively high need for modeling tools. A crucial aspect of exposure modeling and effect assessment in the context of ERA is what is to be considered the "ecotoxicologically relevant metric" (ERM) [120]. The ERM is the "common currency" used in the exposure and the effect assessment, which links these two, such that they can lead to a consistent risk characterization. For soluble chemicals, the ERM always is concentration, which is why ERA for chemicals uses the ecotoxicologically relevant concentration. Effect assessment and risk characterization are beyond the scope of this review, which focuses on freshwater models for fate and exposure (see [121] for details). However, in order to frame models in the context of ERA, here we briefly touch upon the wide variety of adverse outcome pathways (AOPs) that exists for plastic debris. As plastic debris is a complex mixture of sizes, types, and shapes, which also can be associated with chemicals, there will be a multitude of ERMs. Each ERM captures the unique features of a particular type of debris present in a habitat in combination with specific traits of species in that habitat, leading to an AOP describing the preset ecological or human health protection goals. Some reported AOPs are entanglement, ingestion/suffocation, blockage of the gastrointestinal tract, food dilution, chemical toxicity from associated chemicals, and a series of biomarker responses, which have been reviewed recently [6, 9, 122]. ERMs for physical effect of plastic can be defined in the form of a matrix where exposure and effect criteria such as habitat, species, life stage, mode of action, plastic size, plastic shape, and exposure duration are tabulated and scored. Using population models, effects on individuals then can be integrated with those from other stressors and habitat factors and, where needed, scaled to the population level similar to pesticide effect models (e.g., [100]). The ERM then needs to be assessed in space and time, dependent on the protection goal and the aim of the ERA. Ideally, fate models as described in this review should thus be able to simulate or predict all relevant ERMs emerging from the broad suite of species and particles that can be encountered in a habitat that has to be protected. Here, as mentioned before, for relatively simple site or material prioritizations, regional background concentrations as produced by multimedia models may suffice. Multimedia models can also be used in probabilistic ERA where spatial heterogeneity is accounted for by using a probability function that quantifies the spatial variation. For site-specific assessments, ERMs may be predicted by explicit models like NanoDUFLOW [36, 96], INCA-contaminants [93], or similar particle models, as long as aggregation of nanosized particle fractions is accounted for. Exposure then can be combined with effect thresholds in a PEC/NEC (NEC is the no-effect concentration) approach, where the NEC may come from data for different dose response models dependent on the AOP (threshold model, log logistic, Weibull, binary). Due to considerable uncertainty compared to ERA for traditional chemicals, probabilistic approaches are recommended, which can be adopted from recent developments in the ERA of engineered nanoparticles [118].

6 Concluding Thoughts

Contamination of the freshwater environment with plastic debris of all sizes has received increasing attention. In this review we argue that in order to conduct a proper risk assessment of plastic pollutants and their sources, and given the scarcity of data, models are useful complementary methods for exposure assessment. These models can build on existing transport models that simulate other types of particles, only changing the plastics-specific parameters and characteristics. As a material to model, plastic is unique given its wide range of sizes, shapes, and densities. It can aggregate or fragment and obtain a biofilm, all of which influence the hydrodynamics and size distribution of the particles. The first models developed for plastic transport so far range from mass-balance point-emission models to spatiotemporally explicit models. These models, however, have not yet been calibrated because of a lack of data. We recommend that before large measurement campaigns start, units to express abundance of plastics and methods for the analysis of plastics in the environment are standardized, which would increase the usability of the measurements.

Acknowledgment This study was funded by the Dutch Technology Foundation STW, project nr 13940. We acknowledge additional support from KWR; IMARES; NVWA; RIKILT; the Dutch Ministry of Infrastructure and the Environment; the Dutch Ministry of Health, Welfare and Sport; Wageningen Food & Biobased Research; STOWA; RIWA; and water boards Hoogheemraadschap van Delfland, Zuiderzeeland, Rijn en IJssel, Vechtstromen, Scheldestromen, Aa en Maas, de Dommel, and Rivierenland.

References

1. Andrady AL (2011) Microplastics in the marine environment. Mar Pollut Bull 62:1596–1605. doi:10.1016/j.marpolbul.2011.05.030
2. Rochman CM, Cook A-M, Koelmans AA (2016) Plastic debris and policy: using current scientific understanding to invoke positive change. Environ Toxicol Chem 35:1617–1626
3. Eriksen M, Lebreton LCM, Carson HS et al (2014) Plastic pollution in the world's oceans: more than 5 trillion plastic pieces weighing over 250,000 tons afloat at sea. PLoS One:1–15. doi:10.1371/journal.pone.0111913
4. Jambeck JR, Geyer R, Wilcox C et al (2015) Plastic waste input from land into the ocean. Science 80(347):768–771
5. Huerta Lwanga E, Gertsen H, Gooren H et al (2016) Microplastics in the terrestrial ecosystem: implications for *Lumbricus terrestris* (Oligochaeta, Lumbricidae). Environ Sci Technol 50:2685–2691
6. Kühn S, Rebolledo ELB, van Franeker JA (2015) Deleterious effects of litter on marine life. In: Bergmann M, Gutow L, Klages M (eds) Marine anthropogenic litter. Springer, Berlin, pp 75–116
7. Cole M, Lindeque P, Fileman E et al (2013) Microplastic ingestion by zooplankton. Environ Sci Technol 47:6646–6655. doi:10.1021/es400663f
8. Scherer C et al (2017) Interactions of microplastics with freshwater biota. In: Wagner M, Lambert S (eds) Freshwater microplastics: emerging environmental contaminants? Springer Nature, Heidelberg. doi:10.1007/978-3-319-61615-5_8 (in this volume)
9. Koelmans AA, Bakir A, Burton GA, Janssen CR (2016) Microplastic as a vector for chemicals in the aquatic environment: critical review and model-supported reinterpretation of empirical studies. Environ Sci Technol 50:3315–3326
10. Barnes DKA (2002) Biodiversity: invasions by marine life on plastic debris. Nature 416: 808–809

11. Lusher AL, Tirelli V, O'Connor I, Officer R (2015) Microplastics in Arctic polar waters: the first reported values of particles in surface and sub-surface samples. Sci Rep 5:14947
12. Van Cauwenberghe L, Vanreusel A, Mees J, Janssen CR (2013) Microplastic pollution in deep-sea sediments. Environ Pollut 182:495–499. doi:10.1016/j.envpol.2013.08.013
13. Lechner A, Keckeis H, Lumesberger-Loisl F et al (2014) The Danube so colourful: a pot-pourri of plastic litter outnumbers fish larvae in Europe's second largest river. Environ Pollut 188:177–181
14. Rech S, Macaya-Caquilpán V, Pantoja JF et al (2014) Rivers as a source of marine litter-a study from the SE Pacific. Mar Pollut Bull 82:66–75
15. Klein S, Worch E, Knepper TP (2015) Occurrence and spatial distribution of microplastics in river shore sediments of the Rhine-main area in Germany. Environ Sci Technol 49: 6070–6076
16. Wagner M, Scherer C, Alvarez-Muñoz D et al (2014) Microplastics in freshwater ecosys-tems: what we know and what we need to know. Environ Sci Eur 26(1):1–9
17. Eerkes-Medrano D, Thompson RC, Aldridge DC (2015) Microplastics in freshwater systems: a review of the emerging threats, identification of knowledge gaps and prioritisation of research needs. Water Res 75:63–82
18. Free CM, Jensen OP, Mason SA et al (2014) High-levels of microplastic pollution in a large, remote, mountain lake. Mar Pollut Bull 85:156–163
19. Mani T, Hauk A, Walter U, Burkhardt-Holm P (2015) Microplastics profile along the Rhine river. Sci Rep 5:17988
20. Castañeda RA, Avlijas S, Simard MA, Ricciardi A (2014) Microplastic pollution in St. Lawrence river sediments. Can J Fish Aquat Sci 71:1767–1771
21. Dris R, Gasperi J, Saad M et al (2016) Synthetic fibers in atmospheric fallout: a source of microplastics in the environment? Mar Pollut Bull 104:290–293
22. Faure F, Demars C, Wieser O et al (2015) Plastic pollution in Swiss surface waters: nature and concentrations, interaction with pollutants. Environ Chem 12:582–591
23. Morritt D, Stefanoudis PV, Pearce D et al (2014) Plastic in the Thames: a river runs through it. Mar Pollut Bull 78:196–200
24. Koelmans AA, Besseling E, Shim WJ (2015) Nanoplastics in the aquatic environment. Critical review. In: Marine anthropogenic litter. Springer, Berlin, pp 325–340
25. Mattsson K, Hansson L-A, Cedervall T (2015) Nano-plastics in the aquatic environment. Environ Sci Process Impacts 17:1712–1721
26. Cole M, Galloway TS (2015) Ingestion of nanoplastics and microplastics by Pacific oyster larvae. Environ Sci Technol 49:14625–14632
27. Gigault J, Pedrono B, Maxit B, Ter Halle A (2016) Marine plastic litter: the unanalyzed nano-fraction. Environ Sci Nano 3:346–350
28. Lambert S, Wagner M (2016) Characterisation of nanoplastics during the degradation of polystyrene. Chemosphere 145:265–268
29. Rist SE, Hartmann NB (2017) Aquatic ecotoxicity of microplastics and nanoplastics—lessons learned from engineered nanomaterials. In: Wagner M, Lambert S (eds) Freshwater microplastics: emerging environmental contaminants? Springer, Heidelberg. doi:10.1007/978-3-319-61615-5_2 (in this volume)
30. da Costa JP, Santos PSM, Duarte AC, Rocha-Santos T (2016) (Nano) plastics in the environ-ment- sources, fates and effects. Sci Total Environ 566:15–26
31. Sanchez W, Bender C, Porcher J-M (2014) Wild gudgeons (*Gobio gobio*) from French rivers are contaminated by microplastics: preliminary study and first evidence. Environ Res 128: 98–100
32. Phillips MB, Bonner TH (2015) Occurrence and amount of microplastic ingested by fishes in watersheds of the Gulf of Mexico. Mar Pollut Bull 100:264–269
33. Peters CA, Bratton SP (2016) Urbanization is a major influence on microplastic ingestion by sunfish in the Brazos River Basin, Central Texas, USA. Environ Pollut 210:380–387

34. Imhof HK, Laforsch C (2016) Hazardous or not- are adult and juvenile individuals of Pota-mopyrgus antipodarum affected by non-buoyant microplastic particles? Environ Pollut 218: 383–391
35. Besseling E, Quik JTK, Koelmans AA (2014) Modeling the fate of nano- and microplastics in freshwater systems. SETAC Annual Meeting, Basel, Switzerland. doi: 10.13140/RG.2.2. 20991.61602
36. Besseling E, Quik JTK, Sun M, Koelmans AA (2017) Fate of nano- and microplastic in freshwater systems: a modeling study. Environ Pollut 220:540–548
37. Nizzetto L, Bussi G, Futter MN et al (2016) A theoretical assessment of microplastic trans-port in river catchments and their retention by soils and river sediments. Environ Sci Process Impacts 18:1050–1059
38. Gasperi J, Dris R, Bonin T et al (2014) Assessment of floating plastic debris in surface water along the Seine River. Environ Pollut 195:163–166
39. Merritt WS, Letcher RA, Jakeman AJ (2003) A review of erosion and sediment transport models. Environ Model Softw 18:761–799. doi:10.1016/S1364-8152(03)00078-1
40. Aksoy H, Kavvas ML (2005) A review of hillslope and watershed scale erosion and sediment transport models. Catena 64:247–271
41. Drewry JJ, Newham LTH, Greene RSB et al (2006) A review of nitrogen and phosphorus export to waterways: context for catchment modelling. Mar Freshw Res 57:757–774
42. Vanderkruk K, Owen K, Grace M, Thompson R (2009) Review of existing nutrient, Suspended Solid and Metal Models
43. Whitehead PG, Howard A, Arulmani C (1997) Modelling algal growth and transport in rivers: a comparison of time series analysis, dynamic mass balance and neural network tech-niques. Hydrobiologia 349:39–46
44. Vermeulen LC, Hofstra N, Kroeze C, Medema G (2015) Advancing waterborne pathogen modelling: lessons from global nutrient export models. Curr Opin Environ Sustain 14: 109–120
45. Gottschalk F, Sun T, Nowack B (2013) Environmental concentrations of engineered nano-materials: review of modeling and analytical studies. Environ Pollut 181:287–300
46. Baalousha M, Cornelis G, Kuhlbusch TAJ et al (2016) Modeling nanomaterial fate and uptake in the environment: current knowledge and future trends. Environ Sci Nano 3: 323–345
47. Ji Z-G (2008) Hydrodynamics and water quality: modeling rivers, lakes, and estuaries. Wiley, Hoboken
48. Fischer HB, List JE, Koh CR et al (2013) Mixing in inland and coastal waters. Elsevier, Amsterdam
49. Wallis S (2007) The numerical solution of the advection-dispersion equation: a review of some basic principles. Acta Geophys 55:85–94
50. Ani E-C, Wallis S, Kraslawski A, Agachi PS (2009) Development, calibration and evaluation of two mathematical models for pollutant transport in a small river. Environ Model Softw 24:1139–1152
51. Brown PP, Lawler DF (2003) Sphere drag and settling velocity revisited. J Environ Eng 129: 222–231. doi:10.1061/(ASCE)0733-9372(2003)129:3(222)
52. Kowalski N, Reichardt AM, Waniek JJ (2016) Sinking rates of microplastics and potential implications of their alteration by physical, biological, and chemical factors. Mar Pollut Bull 109(1):310–319
53. McNown JS, Malaika J (1950) Effects of particle shape on settling velocity at low Reynolds numbers. EOS Trans Am Geophys Union 31:74–82
54. Johnson CP, Li X, Logan BE (1996) Settling velocities of fractal aggregates. Environ Sci Technol 30:1911–1918

55. Khatmullina L, Isachenko I (2016) Settling velocity of microplastic particles of regular shapes. Mar Pollut Bull 114(2):871–880
56. Farley KJ, Morel FMM (1986) Role of coagulation in the kinetics of sedimentation. Environ Sci Technol 20:187–195
57. Long M, Moriceau B, Gallinari M et al (2015) Interactions between microplastics and phytoplankton aggregates: impact on their respective fates. Mar Chem 175:39–46. doi:10.1016/j.marchem.2015.04.003
58. Chubarenko I, Bagaev A, Zobkov M, Esiukova E (2016) On some physical and dynamical properties of microplastic particles in marine environment. Mar Pollut Bull 108(2):105–112
59. Praetorius A, Scheringer M, Hungerbühler K (2012) Development of environmental fate models for engineered nanoparticles- a case study of TiO2 nanoparticles in the Rhine river. Environ Sci Technol 46:6705–6713
60. Quik JTK, Velzeboer I, Wouterse M et al (2014) Heteroaggregation and sedimentation rates for nanomaterials in natural waters. Water Res 48:269–279
61. Lee DG, Bonner JS, Garton LS et al (2000) Modeling coagulation kinetics incorporating fractal theories: a fractal rectilinear approach. Water Res 34:1987–2000. doi:10.1016/S0043-1354(99)00354-1
62. Fazey FMC, Ryan PG (2016) Biofouling on buoyant marine plastics: an experimental study into the effect of size on surface longevity. Environ Pollut 210:354–360
63. Koelmans AA, Quik JTK, Velzeboer I (2015) Lake retention of manufactured nanoparticles. Environ Pollut 196:171–175
64. Mahmood K (1987) Reservoir sedimentation: impact, extent, and mitigation. Technical paper
65. HJ Q, Kroeze C (2010) Past and future trends in nutrients export by rivers to the coastal waters of China. Sci Total Environ 408:2075–2086
66. HJ Q, Kroeze C (2012) Nutrient export by rivers to the coastal waters of China: management strategies and future trends. Reg Environ Chang 12:153–167
67. FAO (2016) AQUASTAT website. Food and Agriculture Organization of the United Nations (FAO), Rome
68. Weilenmann U, O'Melia CR, Stumm W (1989) Particle transport in lakes: models and measurements. Limnol Oceanogr 34:1–18
69. Papanicolaou A, Thanos N, Elhakeem M, Krallis G et al (2008) Sediment transport modeling review—current and future developments. J Hydraul Eng 134:1–14
70. Cushing CE, Minshall GW, Newbold JD (1993) Transport dynamics of fine particulate organic matter in two Idaho streams. Limnol Oceanogr 38:1101–1115
71. Boling RH, Goodman ED, Van Sickle JA et al (1975) Toward a model of detritus processing in a woodland stream. Ecology 56:141–151
72. Meesters JAJ, Koelmans AA, Quik JTK et al (2014) Multimedia modeling of engineered nanoparticles with simpleBox4nano: model definition and evaluation. Environ Sci Technol 48:5726–5736. doi:10.1021/es500548h
73. Meesters JAJ, Quik JTK, Koelmans AA et al (2016) Multimedia environmental fate and speciation of engineered nanoparticles: a probabilistic modeling approach. Environ Sci Nano 3:715–727
74. Merritt DM, Wohl EE (2002) Processes governing hydrochory along rivers: hydraulics, hydrology, and dispersal phenology. Ecol Appl 12:1071–1087
75. Bridge JS, Bennett SJ (1992) A model for the entrainment and transport of sediment grains of mixed sizes, shapes, and densities. Water Resour Res 28:337–363
76. Michels J, Stippkugel A, Wirtz K, Engel A (2015) Aggregation of microplastics with marine biogenic particles. ASLO Aquatic Sciences Meeting 2015
77. Burd AB, Jackson GA (2009) Particle aggregation. Annu Rev Mar Sci 1:65–90
78. Hotze EM, Phenrat T, Lowry G V (2010) Nanoparticle aggregation: challenges to understanding transport and reactivity in the environment. J Environ Qual 39:1909–1924

79. Zhang K, Gong W, Lv J et al (2015) Accumulation of floating microplastics behind the Three Gorges Dam. Environ Pollut 204:117–123

80. Wright SL, Thompson RC, Galloway TS (2013) The physical impacts of microplastics on marine organisms: a review. Environ Pollut 178:483–492. doi:10.1016/j.envpol.2013.02.031

81. Dris R, Imhof H, Sanchez W et al (2015) Beyond the ocean: contamination of freshwater ecosystems with (micro-) plastic particles. Environ Chem 12:539–550

82. Williams AT, Simmons SL (1996) The degradation of plastic litter in rivers: implications for beaches. J Coast Conserv 2:63–72

83. Eriksen M, Mason S, Wilson S et al (2013) Microplastic pollution in the surface waters of the Laurentian Great Lakes. Mar Pollut Bull 77:177–182

84. Lagarde F, Olivier O, Zanella M et al (2016) Microplastic interactions with freshwater micro-algae: hetero-aggregation and changes in plastic density appear strongly dependent on polymer type. Environ Pollut 215:331–339

85. Van Cauwenberghe L (2016) Occurrence, effects and risks of marine microplastics. Ghent University, Gent

86. Van Rijn LC (1984) Sediment transport, part II: suspended load transport. J Hydraul Eng 110: 1613–1641

87. Barnes DKA, Galgani F, Thompson RC, Barlaz M (2009) Accumulation and fragmentation of plastic debris in global environments. Philos Trans R Soc Lond Ser B Biol Sci 364: 1985–1998. doi:10.1098/rstb.2008.0205

88. Klein S et al (2017) Analysis, occurrence, and degradation of microplastics in the aqueous environment. In: Wagner M, Lambert S (eds) Freshwater microplastics: emerging environmental contaminants? Springer, Heidelberg. doi:10.1007/978-3-319-61615-5_3 (in this volume)

89. Corcoran PL, Norris T, Ceccanese T et al (2015) Hidden plastics of Lake Ontario, Canada and their potential preservation in the sediment record. Environ Pollut 204:17–25

90. van Wezel A, Caris I, Kools S (2015) Release of primary microplastics from consumer products to wastewater in The Netherlands. Environ Toxicol Chem 35:1627–1631

91. Siegfried M, Verburg C, Gabbert S, et al (2016) Modelling river export of micro- and macro plastics from land to sea. EGU General Assembly Conference Poster

92. Siegfried M, Gabbert S, Koelmans AA (2016) River export of plastic from land to sea: a global modeling approach. In: EGU General Assembly Conference Abstract. p 11507

93. Nizzetto L, Butterfield D, Futter M et al (2016) Assessment of contaminant fate in catchments using a novel integrated hydrobiogeochemical-multimedia fate model. Sci Total Environ 544:553–563

94. Futter MN, Erlandsson MA, Butterfield D et al (2014) PERSiST: a flexible rainfall-runoff modelling toolkit for use with the INCA family of models. Hydrol Earth Syst Sci 18:855–873

95. Lazar AN, Butterfield D, Futter MN et al (2010) An assessment of the fine sediment dynamics in an upland river system: INCA-Sed modifications and implications for fisheries. Sci Total Environ 408:2555–2566

96. de Klein JJM, Quik JTK, Bäuerlein PS, Koelmans AA (2016) Towards validation of the NanoDUFLOW nanoparticle fate model for the river Dommel, The Netherlands. Environ Sci Nano 3:434–441

97. Hollander A, Schoorl M, van de Meent D (2016) SimpleBox 4.0: improving the model while keeping it simple. Chemosphere 148:99–107

98. Gottschalk F, Sonderer T, Scholz RW, Nowack B (2009) Modeled environmental concentrations of engineered nanomaterials (TiO2, ZnO, Ag, CNT, fullerenes) for different regions. Environ Sci Technol 43:9216–9222

99. Keller AA, Vosti W, Wang H, Lazareva A (2014) Release of engineered nanomaterials from personal care products throughout their life cycle. J Nanopart Res 16:1–10

100. Gouin T, Avalos J, Brunning I et al (2015) Use of micro-plastic beads in cosmetic products in Europe and their estimated emissions to the North Sea environment. SOFW-J 141:40–46
101. Mayorga E, Seitzinger SP, Harrison JA et al (2010) Global nutrient export from WaterSheds 2 (NEWS 2): model development and implementation. Environ Model Softw 25:837–853
102. Seitzinger SP, Mayorga E, Bouwman AF et al (2010) Global river nutrient export: a scenario analysis of past and future trends. Global Biogeochem Cycles 24:GB0A08
103. Fekete BM, Wisser D, Kroeze C et al (2010) Millennium ecosystem assessment scenario drivers (1970–2050): climate and hydrological alterations. Global Biogeochem Cycles 24(4). doi:10.1029/2009GB003593
104. Bouwman AF, Beusen AHW, Billen G (2009) Human alteration of the global nitrogen and phosphorus soil balances for the period 1970–2050. Global Biogeochem Cycles 23(4). doi:10.1029/2009GB003576
105. Van Drecht G, Bouwman AF, Harrison J, Knoop JM (2009) Global nitrogen and phosphate in urban wastewater for the period 1970 to 2050. Global Biogeochem Cycles 23(4). doi:10. 1029/2009GB003458
106. Liu HH, Cohen Y (2014) Multimedia environmental distribution of engineered nano-materials. Environ Sci Technol 48:3281–3292
107. Sherrington C, Darrah C, Hann S et al (2016) Study to support the development of measures to combat a range of marine litter sources. Report for European commission DG environ-ment. Eunomia Research & Consulting, Bristol
108. Meesters JAJ, Veltman K, Hendriks AJ, van de Meent D (2013) Environmental exposure assessment of engineered nanoparticles: why REACH needs adjustment. Integr Environ Assess Manag 9:e15–e26
109. Quik JTK, de Klein JJM, Koelmans AA (2015) Spatially explicit fate modelling of nano-materials in natural waters. Water Res 80:200–208
110. Clemmens AJ, Holly FM Jr, Schuurmans W (1993) Description and evaluation of program: duflow. J Irrig Drain Eng 119:724–734
111. David I, Beilicci E, Beilicci R (2015) Basics for hydraulic modelling of flood runoff using advanced hydroinformatic tools. In: Extreme weather and impacts of climate change on water resources in the Dobrogea region. IGI Global, pp 205–239
112. Dale AL, Lowry GV, Casman EA (2015) Stream dynamics and chemical transformations control the environmental fate of silver and zinc oxide nanoparticles in a watershed-scale model. Environ Sci Technol 49:7285–7293
113. Klimisch H-J, Andreae M, Tillmann U (1997) A systematic approach for evaluating the quality of experimental toxicological and ecotoxicological data. Regul Toxicol Pharmacol 25:1–5
114. Kase R, Korkaric M, Werner I, Ågerstrand M (2016) Criteria for reporting and evaluating ecotoxicity data (CRED): comparison and perception of the Klimisch and CRED methods for evaluating reliability and relevance of ecotoxicity studies. Environ Sci Eur 28(1):7
115. Rocha-Santos T, Duarte AC (2015) A critical overview of the analytical approaches to the occurrence, the fate and the behavior of microplastics in the environment. TrAC Trends Anal Chem 65:47–53
116. Imhof HK, Ivleva NP, Schmid J et al (2013) Contamination of beach sediments of a sub-alpine lake with microplastic particles. Curr Biol 23:R867–R868
117. Kroeze C, Gabbert S, Hofstra N et al Global modelling of surface water quality: a multi-pollutant approach. Curr Opin Environ Sustain 23:35–45
118. Jacobs R, Meesters JAJ, ter Braak CJF et al (2016) Combining exposure and effect modelling into an integrated probabilistic environmental risk assessment for nanoparticles. Environ Toxicol Chem 35:2958–2967
119. Klaine SJ, Koelmans AA, Horne N et al (2012) Paradigms to assess the environmental impact of manufactured nanomaterials. Environ Toxicol Chem 31:3–14
120. Koelmans AA, Diepens NJ, Velzeboer I et al (2015) Guidance for the prognostic risk assess-ment of nanomaterials in aquatic ecosystems. Sci Total Environ 535:141–149

121. Brennholt N et al (2017) Freshwater microplastics: challenges for regulation and management. In: Wagner M, Lambert S (eds) Freshwater microplastics: emerging environmental contaminants? Springer, Heidelberg. doi:10.1007/978-3-319-61615-5_12 (in this volume)
122. Ziccardi LM, Edgington A, Hentz K et al (2016) Microplastics as vectors for bioaccumulation of hydrophobic organic chemicals in the marine environment: a state-of-the-science review. Environ Toxicol Chem 35:1667–1676

Interactions of Microplastics with Freshwater Biota

Christian Scherer, Annkatrin Weber, Scott Lambert, and Martin Wagner

Abstract The ubiquitous detection of microplastics in aquatic ecosystems promotes the concern for adverse impacts on freshwater ecosystems. The wide variety of material types, sizes, shapes, and physicochemical properties renders interactions with biota via multiple pathways probable.

So far, our knowledge about the uptake and biological effects of microplastics comes from laboratory studies, applying simplified exposure regimes (e.g., one polymer and size, spherical shape, high concentrations) often with limited environmental relevance. However, the available data illustrates species- and material-related interactions and highlights that microplastics represent a multifaceted stressor. Particle-related toxicities will be driven by polymer type, size, and shape. Chemical toxicity is driven by the adsorption-desorption kinetics of additives and pollutants. In addition, microbial colonization, the formation of hetero-aggregates, and the evolutionary adaptations of the biological receptor further increase the complexity of microplastics as stressors. Therefore, the aim of this chapter is to synthesize and critically revisit these aspects based on the state of the science in freshwater research. Where unavailable we supplement this with data on marine biota. This provides an insight into the direction of future research.

In this regard, the challenge is to understand the complex interactions of biota and plastic materials and to identify the toxicologically most relevant characteristics of the plethora of microplastics. Importantly, as the direct biological impacts of

---

This chapter has been externally peer reviewed.

C. Scherer (✉), A. Weber, and S. Lambert
Department Aquatic Ecotoxicology, Goethe University, Max-von-Laue-Str. 13, 60438
Frankfurt am Main, Germany
e-mail: c.scherer@bio.uni-frankfurt.de

M. Wagner
Department of Biology, Norwegian University of Science and Technology (NTNU),
Trondheim, Norway

M. Wagner, S. Lambert (eds.), *Freshwater Microplastics*,
Hdb Env Chem 58, DOI 10.1007/978-3-319-61615-5_8,
© The Author(s) 2018

natural particles may be similar, future research needs to benchmark synthetic against natural materials. Finally, given the scale of the research question, we need a multidisciplinary approach to understand the role of microplastics in a multiple-particle world.

Keywords Autecology, Feeding types, Microplastic-biota interaction, Polymers, Suspended solids, Vector

1 Introduction

Over the past decade, microplastics (MPs) have become a prominent environmental concern, mainly because of their frequent and ubiquitous detection in marine and freshwater ecosystems. Therefore, biota will likely encounter and interact with MPs. In addition, MPs are a heterogeneous class of pollutants with a broad range of individual properties such as material type, particle size, and particle shape. These diverse material characteristics make them potentially available to a broad range of neustonic (buoyant materials, density <1 g cm^{-3}), pelagic (materials in suspension), and benthic species (sedimenting materials, density >1 g cm^{-3}). This enables MPs to penetrate aquatic food webs at multiple trophic levels and ecological niches.

To date, research into MP exposure for freshwater biota is limited. Yet, marine research has shown malnutrition caused by the intensive feeding on MPs replacing parts of the natural diet [1–3]. Additionally, further ingestion-related effects include blockages and injuries to the digestive tract [4], inflammatory response [5], and desorption of xenobiotics [6]. Obviously, all of these responses presuppose feeding and ingestion of MPs. As such, the aim of this chapter is to discuss the diverse interactions between MPs and biota that may occur in the environment. In the first section, we focus on factors influencing the ingestion of MPs considering the impact of the different physical properties of MPs and feeding types of freshwater species. In the second section, we provide an overview and analysis of the observed MP effects in terms of their physical, chemical, and vector-related impacts. This is followed by a comparison of the similarities in the effects caused by exposure to naturally occurring particles and MPs. Finally, we conclude by discussing the wider implications of MPs toward freshwater systems.

2 Factors Influencing Microplastic Ingestion by Freshwater Biota

Species in freshwater ecosystems are part of complex food webs and forage on a wide diversity of food types, utilizing a variety of different feeding strategies. Notwithstanding this diversity, the classification by feeding types or by food types is commonly used to group biota. For instance, suspension feeders obtain nutrients from particles suspended in water, deposit feeders forage for particles in sediments, fluid feeders feed on other biotas fluids, and suction feeders ingest the prey together with the surrounding water. The utilized morphological structures determine further classifications. For example, filter feeders (e.g., daphnids) use specialized filtering structures to strain suspended particles, and raptorial feeders (e.g., copepods) actively capture and process suspended particles by modified appendages. Further typically used classifications are collectors (e.g., chironomids), shredders (e.g., amphipods), scrapers (e.g., gastropods), and predators (e.g., odonates) [7]. Another way to categorize species is based on their diet. For instance, bacterivores feed on bacteria, herbivores feed on plants, carnivores feed on animals (e.g., zooplanktivores, insectivores), and detritivores feed on decomposing materials. These groupings imply clear boundaries, although some species feed on multiple food sources (e.g., generalist, omnivorous) or have the ability to switch between food sources (opportunistic feeders).

Primary producers like unicellular algae or bacteria as well as particulate organic matter (POM) provide nutrients for a broad range of pelagic and benthic species. Thus, small MPs are in a similar size range to the natural food of these consumers. To understand the capacities of different species to feed on specific size classes, limnologists have frequently used polymer beads as tracers [8–10]. Although these studies primarily focus on pelagic zooplankton communities, they illustrate that the intake of food and MPs depend on complex interactions between biotic (e.g., feeding type, physiological state, competition, food size, and availability) and abiotic factors (e.g., temperature). Accordingly, they provide a useful starting point to discuss MP ingestion and effects.

2.1 The Role of Feeding Types

2.1.1 Invertebrates

Suspension and filter feeders like protozoans, rotifers, cladocerans, and mussels are assumed to be especially prone to MP ingestion because they commonly feed on suspended particulate matter (SPM) and ingest a variety of seston components. The ingestion of MPs by these feeding types has been shown in numerous studies (Table 1). For instance, bacterivorous and herbivorous ciliates (e.g., *Halteria* sp.), flagellates (e.g., *Vorticella* sp.), rotifers (e.g., *Anuraeopsis fissa*), and cladocerans (*Daphnia* sp.) can feed readily on plastic beads [9, 10]. While data on MP ingestion by pelagic filter-feeding zooplankton is relatively abundant, one prominent group of filter feeders, the bivalves, is underrepresented. Bivalves are known to feed effectively on SPM, including MP, which is ingested by marine mussels (e.g., *Mytilus edulis*, [24]) and freshwater clams (*Sphaerium corneum*, 1–10 μm polystyrene (PS) beads; *Anodonta cygnea*, 5–90 μm polystyrene (PS) beads and fragments; unpublished data).

In addition to organisms specialized in feeding on SPM, a variety of organisms forage for particles in sediments. Although MP exposure may be as relevant for deposit feeders (feeding on fine particulate matter and associated biota in sediments) as for filter feeders, only a few studies have investigated the ingestion of MPs for this mode of feeding. The blackworm *Lumbriculus variegatus* and the

Table 1 Summary of the results of uptake studies with microplastic particles and freshwater species

Uptake (P Ind^{-1} h^{-1})	Species
Ciliates	
Yes (13.6–1,200)	*Epistylis plicatilis*[a], *Epistylis rotans*[c], *Halteria grandinella*[a], *Halteria* sp.[b], *Pelagohalteria viridis*[b], *Stokesia* sp.[a], *Strombidium viride*[b], *Strombidium* sp.[a,b], *Vorticella microstoma*[a], *Vorticella natans*[a], *Vorticella* sp.[b], Unident. oligotrichs[a], Unident. Scuticociliatida[a]
No	*Askenasia volvox*[b], *Balanion* sp.[b], *Coleps* sp.[a], *Condylostoma* sp.[a], *Cyclidium* sp.[b], *Didinium* sp.[a,b], *Lembadion magnum*[a], *Litonotus* spp.[a], *Mesodinium* spp.[a], *Paradileptus* sp.[a], *Paradileptus elephantinus*[b], *Strobilidium caudatum*[b], *Strombidium viride*[a], *Suctoria*[b], *Tintinnidium fluviatile*[a], *Tintinnopsis lacustris*[a], *Urotricha furcata*[b], Unident. Scuticociliatida[b]
Flagellates	
Yes (2.6–103)	*Chrysostephanospharea globulifera*[a], *Cryptomonas ovata*[b], *Dinobryon bavaricum*[a], *Dinobryon cylindricum*[a,b], *Monas* spp.[a], *Monas-like cells*[b], *Ochromonas* sp.[a], Undetermined Choanoflagellate[a], Unident. heterotrophs[a]
No	*Chrysidalis* sp.[b], *Chrysococcus* sp.[b], Chrysomonadine[b], *Kathablepharis* sp.[b], *Mallomonas* sp.[b], *Pandorina morum*[b], *Peridinium volzii*[b], *Rhodomonas minuta*[b], *Synura* sp.[b], Undetermined choanoflagellate[b], Unident. heterotrophs[b]
Rotifera	
Yes (2.5–3,200)	*Anuraeopsis fissa*[a,d,e], *Brachionus angularis*[e], *Brachionus calyciflorus*[f], *Brachionus koreanus*[s], *Conochilus unicornis*[b,d,e], *Conochilus* sp.[a], *Filinia longiseta*[a,d,e], *Filinia terminalis*[f+], *Gastropus* sp.[a], *Hexarthra mira*[b], *Hexarthra* sp.[a], *Kellicottia bostoniensis*[a], *Keratella cochlearis*[d,e], *Keratella cochlearis tecta*[d], *Keratella quadrata*[e], *Keratella* spp.[a], *Lecane* sp.[e], *Lepadella* sp.[e], *Pompholyx complanata*[d0.5], *Pompholyx sulcata*[e]
No	*Anuraeopsis fissa*[d6], *Ascomorpha saltans*[e], *Asplanchna priodonta*[b], *Asplanchna* sp.[e], *Collotheca* spp.[e], *Conochilus unicornis*[d6], *Filinia longiseta*[d6], *Filinia terminalis*[b,f], *Kellicottia longispina*[b], *Keratella cochlearis*[d6], *Keratella cochlearis tecta*[d6], *Keratella quadrata*[d], *Polyarthra* spp.[a,b,e], *Pompholyx complanata*[d3–6], *Synchaeta* spp.[b,e], *Trichocerca pusilla*[e], *Trichocerca* spp.[a]
Annelida	
Yes (0–1)	*Lumbriculus variegatus*[h,i]
No	–
Crustacea	
Yes (1–28,000)	*Bosmina coregoni*[d,e,g], *Bosmina longirostris*[a,b,d,e,f], *Ceriodaphnia lacustris*[a], *Ceriodaphnia quadrangula*[b,f], *Chydorus sphaericus*[d,e,f,g], *Cyclops bicuspidatus thomasi*[f], *Cyclops bicuspidatus thomasi (nauplii)*[f], *Daphnia cucullata*[d,e,g], *Daphnia galeata mendotae*[f], *Daphnia longispina*[b], *Daphnia magna*[f,h1–10], *Daphnia parvula*[a], *Diaphanosoma birgei*[f], *Diaphanosoma brachyurum*[a,g], *Diaptomus siciloides*[f], *Diaptomus siciloides* (nauplii)[f], *Eubosmina coregoni*[f], *Eudiaptomus gracilis*[b], *Gammarus pulex*[h], *Holopedium amazonicum*[a], *Hyalella azteca*[k], *Notodromas monacha*[i], *Simocephalus vetulus*[f]
No	*Acanthocyclops robustus*[b,e], *Alona* sp.[e], *Chydorus sphaericus*[g19], *Cyclops vicinus*[b], *Daphnia magna*[h90], *Diacyclops bicuspidatus*[e], *Diaptomus mississippiensis*[a], *Eudiaptomus gracilis*[g], *Mesocyclops edax*[a], *Mesocyclops leuckarti*[e], nauplii[a,e], *Tropocyclops prasinus mexicanus*[a]

(continued)

Table 1 (continued)

Uptake (P Ind^{-1} h^{-1})	Species
Insecta	
Yes (0.05–15.6)	*Chironomus riparius*[h]
No	–
Mollusca	
Yes (0.16–104)	*Anodonta cygnea*[w], *Physella acuta*[h], *Potamopyrgus antipodarum*[I], *Sphaerium corneum*[u1–10]
No	*Sphaerium corneum*[u90]
Pisces	
Yes	*Cathorops agassizii*[n], *Cathorops spixii*[n], *Eucinostomus melanopterus*[o], *Eugerres brasilianus*[o], *Diapterus rhombeus*[o], *Dorosoma cepedianum*[q], *Sciades herzbergii*[n], *Stellifer brasiliensis*[m], *Stellifer stellifer*[m]
No	–

[a] [9], carboxylated microspheres 0.57 μm; [b] [10], plain microspheres 0.5 μm; [c] [11], latex beads 0.57–1.05 μm; [d] [12], PS spheres 0.5, 3, and 6 μm; [e] [13], carboxylated PS spheres 0.51 μm; [f] [14], PS spheres 6.5 μm ([f+]) flavored ([f−]) non-flavored; [g] [15], carboxylated spheres 2.1, 6.2, 10.8, and 19.4 μm; [h] [16], PS spheres 1, 10, and 90 μm; [i] [17], polymethyl methacrylat 29.5 ± 26 μm; [k] [18], PE particles and PP fibers 10–75 μm; [m] [19], field study, nylon rope fibers; [n] [20], field study, nylon fragments; [o] [21], field study, nylon fragments; [q] [22], microspheres 10–82 μm; [s] [23], PS beads, 0.05, 0.5, and 6 μm; [u] pers. observation, PS beads 1, 10, and 90 μm; [w] pers. observation, PS beads and fragments 5–90 μm. Superscript numbers indicate particle sizes

aquatic larvae of *Chironomus riparius* ingest a broad size range of MPs implying a relative nonselective feeding on sediment components (Table 1 [16]). Surface-grazing gastropods *Physella acuta* and *Potamopyrgus antipodarum* as well as the shredder *Gammarus pulex* have also been shown to ingest MPs through water-/sediment-borne (*P. acuta* and *G. pulex* [16]) and food-associated (*P. antipodarum* and *G. pulex* [17]) exposure routes. It is unknown if these results are relevant for other benthic deposit feeders considering the diverse ecological niches and feeding types (e.g., collector-gatherer, filter-gatherer, shredders, scrapers).

An analysis of studies on MP ingestion by freshwater species indicates that their general role in the food web (generalist vs. specialized feeders) may determine dietary MP uptake. Generalists (e.g., *Daphnia* sp.) or deposit feeders like the dipteran *C. riparius* frequently ingested MPs in laboratory experiments, while this is not the case for more specialized raptorial and carnivorous feeders like the cyclopoid copepod *Mesocyclops* sp., the rotifer *Asplanchna* sp. as well as the ciliate *Didinium* sp. (Table 1). However, given the potential of MPs to enter complex aquatic food webs at low trophic levels, an indirect ingestion via the prey is also likely for carnivorous predators. For instance, the transfer of MPs via prey was observed in food chain experiments with *D. magna* and *Chaoborus flavicans* (personal observation). While the predator *C. flavicans* did not directly ingest suspended MPs (PS beads, 10 μm), the feeding of MP-containing daphnids (pre-fed on MPs) resulted in an indirect uptake of 10 μm MPs.

Besides these general trends, available studies illustrate that species of the same functional feeding type have species-specific and sometimes highly divergent MP feeding rates. For instance, the filter-feeding cladocerans *Daphnia longispina* and *Ceriodaphnia quadrangula* ingested 230×10^3 P l^{-1} h^{-1} and 176×10^3 P l^{-1} h^{-1}, respectively. In comparison, rotifers (e.g., *Hexarthra mira*, 38.1×10^3 P l^{-1} h^{-1}) and ciliates (e.g., *Halteria* sp., 46.8 P l^{-1} h^{-1}) ingest MPs at a much slower rate [10]. While differences are mainly caused by the species' morphology and autecology, numerous other factors (e.g., appetite, MP type and concentration, quantification methods) may also contribute. Overall, the most commonly studied invertebrate species are zooplankton. However, we still know little about the interactions of MPs with other prominent invertebrate freshwater taxa, e.g., Annelida, Insecta, Decapoda, and Mollusca.

2.1.2 Vertebrates

When considering vertebrate species, MP uptake is documented in laboratory and field studies for several fish species (Table 1). In contrast, no information is available for amphibians. Considering the diversity of vertebrates acting as predators, herbivores, detritivores, or omnivores, we can assume that many species, at least in principle, have the capacity to ingest MPs depending on their feeding strategies. However, predicting MP ingestion by vertebrates solely based on feeding types may be too short sighted. For instance, grouping fishes into specific guilds/feeding groups is an imprecise and difficult task. Indeed, typical terms like detritivores, herbivores, and carnivores as well as generalist, specialist, and opportunist are used, but the variability of feeding (e.g., during development) and the trophic adaptability (ability to switch food sources) impede a precise classification [25]. The ingestion of prey through suction feeding is utilized by the majority of teleosts, which allows this high flexibility to exploit a variety of food sources [26]. Thus, accidental (mistake MPs for prey) and indirect ingestion of MPs (via prey containing MPs) are probable. The documented MPs in several fishes collected in the field (e.g., catfish, perch, drum, Table 1) support this assumption.

2.2   The Role of Particle Size, Shape, and Taste

2.2.1   Size and Shape

The importance of particle size in the acquisition of particulate food has been studied for pelagic protozoans, rotifers, and crustaceans (e.g., [26, 27]). For filter-feeding taxa, a distinct relation between morphology and particle size has been observed. Here, the minimum ingested particle size is mainly determined by the mesh size of the filtering apparatus. The maximum size is determined by the morphology of mouthparts and, in the case of cladocerans, the opening width of the carapace. Additionally, Burns [8] and Fenchel [27] describe a correlation

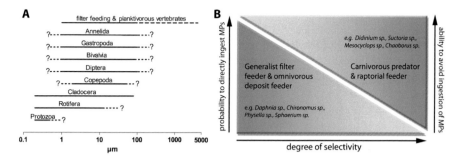

Fig. 1 Estimated feeding size ranges on microplastic particles (**a**). *Dotted lines* and question marks indicate the lack of min to max limits based on ingested size classes. An increasing feeding selectivity decreases the probability to directly ingest microplastics (**b**)

between the maximum ingestible particle size and the overall size of several cladoceran and protozoan species. Studies with the aquatic larvae of the dipteran *C. riparius* confirm this pattern for a benthic deposit feeder. Here, only individuals with a head capsule width larger than 400 μm ingested 90 μm PS spheres ([16], Table 1).

Fine-mesh filter feeders (size range 0.2–75 μm; e.g., *Daphnia magna*) are highly efficient bacteria feeders, whereas coarse mesh filter feeders (macrofiltrators, size range >2 μm; e.g., *Holopedium gibberum*) feed mainly on larger particles [28]. Results from feeding studies with polymer spheres illustrate that several protozoans feed effectively on 0.5 μm particles [9]; several rotifers on 0.5, 3, and 6 μm particles [19]; and cladocerans on 0.5, 3, 6, 10, and 20 μm particles ([13], Table 1). In comparison, calanoid copepods are macrofiltrators and ingest particles >2.1 μm but not 0.5 μm particles (e.g., [10], Fig. 1a). In addition, some species with a broad feeding size range have been shown to selectively forage on specific sizes when exposed to multiple size fractions. For instance, *Bosmina* sp. ingested large algae cells (*Cosmarium* sp.) six times faster than a small algae species (*Chlorella* sp.) [29, 30]. Furthermore, Agasild and Nõges [12] observed higher filtering rates of *Daphnia cucullata* on 3 and 6 μm compared to 0.5 μm MPs, whereas the rotifer *Conochilus unicornis* exhibited an increased filtering rate on 3 μm compared to 0.5 μm MPs.

Particle shape is another important property determining MP-biota interactions. Currently, the majority of the available literature focuses on MP beads, and it remains unclear whether the investigated species have similar feeding rates on non-spherical MPs (e.g., fibers, fragments). Some species (e.g., *G. pulex, D. magna, Notodromas monacha*) feed readily on secondary, irregularly shaped MPs [17, 31] with different toxicological profiles (see Sect. 3.1). As most of the MPs found in aquatic ecosystems are not spherical, more research is needed on irregularly shaped MPs.

2.2.2 Taste Discrimination

Many species are able to identify particles with nutritional value. For example, some bacterivorous and herbivorous protozoan, rotifer, and copepod species do not ingest polymer particles in their preferred size ranges (Table 1). Studies with fluorescently labeled bacteria have shown that some ciliates (estuarine oligotrichs) and flagellates prefer bacteria over MPs, while other species (estuarine scuticociliates; e.g., *Uronema narina*) cannot discriminate between bacteria and MPs [32].

The essential role of "taste" in the feeding of zooplankton [14, 32, 33] was acknowledged when discussing the comparability of feeding studies with synthetic microspheres and labeled bacteria or algae [9, 10, 15]. In rotifers, *Bosmina* (cladoceran), and copepods (calanoid and cyclopoid), DeMott [14] observed significant differences between feeding rates on flavored and non-flavored polymer particles. While *Bosmina* and the rotifer *Filinia terminalis* preferred algal-flavored spheres over untreated ones, *D. magna* and *Brachionus calyciflorus* did not [14]. This degree of selectivity was even higher in feeding trials with copepods. Here, calanoid (e.g., *Diaptomus siciloides*) and cyclopoid (e.g., *Cyclops bicuspidatus thomasi*) species strongly avoided untreated polymer spheres [14].

Despite the abundance of studies that illustrate pelagic zooplankton feeding on MPs, information about benthic invertebrates and vertebrates in general is scarce. Although drawing conclusions for unexamined species is highly speculative, knowledge on zooplankton can be used as a template to a certain extent. The examined species cover a broad spectrum in terms of their autecology (feeding types, selectivity, and food preferences). The same is true for the unexamined species, which inhabit similar niches and have equally diverse autecologies. Therefore, we hypothesize a similar pattern regarding species-specific size and taste discrimination: Some species will directly feed on available MPs in the size range of their food, while more selective feeders will avoid MP ingestion.

2.3   Conclusion

Primary consumers featuring bacterivorous, herbivorous, detritivorous, and deposit-feeding species are commonly specialized in foraging on particulate matter and have the capacity to ingest MP particles. The direct ingestion of MPs might be the major route for primary (e.g., herbivores) and secondary consumers (e.g., zooplanktivores), while apex predators are additionally prone to an indirect ingestion of MPs via prey (food web). The limited literature suggests that generalist and nonselective filter feeders (e.g., daphnids) have higher feeding rates compared to raptorial (e.g., copepods) and deposit feeders. Although studies on benthic invertebrates are scarce, species with detritivorous and omnivorous feeding types (e.g., Annelida, Insecta, Decapoda) may have the potential for ingesting MPs.

However, the feeding type is not a reliable predictor of MP ingestion as several studies on pelagic zooplankton communities highlight a far more complex MP-biota interaction than currently understood.

Overall, the feeding on particulate matter is a sequential process involving the encounter, pursuit, capture, and ingestion of potential prey [30]. Every single stage is determined by species-specific abilities and preferences to distinguish between favored and non-favored food sources (e.g., size, shape, taste, motile, sessile). Additionally, many taxa can adapt their feeding habits (e.g., targeting a preferred size class and/or nutritional value) in response to environmental conditions (optimal foraging). In general, it appears that the capability to directly ingest MPs decreases with an increasing selectivity in feeding (Fig. 1b). Generalist filter feeders will actively and directly ingest MPs from the water column or sediments in the size range of their typical food, whereas more specialized feeders (e.g., fluid feeders, raptorial carnivorous feeders) will indirectly ingest MPs associated with their prey.

The variety of feeding types and degrees of selective feeding present in aquatic fauna complicates generalizing patterns of MP uptake. This is especially true when comparing experimental to the real exposure scenarios. In the laboratory, virgin spherical microbeads are used, whereas in the environment, irregularly shaped MPs are colonized by microbes (see Sect. 3.2), adsorb extracellular proteins (biofilm), and form hetero-aggregates (increasing size). While MP-biota interactions are hard to predict based on the currently available data, feeding selectivity may be a driving factor (see Fig. 1b for a conceptual model).

3 Effects on Freshwater Biota

Studies on the potential adverse effects caused by MP exposures are scarce for freshwater compared to marine species. The few available studies (Table 2) include the filter feeder *D. magna* [34, 35, 41], the amphipods *Hyalella azteca* [18] and *G. pulex* [31], the freshwater snail *P. antipodarum* [38] as well as several fishes [37–39]. In this section, the outcomes of these studies are discussed.

3.1 Physical Impacts

The evaluation of feeding types (Sect. 2.1) suggests that nonselective filter feeders are especially prone to MP exposures. Based on their high rates of MP filtration and ingestion in laboratory studies, adverse effects induced by the particle toxicity may include blockages, reduced dietary intake, and internal injuries.

Table 2 Overview of effect studies with MPs and freshwater fauna

Species and test duration	Material (g cm^{-3})	Size (μm)	Concentration	Effect	Ref.
Cladocera					
Daphnia magna, 96 h	PE (0.69), beads	1–4 90–106	12.5–400 mg L^{-1}	1 μm: LC$_{50}$ 57.43 mg L^{-1} 100 μm: no effect	[34]
Daphnia magna, 21 days	Primary MPs (1.3) Secondary MPs (PE; 1.0) Kaolin (2.6)	1–5 2.6 ± 1.8 4.4 ± 1.1	10^2–10^5 P mL^{-1}	Primary MPs: no effect Secondary MPs: elevated mortality Kaolin: no effect	[35]
Amphipoda					
Gammarus pulex, 48 days	PET, fragments	10–150	0.4–4,000 P mL^{-1}	No effect on growth, behavior, metabolism	[31]
Hyalella azteca, 10 and 42 days	PE (1.13), powder PP, fibers	PE: 10–27 PP: 20–75	PE$_{10}$: 0–100,000 P mL^{-1} PP$_{10}$: 0–90 P mL^{-1} PE$_{42}$: 0–20,000 P mL^{-1}	LC$_{50}$ PE: 4.64 × 10^6 P mL^{-1} LC$_{50}$ PP: 71.4 P mL^{-1} PE$_{42}$: non-dose dependent effect on reproduction and growth	[18]
Gastropoda					
Potamopyrgus antipodarum, 56 days	PA, PET, PC, PS, PVC; fragments	4.64–602	0, 30, 70% MP mixture; dosing via agar slides (food)	No effects on morphology, reproduction, and development	[36]
Osteichthyes					
Clarias gariepinus, 96 h	LDPE (0.91) fragments, +Phenanthrene (Phe)	>95% were <60 μm	MPs: 50 and 500 μg L^{-1} Phe: 10 and 100 μg L^{-1}	MPs affected DTC in the liver and gills, lowered blood HDL levels, decreased plasma globulin, increased albumin/globulin ratio, decreased transcription level of *foxl2* No effect on plasma lipase, total and direct bilirubin, lactate, AST, ALT, ALP, γGT, LDL, triglycerides Significant interactions between MPs and Phe: DTC in gills, cholesterol, HDL, total protein, albumin, globulin, transcription of *ftz-f1*, *GnRH*, *11β-hsd2*, glycogen storages	[37]

Danio rerio, 21 days	PS, beads	0.07, 5	20–2,000 µg L⁻¹	Inflammation and lipid accumulation in liver; oxidative stress (liver cells), metabolic alterations	[38]
Oryzias latipes, 2 months	LDPE, fragments virgin-MPs vs. marine-MPs (pre-conditioned in a marina)	<500	Dosing via food, ~8 ng plastic mL⁻¹	Signs of stress in the liver: glycogen depletion (74% marine-MP fishes, 46% virgin-MP fishes, 0% control), fatty vacuolation (47% marine-MP fishes, 29% virgin-MP fishes, 21% control), single cell necrosis (11% marine-MP fishes, 0% virgin-MP fishes and control)	[39]
Pomatoschistus microps, 96 h	PE, beads +Pyrene	1–5	MP: 18.4–184 µg L⁻¹ Pyrene: 20–200 µg L⁻¹	MPs delayed pyrene-induced mortality, reduced AChE activity, no effects on GST and LPO	[40]

Only studies with MPs >1 µm are considered

3.1.1 Algae

So far, the majority of studies focused on the effects of MPs on consumers of aquatic food webs, and information on primary producers is limited. However, there are some indications that MPs adversely affect algae in a concentration and size-dependent manner [41–43]. For instance, 1 μm PVC fragments inhibited the growth and negatively affected photosynthesis (50 mg L^{-1}) of the marine algae *Skeletonema costatum* [43], while 1 mm PVC fragments did not induce such alterations. The underlying mechanisms are still unknown, whereby the direct interaction between MPs and algae and formation of aggregates seem to be strongly related. Since algae are used as a food source in ecotoxicological experiments, MPs may induce direct and indirect (quality and quantity of the algae) effects in the consumer.

3.1.2 *Daphnia magna*

In contrast to marine studies, only one filter-feeding freshwater species, *D. magna*, has been tested thoroughly in chronic and acute exposure regimes. Acute toxicity testing over 96 h resulted in an elevated immobilization at extremely high concentrations of 1 μm polyethylene (PE) particles [34]. With a median lethal concentration (LC_{50}) of 75.3 mg L^{-1}, these acute effects are (presumably) not environmentally relevant. Compared to this, chronic exposure to nanoscale PS over 21 days (0.22–150 mg L^{-1}, [41]) was not lethal. However, high concentrations of nano-PS (>30 mg L^{-1}) induced neonatal malformations and slightly decreased the reproductive output. Interestingly, the mortality as well as the amount of malformations increased when the daphnids were fed with nano-PS incubated algae (5 days). Since nano-PS particles might be too small for a direct ingestion, the formation of particle-algae aggregates may have resulted in a higher exposure. Furthermore, nano-PS reduced the growth and the chlorophyll a content of algae (*Scenedesmus obliquus*) indicating a reduced nutritional value of algae cultured with polymer particles.

Ogonowski et al. [35] conducted life-history experiments with *D. magna* exposed to primary MPs (spherical beads, 1.3 g cm^{-3}, 4.1 μm), secondary MPs (PE fragments, 1.0 g cm^{-3}, 2.6 μm), and kaolin (2.6 g cm^{-3}, 4.4 μm) under food-limited conditions. They observed an increased mortality and slightly decreased reproduction of daphnids for the highest concentration of secondary MPs (10^5 P mL^{-1}). However, incoherent exposure regimes (different particle sizes, concentrations, and exposure durations, among others) limit a general comparability and conclusion. In fact, the strongest response was driven by the low amount of food (reproduction far below validation criteria, OECD). However, these studies illustrate that (a) adverse effects depend on several factors, e.g., the size and shape of primary vs. secondary MPs, particle concentrations, polymer densities, as well as particle interaction with other stressors, and (b) *D. magna* seems relatively resistant to MP exposures.

The low sensitivity of *D. magna* could be due to its behavioral and morphological adaptations as a generalist filter feeder. *D. magna* feeds nonselectively on seston components encountering multiple particle sizes, shapes, and materials. High concentrations of SPM reduce the filtration rates as daphnids reject collected particles before ingestion or even narrow their carapace opening to avoid large particles [44, 45]. Besides pre-ingestion adaptations to unsuitable SPM, the peritrophic membrane protects the epithelium of the digestive tract from particle-induced injury. It consists of a complex matrix of chitin microfibrils, polysaccharides, as well as proteins and surrounds the food bolus in the digestive tract of many arthropods [46, 47]. Pores of several nanometers in diameter ensure the transport of digestive fluids and nutrients and protect against pathogens and mechanical damage. The packed food particles pass the digestive tract and are egested with the surrounding peritrophic membrane. Therefore, a direct interaction of MPs with epithelial cells in the digestive tract and thus injuries and a transfer of MPs into the surrounding tissue are unlikely. However, Rosenkranz et al. [48] observed 20 and 1,000 nm particles in the oil droplets of *D. magna* implying a translocation through the gut's epithelial cells, whereas the majority of studies with nanomaterials did not confirm this observation [49, 50].

3.1.3 Other Crustaceans

Null effects were found in the amphipod *Gammarus pulex* exposed to irregular polyethylene terephthalate (PET) fragments (0.4–4,000 P mL^{-1}, size 10–150 µm; [31]). After 48 days, MPs did not induce any effects on behavior (feeding activity), metabolism (energy reserves), development (molting), and growth. Au et al. [18] tested weathered polypropylene (PP) fibers (20–75 µm, 0–90 P mL^{-1}) as well as laboratory-made PE fragments (10–27 µm, 0–10^5 P mL^{-1}) in the amphipod *Hyalella azteca*. In a 10-day acute exposure, PP fibers were more toxic than PE fragments with LC$_{50}$ values of 71.43 and 46,400 P mL^{-1}, respectively. This might be related to the longer gut retention times of fibers versus fragments and again highlights the importance of particle shape. In the same study, a 42-day chronic exposure to PE fragments significantly decreased growth and reproduction.

At present, besides the studies with *D. magna* and the amphipods, there is very limited data regarding other freshwater crustaceans as the majority of research focuses on marine species. In addition to the increasing number of laboratory studies, the monitoring of wild populations of the common shrimp *Crangon crangon* [51] and the Norway lobster *Nephrops norvegicus* [52] have shown that field populations in marine environments are exposed to MPs. In both studies, MPs (predominantly fibers) were detected in 63% [51] and 83% [52] of the examined animals. A recent study by Welden and Cowie [1] with *N. norvegicus* confirmed that MP exposure negatively affects feeding, body mass, metabolic activity, and energy reserves. An 8-month exposure to PP fibers via food (0.2–5 mm, five fibers per feeding) resulted in formations of MP aggregates in the gut of the langoustine that might have reduced the uptake of nutrients. Effects on survival and growth as an outcome of reduced feeding have also been shown in the marine calanoid copepod *Calanus helgolandicus* [2]. The presence of 20 μm PS beads (75 P mL^{-1}) reduced the feeding on algae and provoked a feeding preference for smaller algae prey.

Although calanoid copepods are raptorial with strong size and taste discrimination, a study by Lee et al. [53] demonstrated a nonselective ingestion of 0.05, 0.5, and 6 μm PS beads by the marine *Tigriopus japonicus*. While all individuals survived an acute exposure (96 h), a two-generation chronic exposure to 0.05 (>12.5 μg mL^{-1}) and 0.5 μm beads (25 μg mL^{-1}) induced a concentration- and size-dependent mortality and a significant decrease in fecundity by 0.5 and 6 μm PS beads. Again, the observed effects were mainly interpreted as related to an impaired nutritional uptake.

In addition to the presumed nutritional effects, Bundy et al. [54] have shown that calanoid copepods regularly attack, capture, and reject 50 μm PS beads. This pre-ingestion behavior may result in a negative energy budget. Additionally, Cole et al. [55] documented that MPs attach to the external carapace and appendages of marine zooplankton, which then might interfere with locomotion, molting, and feeding. The relevance of adhered particles was also shown in the marine crabs *Uca rapax* and *Carcinus maenas* [56, 57]. Here, MP exposure led to an accumulation in the stomach and hepatopancreas but also to an accumulation in the gills. The respiratory uptake and the following adhesion of MPs to the gills might influence the branchial function. For instance, Watts et al. [58] found a significantly decreased oxygen consumption of MP-exposed crabs after 1 h and observed some adaptation as oxygen consumption returned to normal after 16 h.

3.1.4 Bivalves

The transfer of MPs to tissues induces cellular injuries as well as inflammatory responses in the marine filter-feeding mussel *M. edulis*. After 3 days of exposure to 3.0 and 9.6 μm PS beads, Browne et al. [24] observed a translocation to the circulatory (hemolymph) system where they remained for up to 48 days. Although the exact pathway is yet unknown, the transfer may be due to specialized enterocytes which in humans and rodents transport MPs from the gut into follicles from which they can translocate into the circulatory system. In addition, particles accumulating in the digestive gland were taken up by cells of the lysosomal system, which resulted in an inflammatory response and histological alterations (lysosomal membrane destabilization) [5]. As a consequence of particle interaction with tissue or hemolymph cells, marine bivalves can express an immediate stress and immune response. This results in an increased production of reactive oxygen species as well as anti-oxidant and glutathione-related enzymes but also changes the hemocyte phagocytosis activity and the ratio of granulocytes and hyalinocytes [59, 60].

Rist et al. [61] exposed the marine Asian green mussel *Perna viridis* to 1–50 µm polyvinyl chloride (PVC) fragments. MP exposure reduced the filtration and respiration rates, byssus production, as well as motility, while mortality was enhanced. Regarding life-history parameters, MP significantly reduced the reproductive success of *Crassostrea gigas* and negatively affected larval development of the offspring (PS spheres, 2, 6 µm). Sussarellu et al. [62] linked these effects to a disrupted energy uptake, which resulted in a shift of resources from reproduction to growth. In contrast, studies with *M. edulis* by van Cauwenberghe et al. [63] showed no significant effects of particle exposure to energy reserves (PS spheres, 10, 30, 90 µm).

Behavioral and physiological responses have also been shown for bivalves exposed to suspended solids. For instance, particle exposure damaged the cilia of the gill filaments in *P. viridis* (<500 µm [64]) and significantly reduced the algal ingestion of *M. mercenaria* (3–40 µm, [65]). Therefore, the lack of studies comparing impacts of both MPs and suspended particles hampers a discrimination of MP-associated and more general particle-associated effects.

These studies provide evidence that MP ingestion can affect marine bivalves. As the general feeding strategies are consistent in both marine and freshwater species, the latter may be similarly affected. Still, morphological details of the feeding-associated organs vary in the different bivalve taxa, which can alter feeding-specific characteristics [66].

3.1.5 Gastropods

In comparison to bivalves, fewer studies have examined MP toxicity in gastropods, which also have a high capacity to ingest MPs (discussed in Sect. 2.1). The only currently available study on MP toxicity in gastropods suggests limited impacts [36]. In this study, the omnivorous surface grazer *P. antipodarum* was exposed to a mixture of five different polymers (4.6–603 µm particle size; polyamide (PA), polycarbonate (PC), PET, PS, PVC) mixed with food at a ratio of 30 and 70%. After 8 weeks, MPs neither affected the growth (shell width, length, body weight) nor the reproduction (number of produced embryos and ratio of embryos with and without shell). Additionally, MP had no effect on the development of the consecutive generation of juveniles.

3.1.6 Fish

Several adverse effects by MP exposures have also been observed for freshwater fishes (Table 2). MPs accumulate in the gills of marine crustaceans, and studies with freshwater fishes demonstrate that this pathway is relevant for vertebrate species too. One example is zebrafish (*Danio rerio*) in which PS beads accumulate in the gills (5 and 20 µm), gut (5 and 20 µm), and liver (5 µm) [38]. Indeed, histopathological analysis revealed an inflammatory response and accumulation of lipids in the liver as well as oxidative stress. However, these findings were only significant at high concentrations (2 mg L^{-1}) of 0.07 and 5 µm beads. In comparison, Karami et al. [37] observed histological alterations in the gills (e.g., basal cell hyperplasia and necrosis in connective tissue) and blood biochemistry parameters (e.g., plasma cholesterol levels, blood HDL levels) of the African catfish (*Clarias gariepinus*) at lower concentrations of HDPE fragments (50 µg L^{-1}). More severe changes (epithelial lifting, hyperplasia, extensive cell sloughing) were reported for higher particle concentration (500 µg L^{-1}). Additionally, concentrations of 500 µg L^{-1} significantly affected the degree of tissue change in the liver of exposed individuals. Overall, the authors point toward ethylene monomers (released from HDPE) and internal as well as external abrasions (caused by sharp edges of the fragments) as possible mechanisms for the changes in biomarker responses.

It is well documented that suspended solids can damage organs in several fish species and cause adverse effects similar to those observed for MPs. High concentrations of SPM can accumulate in the gills, disturb the respiratory function, and have been found to translocate into epithelial cells, cause lipid peroxidation, and reduce the tolerance of infection by pathogens [67, 68]. Additionally, studies with gill epithelial cells (rainbow trout, RTgill-W1) and fluvial fine sediment revealed translocation of fine minerals (<2 µm, 10–250 mg L^{-1}) into the cells as well as material-related cytotoxicity [69]. Here, quartz and feldspar only caused sporadic changes in biomarker response, and exposure to mica (silicate minerals) and kaolin induced cytotoxicity as well as free radicals and cell membrane damage. Therefore, Michel et al. [69] conclude that the uptake of fine particles by gill epithelial cells is a common natural event in aquatic species with the material, size, shape, and concentration determining the impacts.

3.2 Chemical Impacts

So far, MPs detected in freshwater environments represent a range of material types (e.g., PE, PS, PET, PVC, PA, and PP), originate from various sources and applications, and represent a plethora of material characteristics. In general, plastic materials are highly functional compounds of synthetic polymers and additives (e.g., plasticizers, flame retardants, colorants). Leachates from diverse plastic products were found to cause chemical toxicity [70, 71] induced by monomers, residues of production processes (e.g., catalyzers, stabilizers), and additives. For instance, some leaching components were classified as endocrine disrupting chemicals (e.g., phthalates, bisphenol A) adversely affecting life-cycle parameters of a broad range of species [72, 73]. Fries et al. [74] extracted several organic (e.g., phthalates) and inorganic additives (e.g., metals) from MP samples in marine sediments highlighting the relevance of these compounds. Besides additives, adsorbed persistent organic pollutants have been found on MPs (e.g., [75, 76]). The capacity of plastic materials to accumulate hydrophobic organic chemicals is thoroughly studied and frequently applied in passive samplings/monitoring (e.g., [76, 77]). For MPs, the large surface-to-volume ratio supports an accumulation of dissolved pollutants (e.g., PAHs, PBTs, metals), and complex adsorption-desorption patterns have been demonstrated [77, 78].

Although a detailed review of the complexity in adsorption-desorption kinetics is beyond the focus of this chapter, the default hypothesis is that MPs readily sorb hydrophobic compounds and therefore act as vectors transferring waterborne contaminants to aquatic organisms (vector hypothesis). However, this idea is controversially discussed. Several laboratory studies illustrate the capacity of MPs to modify adverse effects of chemicals by affecting the bioavailability or acting as an additional stressor. For instance, (1) the exposure to spiked MPs lead to an accumulation of pollutants to the tissues of lugworms (PVC, [6]), mussels (PE and PS, [59]), amphipods (PE, [79]), and fish (LDPE, [39]); (2) Besseling et al. [80] observed a decreased bioaccumulation of polychlorinated biphenyls in lugworms at higher doses of PS particles; (3) Oliveira et al. [40] confirmed a delayed pyrene-induced mortality of juvenile fishes (*Pomatoschistus microps*) in the presence of PE MPs; and (4) Karami et al. [37] as well as Paul-Pont et al. [60] detected modulations of adverse effects by an exposure to phenanthrene-loaded LDPE fragments (African catfish) and PS beads and fluoranthene (*Mytilus* spp.), respectively. However, Gouin et al. [81] and Koelmans et al. [82] highlight the minor influence of MPs as vectors for the bioaccumulation of pollutants considering they are outcompeted by natural occurring matter. These authors emphasize the importance of experimental design and chemical analysis in order to understand the relevance and underlying mechanisms of MPs as vectors of bioaccumulative substances. For instance, the introduction of freshly spiked MPs in clean water can result in desorption, which increases dermal exposure [82]. Furthermore, desorbed chemicals might adsorb to food or sediments and decrease the potential relevance of MPs as vectors. In principle, adsorption and desorption patterns follow the partition equilibrium between the available compartments (e.g., biota, food, MPs, sediment, water). This may confound the analysis of single pathways particularly if analytical information is absent (e.g., exposure via ingestion of MPs, food or sediments vs. dermal uptake).

While studies on the vector hypothesis were mostly performed with marine species and persistent organic pollutants, the situation is likely to be very different in freshwater ecosystems. First and foremost, freshwater compartments are exposed to a completely different and much larger spectrum of chemicals than marine systems. This is because they receive a constant input of chemicals from land-based sources (e.g., pesticides) and wastewater (e.g., pharmaceuticals and chemicals from personal care products). Many of these compounds are pseudo-persistent and biologically highly active but do not occur in marine ecosystems (due to dilution or degradation). Accordingly, freshwater MPs will sorb a completely different set of chemicals than marine ones. In addition, being closer to the source of plastic litter and thus "younger," freshwater MPs might contain higher concentrations of plastic additives. With regard to desorption, physical water properties will affect the transfer of pollutants. The adsorption equilibrium of chemicals to organic materials is highly dependent on water temperature, quantity of organic matter, and the content of inorganic salts [83]. Therefore, the partition equilibrium will be different in salt- and freshwater.

Besides the capacity of MPs to influence the bioavailability of toxic compounds, Besseling et al. [41] suggested that MPs can interfere with intra- and interspecies signaling (e.g., phero- and kairomones) as an integral component of aquatic biocoenosis regulating predator-prey interactions as well as population and community structures [84]. Although they found significant interactions between kairomones and nano-PS when investigating the growth of the water flea *D. magna*, it remains unclear whether the nano-PS beads increased the bioavailability of kairomones or they observed an additive effect of both stressors [41]. Any disturbance of this inter- and intraspecies communication can lead to maladaptive responses in both signaler and receiver [85]. So far, it is unclear whether MPs act as info-disruptors as is the case for several metals and pesticides (reviewed in [85]), especially when considering the abundance of additional particulate organic and inorganic matter in aquatic ecosystems.

3.3   Biofilm-Related Impacts

Apart from the potential of MPs to act as carriers for chemicals, MPs can serve as substrates for microorganisms. The formation of biofilms [86] can affect the interaction of MPs with biota on multiple levels. For example, the colonization of MPs with microbes and the adsorption of biopolymers increase the nutritional value and improve the "taste" making them more attractive for biota. In contrast, the colonization of MPs with pathogens [87] and toxic algae/bacteria might induce infections/chemical toxicity or avoidance of "bad tasting" MPs. Additionally, biofouling was shown to affect the fate of MPs by changing the particle properties (e.g., density). The formation of biofilms increases the density of floating or buoyant MPs and leads to sedimentation of these low-density particles (reviewed in [88]). Furthermore, in the environment, MPs are most likely incorporated in so-called hetero-aggregates. These aggregates consist of particulate matter (MPs as well as other suspended solids) and microbes (e.g., protozoans, algae) with biopolymers acting as binders. A laboratory study by Lagarde et al. [89] confirmed polymer-dependent (PP vs. HDPE) aggregations with the algae *Chlamydomonas reinhardtii*. While rapid colonization of the surfaces of both HDPE and PP was observed, expanding hetero-aggregates consisting of polymer particles, algae cells, and exopolysaccharides were solely formed by PP. The upscaling of microscopic particles via aggregation can modify their potential for being ingested. While the abundance of microscopic particles and thereby the availability to micro-feeders (e.g., protozoans, planktonic crustaceans) decreases, large hetero-aggregates are accessible to macro-feeders (e.g., planktivorous fishes). Thus, the uptake of one aggregate by macro-feeders might lead to an internal release and exposure to multiple particles of different sizes as digestive fluids digest the biopolymer matrix. However, the sample preparation needed to separate MPs from environmental

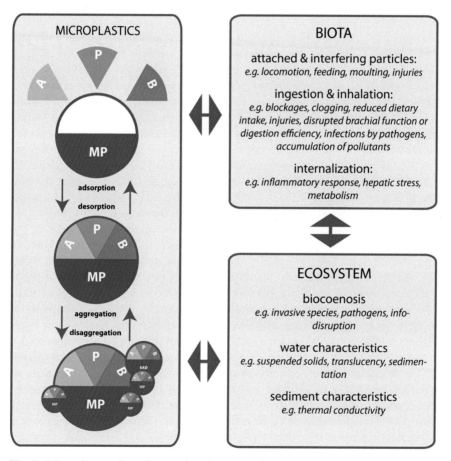

Fig. 2 Schematic overview of interactions between microplastics, biota, and ecosystems. The term microplastics comprises the following interdependent factors: *A* additives (e.g., polymer monomers, production residues), *P* pollutants (e.g., HOCs), *B* biofilm and biopolymers, *MP* microplastics including varieties of material, density, shape, size, and surface characteristics

samples destroys hetero-aggregates and makes it almost impossible to investigate them in their natural state.

Overall, MP-associated risks are multifaceted in their nature and the following must be considered: MP-biota interactions, toxicity of polymer-related leachates, adsorption-desorption kinetics of co-occurring compounds, biofilm-related effects, and the formation of hetero-aggregates. Thus, understanding the interaction of all these factors in real-world situations is necessary to evaluate the environmental risk of MP exposures (Fig. 2).

4 Natural vs. Synthetic Particle Toxicity

The similarities in the effects caused by exposure to natural fine particles and MPs (see Sect. 3.1) provoke the legitimate question whether MPs have a different toxicological profile compared to natural solids. In general, organisms interact with a variety of particulate matter in freshwater ecosystems and possess adaptations to this potential stressor (e.g., peritrophic membrane, mucus). Species occupying turbid waters might be less sensitive to high concentrations of SPM than species inhabiting clear water. Species-specific effects of exposures to suspended solids were highlighted in numerous studies investigating the anthropogenic introduction of particulate matter (e.g., arising from erosion, dredging; reviewed in [90, 91]). Suspended particles or fine sediments can reduce feeding rates, decrease reaction distance to prey, influence embryo development, increase mortality, reduce primary production, reduce species diversity, and decrease population size [90–94]. Bilotta and Brazier [90] conclude that the magnitude of adverse effects depends on concentration, exposure duration, chemical composition, and particle size distribution. Tolerant species suffered moderately negative effects, while strong effects mainly occurred in intolerant species (see a review on fish in [91]).

These outcomes are also applicable for the effect studies with MPs and, thus, illustrate the importance of benchmarking the toxicity of MPs in comparison to naturally occurring particles. Considering the available literature, we can hypothesize a higher particle toxicity of MPs since adverse effects were observed at lower concentrations compared to fine sediments. However, studies with suspended solids have used a variety of units (particle per volume, mass per volume, parts per million), size classes, densities, and experimental conditions, which impedes a direct comparison. Accordingly, to answer the question whether the particle toxicity of MPs is indeed different from natural materials, ecotoxicological studies need to include reference treatments with natural particles (e.g., minerals, charcoal). However, investigating particle toxicity necessitates a highly complex approach featuring multiple factors, e.g., concentration, material, size, shape, density, and surface characteristics.

In principle, aquatic species interact with MPs through a variety of pathways featuring direct or indirect ingestion, respiration, or attachment to the body surface. Therefore, a single stressor (e.g., inert particle) influences life-cycle parameters on multiple levels. For instance, the presence of MPs can limit the nutrient assimilation by reducing the proportion of available food particles or by interfering with feeding mechanisms and locomotion, influencing digestion efficiency, and driving behavioral adaptations (e.g., avoidance, foraging). This implies that effect studies with MPs should focus on multiple endpoints including typical life-cycle parameters (e.g., reproduction, growth, nutritional state), histological analyses, and biomarker responses. Furthermore, the implementation of time-course, chronic, and multigenerational test designs might help uncover adaptive responses as well as cascading effects in populations. Only the simultaneous investigation and direct comparison of the toxicity of natural and polymeric particles will enable discovering specific MP-associated risks in the diversity of particulate matter. In the absence of this reference, adverse effects of MPs observed in the laboratory could be nothing but a representation of the (normal) biological response and physiological condition induced by natural particles. However, species in freshwater systems are adapted to naturally occurring particles, and it remains relatively unclear whether polymer particles act differently or have the potential to bypass protective adaptations.

5 Implications for Freshwater Ecosystems

Although plastics have been released into the environment for many years, researchers have barely begun to understand the extent of MP distribution in freshwater systems. As such, the environmental impacts of MPs have not been thoroughly evaluated. Importantly, the term "microplastics" encompasses a tremendous variety of polymers that in turn spans a very wide range of sizes, shapes, and chemical compositions. In this sense, MPs do not represent one stressor, whose impacts can be evaluated relatively easily, but a very large number of stressors that potentially act jointly. The use of copolymers, product-specific mixtures of additives, and source- and pathway-specific sorbed pollutants further complicates the situation.

In physical terms, MPs can influence water (e.g., translucency [42]), sedimentation (e.g., feces [95]), and sediment (e.g., thermal conductivity [96]) characteristics. In ecological terms, MPs can affect the aquatic biocoenosis on a large scale (Fig. 2), for instance, as vectors for invasive species and pathogens [97–99]. The existing toxicological studies mostly focus on the interaction of MPs and biota in simplified exposure regimes, commonly using spherical microbeads composed of a single polymer. Here, there is a tendency for elevated adverse effects (e.g., reduced reproduction, inflammatory response) with decreasing particle sizes. At the current state of research, MP toxicity has been studied and in some cases demonstrated at relatively high concentrations. This has been criticized as lacking ecological relevance. However, the environmental concentrations of very small, biologically relevant MPs (<100 μm) remain unknown but may be higher than predicted based on analyzing larger MPs. In addition, species-specific responses may be incorrectly estimated by using microspheres alone. The use of multiple polymer types, shapes, and sizes may establish that some species are more sensitive than originally predicted.

It is already established that high concentrations of suspended solids affect community structure through changes in growth, reproduction, and species interactions. Accordingly, evolutionary adaptations (e.g., peritrophic membrane, mucus, avoidance) might explain the species-dependent resistance to high concentrations of MPs (e.g., *D. magna*, *G. pulex*). However, MPs can infiltrate habitats normally low in suspended solid and thereby affect more sensitive species.

The continuing release of MPs through the breakdown of littered plastics that are already present in the environment means that MPs may become an increasingly important freshwater pollutant in the future. In addition, the high demand of plastic materials/products will not decrease if continuing the business-as-usual mode. Accordingly, without rethinking and restructuring our resource production and use (e.g., within the framework of a circular economy, [100]), plastic waste will further accumulate in the biosphere.

Overall, traditional approaches for toxicity testing may not be appropriate for a multifaceted stressor such as MPs. The default assumption that standard model organisms act as appropriate surrogates for aquatic biocoenoses may ignore species-specific responses of more sensitive species. In addition, consideration of future scenarios may render vector-related impacts (e.g., biofilms, transfer of additives, and hydrophobic persistent pollutant) more prominent.

Our knowledge regarding the impacts of MPs on freshwater species is limited at the present time, although we are beginning to appreciate some of the complexities as more laboratory and field data becomes available. First and foremost, we need to prioritize which physical and chemical MP characteristics are toxicologically and ecologically most important. In this context, there is also a lot to learn from other disciplines with important data already abundant (e.g., ecological feeding studies, suspended solids, medicine, nanomaterials; see e.g., chapter "Freshwater Microplastics: Challenges for Regulation and Management"). Ecological knowledge regarding the adaptations of specific species as well as factors driving species compositions might help to identify especially sensitive biota. In addition, understanding the role of MPs relative to other stressors will require a multidisciplinary approach. Overall, understanding the complex interactions of plastics and the environment can only be achieved by a joint effort. The upcoming challenge will be to unravel the role that MPs play in a multiple-particle world.

References

1. Welden NAC, Cowie PR (2016) Environment and gut morphology influence microplastic retention in langoustine, *Nephrops norvegicus*. Environ Pollut 214:859–865. doi:10.1016/j.envpol. 2016.03.067
2. Cole M, Lindeque P, Fileman E et al (2015) The impact of polystyrene microplastics on feeding, function and fecundity in the marine copepod *Calanus helgolandicus*. Environ Sci Technol 49:1130–1137. doi:10.1021/es504525u
3. Phuong NN, Zalouk-Vergnoux A, Poirier L et al (2016) Is there any consistency between the microplastics found in the field and those used in laboratory experiments? Environ Pollut 211:111–123. doi:10.1016/j.envpol.2015.12.035
4. Wright SL, Thompson RC, Galloway TS (2013) The physical impacts of microplastics on marine organisms: a review. Environ Pollut 178:483–492. doi:10.1016/j.envpol.2013.02.031
5. von Moos N, Burkhardt-Holm P, Koehler A (2012) Uptake and effects of microplastics on cells and tissue of the Blue mussel *Mytilus edulis* L. after an experimental exposure. Environ Sci Technol 46:327–335. doi:10.1021/es302332w
6. Browne MA, Niven SJ, Galloway TS et al (2013) Microplastic moves pollutants and additives to worms, reducing functions linked to health and biodiversity. Curr Biol 23:2388–2392. doi:10.1016/j.cub.2013.10.012
7. Cummins KW, Klug MJ (1979) Feeding ecology of stream invertebrates. Annu Rev Ecol Syst 10:147–172. doi:10.1146/annurev.es.10.110179.001051
8. Burns CW (1968) The relationship between body size of filter-feeding Cladocera and the maximum size of particle ingested. Limnol Oceanogr 13:675–678. doi:10.4319/lo.1968.13.4.0675
9. Sanders RW, Porter KG, Bennett SJ, Debiase AE (1989) Seasonal patterns of bacterivory by flagellates, ciliates, rotifers, and cladocerans in a freshwater planktonic community. Limnol Oceanogr 34:673–687. doi:10.4319/lo.1989.34.4.0673
10. Thouvenot A, Richardot M, Debroas D, Devaux J (1999) Bacterivory of metazooplankton, ciliates and flagellates in a newly flooded reservoir. J Plankton Res 21:1659–1679. doi:10.1093/plankt/21.9.1659
11. Børsheim KY (1984) Clearance rates of bacteria-sized particles by freshwater ciliates, measured with monodisperse fluorescent latex beads. Oecologia 63:286–288. doi:10.1007/BF00379891
12. Agasild H, Nõges T (2005) Cladoceran and rotifer grazing on bacteria and phytoplankton in two shallow eutrophic lakes: in situ measurement with fluorescent microspheres. J Plankton Res 27:1155–1174. doi:10.1093/plankt/fbi080
13. Ooms-Wilms AL, Postema G, Gulati RD (1995) Evaluation of bacterivory of Rotifera based on measurements of in situ ingestion of fluorescent particles, including some comparisons with Cladocera. J Plankton Res 17:1057–1077. doi:10.1093/plankt/17.5.1057
14. DeMott WR (1986) The role of taste in food selection by freshwater zooplankton. Oecologia 69:334–340. doi:10.1007/BF00377053
15. Bern L (1990) Size-related discrimination of nutritive and inert particles by freshwater zooplankton. J Plankton Res 12:1059–1067. doi:10.1093/plankt/12.5.1059
16. Scherer C, Brennholt N, Reifferscheid G, Wagner M (2017) Feeding strategy and development drive the ingestion of microplastics by freshwater invertebrates, in preparation
17. Imhof HK, Ivleva NP, Schmid J et al (2013) Contamination of beach sediments of a subalpine lake with microplastic particles. Curr Biol 23:1–15. doi:10.1016/j.cub.2013.09.001
18. Au SY, Bruce TF, Bridges WC, Klaine SJ (2015) Responses of *Hyalella azteca* to acute and chronic microplastic exposures. Environ Toxicol Chem 34:2564–2572. doi:10.1002/etc.3093
19. Dantas DV, Barletta M, da Costa MF (2012) The seasonal and spatial patterns of ingestion of polyfilament nylon fragments by estuarine drums (Sciaenidae). Environ Sci Pollut Res 19:600–606. doi:10.1007/s11356-011-0579-0

20. Possatto FE, Barletta M, Costa MF et al (2011) Plastic debris ingestion by marine catfish: an unexpected fisheries impact. Mar Pollut Bull 62:1098–1102. doi:10.1016/j.marpolbul.2011. 01.036
21. Ramos JAA, Barletta M, Costa MF (2012) Ingestion of nylon threads by gerreidae while using a tropical estuary as foraging grounds. Aquat Biol 17:29–34. https://doi.org/10.3354/ ab00461
22. Drenner RW, Mummert JR, Denoyelles F, Kettle D (1984) Selective particle ingestion by a filter-feeding fish and its impact on phytoplankton community structure. Limnol Oceanogr 29:941–948. doi:10.4319/lo.1984.29.5.0941
23. Jeong C-B, Won E-J, Kang H-M et al (2016) Microplastic size-dependent toxicity, oxidative stress induction, and p-JNK and p-P38 activation in the monogonont rotifer (Brachionus koreanus). Environ Sci Technol 50:8849–8857. doi:10.1021/acs.est.6b01441
24. Browne MA, Dissanayake A, Galloway TS et al (2008) Ingested microscopic plastic translocates to the circulatory system of the mussel, Mytilus edulis (L.) Environ Sci Technol 42: 5026–5031. doi:10.1021/es800249a
25. Wootton RJ (1989) Ecology of teleost fishes. Springer, Dordrecht. doi:10.1007/978-94-009-0829-1
26. Gerking SD (1994) Feeding ecology of fish. Academic Press, San Diego
27. Fenchel T (1980) Suspension feeding in ciliated protozoa: feeding rates and their ecological significance. Microb Ecol 6:13–25. doi:10.2307/4250595
28. Geller W, Müller H (1981) The filtration apparatus of cladocerans: filter mesh-size and their implications on food selectivity. Oecologia 2:316–321
29. Bleiwas AH, Stokes PM (1985) Collection of large and small food particles by Bosmina. Limnol Oceanogr 30:1090–1092. doi:10.4319/lo.1985.30.5.1090
30. Price HJ (1988) Feeding mechanisms in marine and freshwater zooplankton. Bull Mar Sci 43: 327–343
31. Weber A, Scherer C, Brennholt N et al (2017) Microplastics do not negatively affect the freshwater invertebrate Gammarus pulex, in preparation
32. Sherr BF, Sherr EB, Fallon RD (1987) Use of monodispersed, fluorescently labeled bacteria to estimate in situ protozoan bacterivory. Appl Environ Microbiol 53:958–965. doi:10.1016/ 0198-0254(87)90950-2
33. DeMott WR (1988) Discrimination between algae and artificial particles by freshwater and marine copepods. Limnol Oceanogr 33:397–408. doi:10.4319/lo.1988.33.3.0397
34. Rehse S, Kloas W, Zarfl C (2016) Short-term exposure with high concentrations of pristine microplastic particles leads to immobilisation of Daphnia magna. Chemosphere 153: 91–99. doi:10.1016/j.chemosphere.2016.02.133
35. Ogonowski M, Schür C, Jarsén Å, Gorokhova E (2016) The effects of natural and anthropogenic microparticles on individual fitness in Daphnia magna. PLoS One 11:e0155063. doi:10.1371/journal.pone.0155063
36. Imhof HK, Laforsch C (2016) Hazardous or not – are adult and juvenile individuals of Potamopyrgus antipodarum affected by non-buoyant microplastic particles? Environ Pollut 218:383–391. doi:10.1016/j.envpol.2016.07.017
37. Karami A, Romano N, Galloway T, Hamzah H (2016) Virgin microplastics cause toxicity and modulate the impacts of phenanthrene on biomarker responses in African catfish (Clarias gariepinus). Environ Res 151:58–70. doi:10.1016/j.envres.2016.07.024
38. Lu Y, Zhang Y, Deng Y et al (2016) Uptake and accumulation of polystyrene microplastics in zebra fish (Danio rerio) and toxic effects in liver. Environ Sci Technol 50:4054–4060. https://doi.org/10.1021/acs.est.6b00183
39. Rochman CM, Hoh E, Kurobe T, Teh SJ (2013) Ingested plastic transfers hazardous chemicals to fish and induces hepatic stress. Sci Rep 3:3263. doi:10.1038/srep03263

40. Oliveira M, Ribeiro A, Hylland K, Guilhermino L (2013) Single and combined effects of microplastics and pyrene on juveniles (0+ group) of the common goby *Pomatoschistus microps* (Teleostei, Gobiidae). Ecol Indic 34:641–647. doi:10.1016/j.ecolind.2013.06.019

41. Besseling E, Wang B, Lurling M, Koelmans AA (2014) Nanoplastic affects growth of *S. obliquus* and reproduction of *D. magna*. Environ Sci Technol 48:12336–12343. doi:10.1021/es503001d

42. Sjollema SB, Redondo-Hasselerharm P, Leslie HA et al (2016) Do plastic particles affect microalgal photosynthesis and growth? Aquat Toxicol 170:259–261. doi:10.1016/j.aquatox.2015.12.002

43. Zhang C, Chen X, Wang J, Tan L (2017) Toxic effects of microplastic on marine microalgae *Skeletonema costatum*: interactions between microplastic and algae. Environ Pollut 220:1282–1288. doi:10.1016/j.envpol.2016.11.005

44. Gliwicz ZM, Siedlar E (1980) Food size limitation and algae interfering with food collection in Daphnia. Arch Hydrobiol 88:155–177

45. Kirk KL (1991) Inorganic particles alter competition in grazing plankton: the role of selective feeding. Ecology 72:915–923. doi:10.2307/1940593

46. Peters W (1968) Occurrence, composition and fine structure of peritrophic membranes in animals. Z Morphol Tiere 62:9–57. doi:10.1007/BF00401488

47. Georgi O (1969) Fine structure of peritrophic membranes in Crustacea. Z Morphol Tiere 65:225–273

48. Rosenkranz P, Chaudhry Q, Stone V, Fernandes TF (2009) A comparison of nanoparticle and fine particle uptake by *Daphnia magna*. Environ Toxicol Chem 28:2142–2149

49. Tervonen K, Waissi G, Petersen EJ et al (2010) Analysis of fullerene-c60 and kinetic measurements for its accumulation and depuration in *Daphnia magna*. Environ Toxicol Chem 29:1072–1078. doi:10.1002/etc.124

50. Khan FR, Kennaway GM, Croteau M-N et al (2014) In vivo retention of ingested Au NPs by *Daphnia magna*: no evidence for trans-epithelial alimentary uptake. Chemosphere 100:97–104. doi:10.1016/j.chemosphere.2013.12.051

51. Devriese LI, van der Meulen MD, Maes T et al (2015) Microplastic contamination in brown shrimp (*Crangon crangon*, Linnaeus 1758) from coastal waters of the Southern North Sea and Channel area. Mar Pollut Bull 98:179–187. doi:10.1016/j.marpolbul.2015.06.051

52. Murray F, Cowie PR (2011) Plastic contamination in the decapod crustacean *Nephrops norvegicus* (Linnaeus, 1758). Mar Pollut Bull 62:1207–1217. doi:10.1016/j.marpolbul.2011.03.032

53. Lee K, Shim WJ, Kwon OY, Kang J (2013) Size-dependent effects of micro polystyrene particles in the marine copepod *Tigriopus japonicus*. Environ Sci Technol 47:11278–11283. doi:10.1021/es401932b

54. Bundy MH, Gross TF, Vanderploeg HA, Strickler JR (1998) Perception of inert particles by calanoid copepods: behavioral observations and a numerical model. J Plankton Res 20:2129–2152. doi:10.1093/plankt/20.11.2129

55. Cole M, Lindeque P, Fileman E et al (2013) Microplastic ingestion by zooplankton. Environ Sci Technol 47:6646–6655. doi:10.1021/es400663f

56. Brennecke D, Ferreira EC, Costa TMM et al (2014) Ingested microplastics (<100 μm) are translocated to organs of the tropical fiddler crab *Uca rapax*. Mar Pollut Bull 96:491–495. doi:10.1016/j.marpolbul.2015.05.001

57. Watts AJR, Lewis C, Goodhead RM et al (2014) Uptake and retention of microplastics by the shore crab *Carcinus maenas*. Environ Sci Technol 48:8823–8830. doi:10.1021/es501090e

58. Watts AJR, Urbina MA, Goodhead RM et al (2016) Effect of microplastic on the gills of the shore crab *Carcinus maenas*. Environ Sci Technol 50:5364–5369. doi:10.1021/acs.est.6b01187

59. Avio CG, Gorbi S, Milan M et al (2015) Pollutants bioavailability and toxicological risk from microplastics to marine mussels. Environ Pollut 198:211–222. doi:10.1016/j.envpol.2014.12.021

60. Paul-Pont I, Lacroix C, González Fernández C et al (2016) Exposure of marine mussels *Mytilus* spp. to polystyrene microplastics: toxicity and influence on fluoranthene bioaccumulation. Environ Pollut 216:724–737. doi:10.1016/j.envpol.2016.06.039

61. Rist SE, Assidqi K, Zamani NP et al (2016) Suspended micro-sized {PVC} particles impair the performance and decrease survival in the Asian green mussel *Perna viridis*. Mar Pollut Bull 111:213–220. doi:10.1016/j.marpolbul.2016.07.006

62. Sussarellu R, Suquet M, Thomas Y et al (2016) Oyster reproduction is affected by exposure to polystyrene microplastics. Proc Natl Acad Sci 113:2430–2435. doi:10.1073/pnas.1519019113

63. Van Cauwenberghe L, Claessens M, Vandegehuchte MB, Janssen CR (2015) Microplastics are taken up by mussels (*Mytilus edulis*) and lugworms (*Arenicola marina*) living in natural habitats. Environ Pollut 199:10–17. doi:10.1016/j.envpol.2015.01.008

64. Cheung SG, Shin PKS (2005) Size effects of suspended particles on gill damage in green-lipped mussel *Perna viridis*. Mar Pollut Bull 51:801–810

65. Bricelj VM, Malouf RE (1984) Influence of algal and suspended sediment concentrations on the feeding physiology of the hard clam *Mercenaria mercenaria*. Mar Biol 84:155–165. doi:10.1007/BF00393000

66. Dillon RTJ (2000) The ecology of freshwater molluscs. Cambridge University Press, Cambridge, p 499. doi:10.1017/CBO9781107415324.004

67. Redding JM, Schreck CB, Everest FH (1987) Physiological effects on coho salmon and steelhead of exposure to suspended solids. Trans Am Fish Soc 116:737–744. doi:10.1577/1548-8659(1987)116<737:PEOCSA>2.0.CO;2

68. Michel C, Schmidt-Posthaus H, Burkhardt-Holm P, Richardson J (2013) Suspended sediment pulse effects in rainbow trout (*Oncorhynchus mykiss*) – relating apical and systemic responses. Can J Fish Aquat Sci 70:630–641. doi:10.1139/cjfas-2012-0376

69. Michel C, Herzog S, De Capitani C et al (2014) Natural mineral particles are cytotoxic to rainbow trout gill epithelial cells in vitro. PLoS One 9:e100856. doi:10.1371/journal.pone.0100856

70. Lithner D, Larsson A, Dave G (2011) Environmental and health hazard ranking and assessment of plastic polymers based on chemical composition. Sci Total Environ 409:3309–3324. doi:10.1016/j.scitotenv.2011.04.038

71. Lithner D, Damberg J, Dave G, Larsson Å (2009) Leachates from plastic consumer products - Screening for toxicity with *Daphnia magna*. Chemosphere 74:1195–1200. doi:10.1016/j.chemosphere.2008.11.022

72. Oehlmann J, Schulte-Oehlmann U, Kloas W et al (2009) A critical analysis of the biological impacts of plasticizers on wildlife. Philos Trans R Soc Lond Ser B Biol Sci 364:2047–2026

73. Lambert S, Sinclair C, Boxall A (2014) Occurrence, degradation, and effect of polymer-based materials in the environment. Rev Environ Contam Toxicol 227:1–53

74. Fries E, Dekiff JH, Willmeyer J et al (2013) Identification of polymer types and additives in marine microplastic particles using pyrolysis-GC/MS and scanning electron microscopy. Environ Sci: Processes Impacts 15:1949–1956. doi:10.1039/c3em00214d

75. Antunes JC, Frias JGL, Micaelo AC, Sobral P (2013) Resin pellets from beaches of the Portuguese coast and adsorbed persistent organic pollutants. Estuar Coast Shelf Sci 130:62–69. doi:10.1016/j.ecss.2013.06.016

76. Mizukawa K, Takada H, Ito M et al (2013) Monitoring of a wide range of organic micro-pollutants on the Portuguese coast using plastic resin pellets. Mar Pollut Bull 70: 296–302. doi:10.1016/j.marpolbul.2013.02.008

77. Bakir A, Rowland SJ, Thompson RC (2012) Competitive sorption of persistent organic pollutants onto microplastics in the marine environment. Mar Pollut Bull 64:2782–2789. doi:10.1016/j.marpolbul.2012.09.010

78. Bakir A, Rowland SJ, Thompson RC (2014) Enhanced desorption of persistent organic pollutants from microplastics under simulated physiological conditions. Environ Pollut 185: 16–23. doi:10.1016/j.envpol.2013.10.007

79. Chua EM, Shimeta J, Nugegoda D et al (2014) Assimilation of polybrominated diphenyl ethers from microplastics by the marine amphipod, *Allorchestes compressa*. Environ Sci Technol 48:8127–8134. https://doi.org/10.1021/es405717z

80. Besseling E, Wegner A, Foekema EM et al (2013) Effects of microplastic on fitness and PCB bioaccumulation by the lugworm *Arenicola marina* (L.) Environ Sci Technol 47:593–600. doi:10.1021/es302763x

81. Gouin T, Roche N, Lohmann R, Hodges G (2011) A thermodynamic approach for assessing the environmental exposure of chemicals absorbed to microplastic. Environ Sci Technol 45: 1466–1472. doi:10.1021/es1032025

82. Koelmans AA, Bakir A, Burton GA, Janssen CR (2016) Microplastic as a vector for chemicals in the aquatic environment: critical review and model-supported reinterpretation of empirical studies. Environ Sci Technol 50:3315–3326. doi:10.1021/acs.est.5b06069

83. Endo S, Koelmans AA (2010) Sorption of hydrophobic organic compounds to plastics in the marine environment: equilibrium. In: The handbook of environmental chemistry. Springer, Berlin, pp 41–53

84. Pohnert G, Steinke M, Tollrian R (2007) Chemical cues, defence metabolites and the shaping of pelagic interspecific interactions. Trends Ecol Evol 22:198–204. doi:10.1016/j.tree.2007.01.005

85. Lürling M, Scheffer M (2007) Info-disruption: pollution and the transfer of chemical information between organisms. Trends Ecol Evol 22:374–379. doi:10.1016/j.tree.2007.04.002

86. Harrison JP, Hoellein TJ, Sapp M, Tagg AS, Ju-Nam Y, Ojeda JJ (2017) Microplastic-associated biofilms: a comparison of freshwater and marine environments. In: Wagner M, Lambert S (eds) Freshwater microplastics: emerging environmental contaminants? Springer, Heidelberg. doi:10.1007/978-3-319-61615-5_9 (in this volume)

87. Kirstein IV, Kirmizi S, Wichels A et al (2016) Dangerous hitchhikers? Evidence for potentially pathogenic *Vibrio* spp. on microplastic particles. Mar Environ Res 120:1–8. doi:10.1016/j.marenvres.2016.07.004

88. Oberbeckmann S, Löder MGJ, Labrenz M (2015) Marine microplastic-associated biofilms – a review. Environ Chem 12:551–562. doi:10.1071/EN15069

89. Lagarde F, Olivier O, Zanella M et al (2016) Microplastic interactions with freshwater microalgae: hetero-aggregation and changes in plastic density appear strongly dependent on polymer type. Environ Pollut 215:331–339. doi:10.1016/j.envpol.2016.05.006

90. Bilotta GS, Brazier RE (2008) Understanding the influence of suspended solids on water quality and aquatic biota. Water Res 42:2849–2861. doi:10.1016/j.watres.2008.03.018

91. Chapman JM, Proulx CL, Veilleux MAN et al (2014) Clear as mud: a meta-analysis on the effects of sedimentation on freshwater fish and the effectiveness of sediment-control measures. Water Res 56:190–202. doi:10.1016/j.watres.2014.02.047

92. Cordone A, Kelley D (1961) The influence of inorganic sediment on the aquatic life of streams. Calif Fish Game 47:189–228

93. Newcombe CP, MacDonald D (1991) Effects of suspended sediments on aquatic ecosystems. North Am J Fish Manag 11:72–82. doi:10.1577/1548-8675(1991)011<0072

94. Kjelland ME, Woodley CM, Swannack TM, Smith DL (2015) A review of the potential effects of suspended sediment on fishes: potential dredging-related physiological, behavioral, and

transgenerational implications. Environ Syst Decis 35:334–350. doi:10.1007/s10669-015-9557-2

95. Cole M, Lindeque PK, Fileman E et al (2016) Microplastics alter the properties and sinking rates of zooplankton faecal pellets. Environ Sci Technol 50:3239–3246. doi:10.1021/acs.est.5b05905

96. Carson HS, Colbert SL, Kaylor MJ, McDermid KJ (2011) Small plastic debris changes water movement and heat transfer through beach sediments. Mar Pollut Bull 62:1708–1713. doi:10.1016/j.marpolbul.2011.05.032

97. Zettler ER, Mincer TJ, Amaral-Zettler LA (2013) Life in the "plastisphere": microbial communities on plastic marine debris. Environ Sci Technol 47:7137–7146. doi:10.1021/es401288x

98. Goldstein MC, Rosenberg M, Cheng L (2012) Increased oceanic microplastic debris enhances oviposition in an endemic pelagic insect. Biol Lett 8:817–820. doi:10.1098/rsbl.2012.0298

99. Barnes DKA, Milner P (2005) Drifting plastic and its consequences for sessile organism dispersal in the Atlantic Ocean. Mar Biol 146:815–825. doi:10.1007/s00227-004-1474-8

100. Eriksen M, Thiel M, Prindiville M, Kiessling T (2017) Microplastic: what are the solutions?. In: Wagner M, Lambert S (eds) Freshwater microplastics: emerging environmental contaminants? Springer, Heidelberg. doi:10.1007/978-3-319-61615-5_13 (in this volume)

101. Lohmann R, Booij K, Smedes F, Vrana B (2012) Use of passive sampling devices for monitoring and compliance checking of POP concentrations in water. Environ Sci Pollut Res 19:1885–1895. doi:10.1007/s11356-012-0748-9

Microplastic-Associated Biofilms: A Comparison of Freshwater and Marine Environments

Jesse P. Harrison, Timothy J. Hoellein, Melanie Sapp, Alexander S. Tagg, Yon Ju-Nam, and Jesús J. Ojeda

Abstract Microplastics (<5 mm particles) occur within both engineered and natural freshwater ecosystems, including wastewater treatment plants, lakes, rivers, and estuaries. While a significant proportion of microplastic pollution is likely sequestered within freshwater environments, these habitats also constitute an important conduit of microscopic polymer particles to oceans worldwide. The quantity of aquatic microplastic waste is predicted to dramatically increase over the next decade, but the fate and biological implications of this pollution are still poorly understood. A growing body of research has aimed to characterize the formation, composition, and spatiotemporal distribution of microplastic-associated ("plastisphere") microbial biofilms. Plastisphere microorganisms have been suggested to play significant roles in pathogen transfer, modulation of particle buoyancy, and biodegradation of plastic polymers and co-contaminants, yet investigation of these topics within freshwater environments is at a very early stage. Here, what is known about marine plastisphere assemblages is systematically compared with up-to-date findings from freshwater habitats. Through analysis of key differences and likely commonalities between environments, we discuss how

J.P. Harrison (✉)
Division of Microbial Ecology, Department of Microbiology and Ecosystem Science, Research Network "Chemistry Meets Microbiology", University of Vienna, Vienna A-1090, Austria
e-mail: jesse.p.harrison@gmail.com

T.J. Hoellein
Department of Biology, Loyola University Chicago, Chicago, IL 60660, USA

M. Sapp
Institute of Population Genetics, Heinrich Heine University Düsseldorf, Düsseldorf 40225, Germany

A.S. Tagg, Y. Ju-Nam, and J.J. Ojeda
Systems and Process Engineering Centre, College of Engineering, Swansea University, Swansea SA1 8EN, UK

M. Wagner, S. Lambert (eds.), *Freshwater Microplastics*,
Hdb Env Chem 58, DOI 10.1007/978-3-319-61615-5_9,
© The Author(s) 2018

an integrated view of these fields of research will enhance our knowledge of the complex behavior and ecological impacts of microplastic pollutants.

Keywords Biodegradation, Biofilms, Microorganisms, Pathogens, Plastisphere

Abbreviations

BONCAT	Bioorthogonal noncanonical amino acid tagging
FACS	Fluorescence-activated cell sorting
FISH	Fluorescence in situ hybridization
FT-IR	Fourier-transform infrared
HDPE	High-density polyethylene
LDPE	Low-density polyethylene
MALDI-ToF MS	Matrix-assisted laser desorption/ionization time-of-flight mass spectrometry
MDA	Multiple displacement amplification
PET	Polyethylene terephthalate
PHBV	Polyhydroxybutyrate-polyhydroxyvalerate
PP	Polypropylene
PS	Polystyrene
(r)DNA	(Ribosomal) deoxyribonucleic acid
(r)RNA	(Ribosomal) ribonucleic acid
SIMS	Secondary ion mass spectrometry
SNP	Single-nucleotide polymorphism
UV	Ultraviolet
WWTP	Wastewater treatment plant
XPS	X-ray photoelectron spectroscopy
XRD	X-ray diffraction

1 Introduction

Microplastics (particles with an upper size limit of <5 mm) are globally distributed within aquatic environments, with up to 51 trillion pieces estimated to float at sea alone [1, 2]. They are encountered within the water column and sediments, with the latter functioning as a sink for the accumulation of plastic waste [3–5]. Most plastic litter originates from land-based activities, with wastewater treatment plant (WWTP) and inland waters comprising an important route through which this pollution reaches marine environments [6, 7]. While a substantial proportion of microplastic is likely to become sequestered within freshwaters, the amount of plastic entering the sea is predicted to increase by an order of magnitude by 2025 (corresponding to an input of up to 250 million metric tons) [7]. Legislation for phasing out microplastics in cosmetic products (e.g., the Microbead-Free Waters Act of 2015 in the USA) can be expected to achieve only a limited reduction in the quantity of environmental plastic debris.

A growing body of research has investigated the impacts of microplastics on biota, which may involve direct and indirect processes (e.g., physical blockage caused by ingested particles, as well as their ability to transport harmful compounds, pathogens, and algae) [2, 8–10]. Even so, little is known about the ecological effects of microplastics within freshwaters [10]. For example, while microplastic-associated microbial (bacterial, archaeal, and picoeukaryotic) assemblages are likely to profoundly influence the distribution, impacts, and fate of these pollutants, research into this topic has focused on marine environments [11–13]. In streams and other habitats, biofilms[1] are primary sites for carbon and nutrient transformations and form the base of food webs, contributing to local and global ecosystem functioning [14]. As they are also essential to pollutant biodegradation, an improved knowledge of microbial-microplastic interactions is required to predict the environmental impacts of plastic debris [15]. Investigating this topic could inform the development of solutions to manage plastic pollution by determining how it affects processes including microbially mediated primary production and interactions between plastic-associated ("plastisphere") taxa and other organisms [11, 12, 16, 17]. It could also lead to insights concerning the biodegradability of plastic litter and facilitate the development of new approaches to plastic disposal and/or recycling [18].

Freshwater and marine habitats share a number of features, but there are also differences between them that may affect the development and activities of plastisphere consortia. To facilitate investigation of this topic, findings based on marine plastisphere research are compared with those available for freshwaters. Following an assessment of recent discoveries concerning the formation and distribution of plastic-associated biofilms, our knowledge concerning their ecological roles and ability to drive processes including polymer biodegradation is considered. Finally, some of the main knowledge gaps in plastisphere research are discussed and used to highlight methodological advances in microbial ecology that could be used to improve our understanding of microbial-microplastic interactions.

2 Freshwater Plastisphere Assemblages: State of the Science

2.1 Factors Contributing to Biofilm Formation and Composition

Fundamental processes involved in biofilm formation are well established, with initial attachment followed by maturation and the eventual detachment of cells [19]. There are also further factors that may influence the formation, composition,

---

[1]Surface-associated aggregates of microbial cells encased in a matrix of extracellular polymeric substances.

Factors driving biofilm formation and composition on plastic

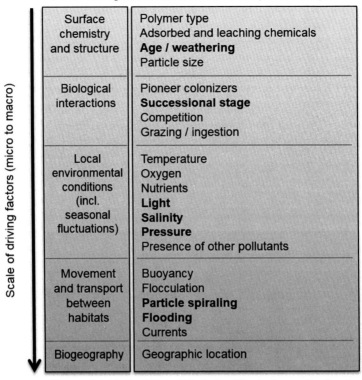

*Parameters **in bold** most likely to differ between freshwater and marine habitats*

Fig. 1 Physical, chemical, and biological factors likely to affect the formation and composition of plastisphere microbial assemblages. Only a limited selection of these parameters has been investigated with specific reference to microplastics

and activities of plastic-associated biofilms (Fig. 1). Only some of the parameters shown in Fig. 1 have been investigated with reference to microplastics. However, efforts to identify factors driving the formation of these assemblages in marine habitats have recently been reviewed [12, 13, 20].

Microplastics are rapidly colonized by environmental microorganisms (within hours; [21]). Many factors driving the development of plastisphere communities are likely to be similar between freshwater and marine habitats. For example, in agreement with research into biofilm formation on other artificial substrata [19, 22], there is evidence for the importance of surface properties (including roughness and hydrophobicity) during early colonization of microplastics [12, 23]. Exposure to ultraviolet (UV) radiation and waves can modify the surface chemistry and structure of plastics (e.g., via the formation of cracks and pits, a reduction in molecular weight, and an increase in surface oxidation), which may facilitate biofilm formation [24, 25]. Plastic-colonizing microorganisms have also been found to influence the surface properties and buoyancy of polymers [12, 20, 26]. Since microplastics are likely to be transported into marine environments via WWTP, rivers, and streams [6, 7], factors contributing to initial colonization (such as surface roughness and attachment by pioneering colonizers) can be hypothesized to be particularly important within freshwaters. The impacts of particle age and/or weathering on plastisphere consortia may be comparatively pronounced within marine ecosystems where the residence times of plastic often exceed those within rivers and streams [24]. However, microplastics additionally accumulate within environments such as lakes, where they may persist for decades (similar to time-scales predicted for marine habitats) and can be exposed to high levels of UV radiation [2, 27, 28]. Local-scale differences in the composition of plastisphere assemblages between polymer types have been found [12, 29, 30], but it is unknown whether there are any general differences in the dominant types of plastic within freshwater and marine ecosystems. Moreover, although it is possible that the ingestion of plastics by higher organisms could have an impact on plastisphere colonization processes, this topic has not been investigated [11, 20, 30].

Ambient conditions such as temperature, salinity, pressure, and the availabilities of light and oxygen are likely to influence the development of plastic-associated biofilms (Fig. 1) [29, 31]. Many of these conditions differ between freshwater and marine ecosystems, and WWTP and unmanaged freshwaters. For example, the low temperatures ($<5°C$), absence of light, and elevated pressure within deep waters are likely to impose selective forces on plastisphere assemblages that differ from those within shallow habitats. In contrast with the frequently nutrient-poor conditions present within the open ocean, inland and coastal waters receive high fluxes of nutrients from the surrounding environment [14]. In addition to contributions from organic matter input and upwelling, high concentrations of nutrients (e.g., nitrogen and phosphorus) are released by agriculture and other human activities. Many plastisphere members have been affiliated with pollutant degradation [12, 13, 20, 21], and it is probable that several contaminants play a role in shaping biofilm formation and activities on polymers (Fig. 1). Indeed, multiple types of pollutants, as well as heavy metals, are known to become adsorbed onto microplastics [2, 8, 10].

Further to these factors, physical processes contributing to the movement of suspended particles differ between freshwater and marine habitats [2]. Continuous downstream movement of water is a key distinction between freshwater and marine ecosystems. In rivers, sediment movement is characterized using the concept of "spiraling" [32, 33]. The components of one spiral include downstream transport, deposition, bed load transport, and resuspension. This concept is a well-developed approach for modeling particle movement and is quantified using measurements of deposition length and velocity, turnover time, and the retention-export ratio [34]. To date, direct measurements of spiraling metrics have not been applied to microplastic (but see Kowalski et al. [26], Long et al. [35], and Nizzetto et al. [36]).

Each step in a spiral is likely to have implications for plastic-associated biofilm composition and activity, due to accompanying shifts in the surrounding environmental conditions (Fig. 1) [29, 31]. Studies of microplastic spiraling metrics will help estimate the spatial scales over which plastic particles move within lotic environments, informing how the associated microbial communities can be expected to change across multiple downstream spirals. Rivers are also characterized by flooding, which redistributes materials between riparian and aquatic components of the fluvial landscape [37, 38]. Flooding moves plastic from the riparian zone into aquatic habitats and increases stranding of plastic in debris dams [39]. Analogous processes in marine environments include tidal movements and storm surges which strand plastic on intertidal or wrack zones [2]. Despite their likely impacts on plastisphere communities (Fig. 1), the effects of movement between aquatic and terrestrial habitats on plastic-associated biofilms have not been studied.

Hydrology in most lakes includes at least a single upstream inlet and downstream outlet, with water and particle residence times depending on water volume and currents. Little is known about plastisphere communities in lakes (Sect. 2.2), but research into this topic can be expected to benefit from a budgetary approach which measures rates of microplastic inflow, outflow, and retention. These metrics will determine microplastic residence times, which are likely to influence microbial-plastic associations within several habitats, including the epilimnion, littoral, and benthic zones (Sect. 3.1). Wind and wave action are likely to further influence the distribution of microplastics within lakes [2].

It is unclear how transport of microplastics from freshwater to marine environments affects plastisphere assemblages, but they may undergo a variety of taxonomic and physiological shifts during this transition (Sects. 2.2 and 2.3) [20, 40]. For example, subjecting *Pseudomonas aeruginosa* to salt stress (0.5 M NaCl) was found to inhibit biofilm formation and reduce rates of benzoate degradation by this strain [41]. Geographic and seasonal differences in the structure and composition of freshwater plastisphere communities are yet to be investigated. However, the spatiotemporal distribution of marine plastic-colonizing microbial consortia has recently been studied [29, 30, 42]. Based on 6-week in situ exposures of polyethylene terephthalate (PET) bottles in the North Sea, Oberbeckmann et al. [29, 42] found location-dependent and seasonal differences in the structure and composition of plastisphere communities. Similar differences were also reported by Amaral-Zettler et al. [30]. Further to distinct communities being discovered in the North Atlantic and North Pacific subtropical gyres, the authors reported latitudinal gradients in the species richness of plastic-colonizing assemblages [30]. While taxonomic differences were also observed between polymer types, the data suggested that geography is likely to be a stronger predictor of plastisphere community composition at the scale of ocean basins [29, 30, 42].

2.2 Examples of Microbial-Microplastic Interactions in Freshwater Habitats

Despite measurements of plastic density and composition in freshwater ecosystems [10, 43], little is known about microbial associations with plastic in unmanaged freshwaters. A limited number of publications have investigated polymer biodegradation in lakes and rivers (Sect. 2.3), and there are at least three studies that have experimentally characterized the structure, composition, and/or activities of plastic-associated biofilms in these environments [44–46]. Because of differences in the study design and sites and the response parameters that were examined, there are few findings in common among these three studies. Thus, some of the major results of each study are discussed and compared with insights into marine microbial-microplastic interactions.

Hoellein et al. [44] compared bacterial community composition and activity on six substrate types (5 × 5 cm pieces of ceramic tile, glass, aluminum, PET, leaf litter, and cardboard) in a river, a pond, and recirculating laboratory streams. In contrast with McCormick et al. [45] and several studies of marine plastisphere communities [21, 29, 47], the authors found no differences in the composition of plastic-colonizing biofilms relative to those on other solid substrates. The plastic, tile, and glass samples also showed similar rates of gross primary production and respiration. The primary factors for determining bacterial community composition and metabolic rates were the study site (river, pond, or artificial stream) and whether the substrate was hard (tile, glass, aluminum, and PET) or soft (leaf litter and cardboard). While the surface-colonizing assemblages on PET were compositionally similar to those on other surfaces, it was suggested that differences between substrate types may be stronger during early stages of biofilm formation. Similarly, Oberbeckmann et al. [42] found PET- and glass-colonizing communities to be compositionally similar following up to 6 weeks of exposure to seawater; the authors noted that higher-resolution studies may be required to distinguish "plastic-specific" taxa from other biofilm members. Taken together, these studies emphasize how investigating the early-stage development of plastisphere communities in more detail will be necessary not only in marine ecosystems [21] but also in freshwater habitats.

McCormick et al. [45] compared bacterial communities on microplastic, suspended organic matter (i.e., seston) and the water column downstream and upstream of a WWTP. All habitats differed from each other, and the microplastic community had a lower taxon diversity relative to seston and downstream water samples. In marine environments, plastic-associated microbial communities have also been found to be taxonomically distinct from those in the surrounding water [30, 47–49]. Genera selected for on plastic (relative to nonplastic habitats) in the study by McCormick et al. [45] included *Pseudomonas, Arcobacter, Aeromonas, Zymophilus*, and *Aquabacterium*. These genera contain species with the potential for plastic degradation and pathogenesis (Sect. 2.3). *Aquabacterium commune* is a common member of drinking water biofilms [50], and colonization of low-density

Fig. 2 Scanning electron micrograph showing a biofilm attached to a HDPE fragment incubated in aerobic wastewater for 6 months. Microplastics are likely to function as vectors for the transport of microbial taxa from WWTP to other environments. The scale bar is 2 μm (Credit: Alexander S. Tagg)

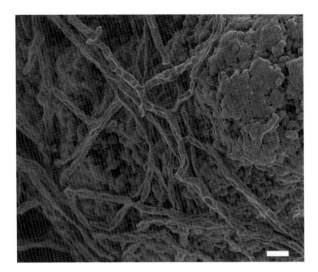

polyethylene (LDPE) by *Arcobacter* spp. has also been shown to occur in coastal marine sediments [21]. The study by McCormick et al. [45] was conducted immediately below a WWTP outfall, and it is unknown if wastewater-affiliated microbial communities will persist further downstream. However, the presence of plastic-colonizing *Arcobacter* spp. in both freshwater and marine habitats [21, 45] implies that certain genera could survive on polymers as they are transported from WWTP to other ecosystems (Fig. 2 and Sect. 2.1). Indeed, *Arcobacter* spp. have been found to be prevalent members of the "landfill microbiome" in the USA [51] and have also been detected in sewage [52].

The objective of Lagarde et al. [46] was to examine the growth of a microalga (*Chlamydomonas reinhardtii*) on plastic particles over time, determine the effect of plastic type on algal growth, and measure particle aggregation. The authors found little effect of plastic (high-density polyethylene [HDPE] or polypropylene [PP]) on algal growth, but contact with polymer particles altered the expression of genes for some sugars used in extracellular polysaccharides. On PP, algal biofilms increased particle aggregation, which was not observed for HDPE. Research has recently been aimed at characterizing the sedimentation rates of microplastics in freshwater and marine environments [26, 35, 36]. Lagarde et al. [46] add to our understanding of microplastic movement by showing that aggregation of plastic particles via biofilm attachment occurs differently among polymer types, which will affect their suspension or deposition. Future studies will benefit from extrapolating this approach to in situ analyses, as well as comparing findings between marine and freshwater environments. For example, the types and sinking rates of algal microplastic aggregates within marine environments are known to be species specific [35], and similar interactions could affect the distribution of microplastics in rivers and lakes.

2.3 Potential for Pathogenesis, Toxicant Transfer, and Biodegradation

2.3.1 Microplastics as Vectors for Pathogen Transfer and Biotoxins

Gene sequencing analyses initially highlighted how microplastics may function as vectors for the transport of potential pathogens including *Vibrio* and *Arcobacter* spp. [21, 30, 45, 48, 53]. A high proportion of 16S rDNA reads (24%) could be attributed to *Vibrio* spp. detected on PP and, to a lesser extent, on polyethylene (PE) collected at a station in North Atlantic waters [48]. Unfortunately, the widely used bacterial metabarcoding technique based on sequencing fragments of the 16S rRNA gene is limited in its ability to provide the required taxonomic resolution for detecting human pathogens [53]. Using oligotyping of 16S rRNA gene data, Schmidt et al. [54] obtained more specific results for taxa within the genus *Vibrio* indicating the presence of potential pathogens affecting animals including fishes, corals, and bivalves in marine or mixed saline plastic samples. The presence of pathogens on plastics sampled from seawater was also implied by increased abundances of genes involved in type IV and type VI secretion systems [49]. However, genes involved in these systems can be involved not only in virulence and infection [55] but also in conjugation [56] and interbacterial interactions [57] that are important in biofilms [58]. *Vibrio* spp. were additionally isolated from plastic collected from a Scottish beach [59], but no further characterization of the isolates was performed. Only recently was the presence of *Vibrio* spp. on marine plastics conclusively confirmed by matrix-assisted laser desorption/ionization time-of-flight mass spectrometry (MALDI-ToF MS) [60]. In their study, Kirstein et al. [60] identified *V. parahaemolyticus*, *V. fluviales*, and *V. alginolyticus* on microplastics from the North Sea. Apart from *V. alginolyticus*, these species were also found on plastics collected in the brackish Baltic Sea. In addition to bacteria, microplastics may transport microbial eukaryotes involved in disease transmission [12]. Potentially harmful algae, including *Ostreopsis* and *Coolia* spp., have been discovered on plastic in the Mediterranean Sea [61]. To date, the only in situ evidence for microplastic-associated pathogens in unmanaged freshwaters identified an increase in *Campylobacteraceae* attached to microplastics sourced from an urban river [45]. Specifically, 16S rRNA gene sequences related to *Arcobacter* and *Pseudomonas* spp. were enriched on plastic in comparison with other suspended matter and the surrounding water.

In summary, current evidence indicates an important role of microplastics as vectors for opportunistic animal and human pathogens. Methodological advances are required to reliably detect viable pathogenic species, so that realistic distribution patterns can be obtained and potential sources can be identified. This is particularly relevant with regard to waters used for recreational [13] but also for industrial purposes such as aquaculture. Relative abundances of *Aeromonas* spp. (a genus harboring fish pathogens) were increased on riverine plastics [45], implying that such species could take advantage of microplastics as vectors. This possibility is reinforced by the presence of *Aeromonas salmonicida*, causing furunculosis in hatcheries, on several plastic types [62]. Recently, 16S rRNA gene sequences affiliated to *Tenacibaculum* spp. (another genus including fish pathogens) were detected on PET in seawater [42]. Research has only started to shed light on this issue, as well as the ability of polymers to transport biologically produced toxins.

2.3.2 Biodegradation and Pollutant Transport

Several reviews of research into plastic biodegradation have been published (e.g., see [11–13, 24, 63–65]). Therefore, only a brief overview of this topic is provided. Plastic biodegradation involves several steps during which the polymer is enzymatically cleaved into oligomers and monomers that can be assimilated by microorganisms [65]. Many microbial taxa can degrade biopolymers[2] including polyhydroxybutyrate (PHB) and polyhydroxybutyrate-polyhydroxyvalerate (PHBV). The biodegradation rates of biopolymers in freshwater have been found to exceed those in marine environments, and higher rates have also been observed in sewage than within natural freshwaters [63, 66, 67]. Even so, these materials can still persist for considerable periods of time in freshwaters, with a lifespan of ~10 years having been estimated for PHBV bottles deposited onto lake sediments at a depth of 85 m [68].

In comparison with biopolymers, traditional plastics (such as PE, PET, and PP) will persist for even longer within aquatic environments (decades or centuries; [11, 63, 64]), with biodegradation typically preceded by abiotic weathering [24, 65]. Although it has been unclear whether plastisphere members can biodegrade conventional plastics [11, 69, 70], a bacterial strain isolated from sediment near a Japanese bottle recycling facility (*Ideonella sakaiensis*) was recently found to assimilate PET [18]. The strain was shown to employ two enzymes to degrade PET at a daily rate of 0.13 mg cm^{-2} when incubated at 30°C [18]. This finding implies that other synthetic plastic-degrading taxa are likely to be present within aquatic environments. Indeed, colonization of plastics by potentially hydrocarbonoclastic bacteria has been observed in both marine and freshwater habitats [21, 45, 47–49]. However, due to a lack of research into plastisphere physiology, the long residence times of plastic waste, and the ability of polymers to adsorb polyaromatic hydrocarbons [11, 12], the mechanisms underlying recruitment of hydrocarbon degraders on microplastics are unknown. These and other taxa could mediate desorption and/or degradation of several plastic-associated compounds, including additives and diverse pollutants, with implications for the ecological impacts of microplastics. Indeed, Bryant et al. [49] already reported the presence of diverse xenobiotic degradation genes in association with marine plastic debris. Since organic contaminants and metals rapidly partition into biofilms

---

[2]Polymers derived from renewable biomass (as opposed to nonrenewable fossil fuels).

[71, 72], plastisphere communities may alternatively be hypothesized to facilitate transport of pollutants between ecosystems and to biota (Sect. 3.2).

3   Knowledge Gaps and Research Needs

3.1   Sources and Transport Between Habitats

Processes contributing to microplastic transport differ between freshwater and marine ecosystems (Sect. 2.1). Conditions encountered within WWTP and unmanaged freshwaters also differ from one another. A priority for research involves determining the extent to which plastic-colonizing taxa associated with wastewater and other sources of plastic (such as landfills) are transported downstream along rivers and streams and whether they remain viable and active upon entering marine habitats [12, 40]. As part of this work, research is required to characterize the residence times of polymer particles within several environments, including different stages of the wastewater treatment process. Most WWTPs are based on three main treatment stages, although slight differences in their configuration can be found. During primary treatment, large debris fragments are removed by using a 6 mm (or larger) screen mesh. During secondary treatment, large aeration tanks are used to remove suspended and dissolved organic material and nutrients through microbial activity. Subsequently, flocculates and settling tanks are used to facilitate separation of sewage sludge from the post-processing effluent prior to a potential disinfection step, also known as advanced tertiary treatment. Studies reporting pathways of microplastics through different wastewater treatment stages are only beginning to emerge [73–75], and little is still known about how these stages influence the development of plastisphere microbial communities.

Overall, studies of microplastic movement and associated biofilms should be based on well-established principles of ecosystem and community ecology [39] and are prerequisite to estimating the spatial scales over which plastics are distributed within a watershed. This approach will best inform how plastic-associated microbial communities can be expected to change with movement from freshwater to marine habitats. There is also a need to compare plastisphere communities in managed and natural environments, within several locations along the water column, as well as between pelagic and benthic habitats. Research into plastic-associated biofilms has focused on surface waters (despite the long-term accumulation of microplastics in sediments; [8, 27]), and investigations of benthic plastisphere assemblages have been restricted to marine habitats [21, 47]. In several environments, no information is available on plastic-associated microbial assemblages. For example, no data have yet been published on plastisphere consortia within WWTP, and although the buildup of plastic debris in deep-sea environments has been reported [76], biofilms associated with this debris have not been studied.

This lack of data limits our ability to predict the ecological consequences and lifetimes of plastic pollution (Sects. 3.2 and 3.3).

3.2 Interactions with Higher Organisms and the Wider Environment

Interactions between plastisphere communities and higher organisms have been recommended as a topic for research in marine environments [11, 12], but they also require investigation within freshwaters. Many organisms including fishes, gastropods, and zooplankton (e.g., *Daphnia magna*) ingest microplastics [2]. Indeed, nanopolystyrene has been found to negatively affect reproduction in *D. magna*, as well as population growth in the primary producer *Scenedesmus obliquus* [77]. Effects of plastic-sorbed chemicals have been rarely studied, but liver toxicity was observed in Japanese medaka [78]. A significant knowledge gap is the in situ analysis of microplastic present within freshwater organisms. Such analyses will need to consider how plastic-associated biofilms may amend the buoyancy of polymer particles and/or influence organismal behavior (e.g., selective feeding). Additionally, research is needed to investigate the pathogenicity of plastic-colonizing microbial taxa, as well as their ability to produce toxins. Oberbeckmann et al. [12] suggested that microplastics could carry pathogens encountered in the feces of marine organisms, and transport of human fecal bacteria on plastics has also been discussed [13]. There is a particular requirement to determine how this debris affects organisms at low trophic levels, such as invertebrates used for biomonitoring purposes [79, 80]. Impacts of plastisphere assemblages on processes such as nutrient cycling and primary production should also be investigated. Indeed, Bryant et al. [49] reported high densities of chlorophyll *a* and an increased abundance of nitrogen fixation genes (*nifH*, *nifD*, and *nifK*) on polymers in comparison with other sample types, leading the authors to suggest that plastic particles may constitute autotrophic "hot spots" in seawater.

Further to impacts on the fitness of plastic-ingesting taxa and processes including elemental cycling, interactions between plastisphere assemblages and other organisms may influence the distribution and fate of plastic waste. For example, microplastics may become transported away from surface waters via encapsulation within fecal pellets [81]. Although this topic has not been investigated in freshwater or marine environments, the gut bacteria of mealworms (larvae of *Tenebrio molitor* Linnaeus) can degrade polystyrene [82], and certain aquatic organisms could harbor microorganisms capable of modifying the surface properties of plastics and/or biodegrading them. Thus, investigating the interactions between plastisphere communities and other organisms is closely connected to research into the transport of plastics between habitats (Sect. 3.1) and the environmental lifetime of this debris (Sect. 3.3).

While this chapter focuses on freshwater and marine environments, plastisphere communities may also be of significance to human health. Risks associated with the human ingestion of microscopic plastics have been identified [83], and investigations of this topic could also be approached from a microbiological viewpoint. In particular, the human health implications of putative pathogens within plastic-associated biofilms (Sect. 2.3.1 and [13]) merit further study.

3.3   In Situ Biodegradability of Plastics and Plastic-Associated Compounds

The recent evidence for PET assimilation by *I. sakaiensis* [18] suggests that, although rates of plastic breakdown in the environment are extremely low (Sect. 2.3.2), several novel polymer-degrading taxa are likely to be present within freshwater and marine ecosystems. Identifying such taxa and investigating their ability to biodegrade different plastic types, additives, and polymer-sorbed compounds are of primary importance to understanding the environmental residence times of plastic waste. Research in this area should focus on habitats functioning as sinks for the accumulation of plastic, including sediments [3–5, 27]. To obtain a complete understanding of the biodegradability of different materials and compounds, there is a need to combine laboratory-based experiments with field-based measurements of plastic degradation in both freshwater and marine environments. Moreover, as nanometer-sized plastic particles become released from the parent polymer as a result of weathering [84], their biodegradation behavior will need to be compared with that of larger fragments that may support a comparatively complex biofilm community. Most research into plastic biodegradation has been based on indirect measurements such as mass loss [11], and a key challenge will be to conclusively demonstrate in situ assimilation of carbon from a given plastic type (or plastic-associated compound) [18]. The toxicity of any degradation products, or of compounds released from the polymer, will also require investigation (Sect. 3.2).

3.4   Analytical and Experimental Advances in Plastisphere Research

Research into plastisphere assemblages has focused on bacterial communities [44, 45]. Little is known about plastic-associated microbial eukaryotes in freshwaters, and there is a need for analyses targeting these organisms, not the least as they are known to occur on marine plastics [48, 49]. Several advances have improved the suitability of metabarcoding for analyzing fungi, diatoms, and protists

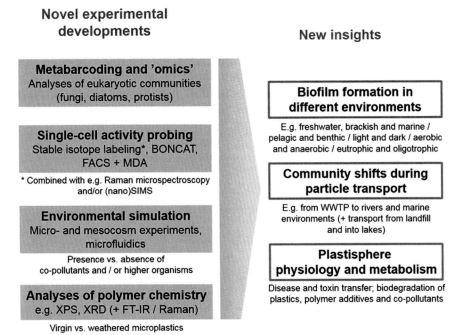

Fig. 3 Moving beyond initial research into the taxonomy and formation of plastisphere microbial assemblages. As investigations of this topic mature, new types of experiments and analytical tools are anticipated to improve our knowledge of topics including how plastisphere communities develop in several types of habitat, how they are affected by transport from freshwater to marine environments, and the metabolic functions of plastic-colonizing microorganisms

[85–87]. This approach is based on taxonomically informative markers and provides no direct information on metabolic activities. Overcoming this limitation could involve using metagenomics or metatranscriptomics, with the former providing information on metabolic capability [49] and the latter enabling investigations of functional gene expression [20] (Fig. 3). The origin of plastic-colonizing pathogens could be determined by whole genome sequencing followed by genome comparisons or identification of single-nucleotide polymorphism (SNP), approaches widely used in bacterial epidemiology. This would result in important insights into the transfer of pathogens on plastics, provided that suitable databases are available for comparison [88, 89].

Several further developments could enable us to move beyond initial studies of biofilm formation on microplastics (Fig. 3). Stable isotope labeling is increasingly used to characterize microbial activity at the single-cell level, including methods such as heavy water labeling [90] and bioorthogonal noncanonical amino acid tagging (BONCAT) [91]. Heavy water labeling is compatible with Raman spectroscopy and cell sorting using optical tweezers [90], and BONCAT has been combined with fluorescence-activated cell sorting (FACS) [91]. These approaches could be followed by multiple displacement amplification (MDA)[3], enabling identification of taxa that are metabolically active under in situ conditions. Raman spectroscopy has been combined with techniques such as fluorescence in situ hybridization (FISH), which can be used to further investigate the presence and activities of specific microbial taxa [92]. Fourier-transform infrared (FT-IR) spectroscopy has additionally been employed to characterize the chemical composition of biofilms, providing a convenient and low-cost method for analyzing microorganisms adhering to opaque materials [93]. Such methods could be used in conjunction with biological rate measurements (e.g., gas evolution) [44, 49]. This, in turn, could advance our understanding of how plastisphere taxa contribute to disease transmission, nutrient fixation, and pollutant degradation.

Research into microplastic-associated biofilms has relied on samples that were collected in situ or exposed to seawater, with only a small selection of studies involving microcosm experiments under controlled conditions [21, 46, 59]. Mesocosm experiments could be used to bridge the current gap between microcosm studies and field-based research into microplastic-associated biofilms (Fig. 3). Microfluidics is also increasingly used as a tool in microbial ecology and could be employed to obtain insights into microbial-microplastic interactions under selected conditions (e.g., in the presence of fluid flow and chemical gradients) [94, 95]. To improve our knowledge of the biodegradation of plastics and plastic-sorbed pollutants, such approaches could be supplemented by advanced surface analysis techniques. X-ray photoelectron spectroscopy (XPS) and secondary ion mass spectrometry (SIMS) have been used to investigate abiotic weathering of plastics [96–98] and could be valuable for monitoring polymer biodegradation (Fig. 3). Indeed, XPS can detect chemical signatures at the parts-per-thousand (‰) range [96], and SIMS (including nanoscale SIMS) has been used to trace microbial uptake of ^{13}C-labeled substrates in environmental samples [99, 100]. While these techniques are suitable for analyzing organic compounds, X-ray diffraction (XRD) analyses are particularly useful for measurements of inorganic materials, including metals. Although microbial interactions with plastic-associated metals (e.g., metal solubilization or precipitation) have not been previously studied, this could be achieved using XRD (e.g., see Roh et al. [101]).

4 Concluding Remarks

Over the past 5 years, several studies have improved our understanding of the taxonomy and potential activities of microbial consortia associated with microplastic particles in the environment. Due to most of these studies focusing on marine ecosystems, there remains a particular lack of information concerning plastisphere assemblages within freshwaters. However, as highlighted in this

---

[3]A method for amplifying very low concentrations of DNA for genomic analysis.

chapter, many of the fundamental processes that underpin the formation and activities of plastic-colonizing biofilms remain poorly understood within both freshwater and marine environments. Establishing an understanding of the implications of microplastic-associated microorganisms for ecosystem and human health, therefore, will require research spanning the entire diversity of environments encountered by these pollutants following their release by industrial and domestic activities.

Acknowledgments We thank Buck Hanson, Toby Samuels, and William Southwell-Wright for their feedback and helpful suggestions.

References

1. van Sebille E, Wilcox C, Lebreton L et al (2015) A global inventory of small floating plastic debris. Environ Res Lett 10(12):124006. doi:10.1088/1748-9326/10/12/124006
2. Eerkes-Medrano D, Thompson RC, Aldridge DC (2015) Microplastics in freshwater systems: a review of the emerging threats, identification of knowledge gaps and prioritisation of research needs. Water Res 75(15):63–82. doi:10.1016/j.watres.2015.02.012
3. Barnes DKA, Galgani F, Thompson RC et al (2009) Accumulation and fragmentation of plastic debris in global environments. Philos Trans R Soc B 364(1526):1985–1998. doi:10.1098/rstb.2008.0205
4. Browne MA, Crump P, Niven SJ et al (2011) Accumulation of microplastic on shorelines worldwide: sources and sinks. Environ Sci Technol 45(21):9175–9179. doi:10.1021/es201811s
5. Corcoran PL (2015) Benthic plastic debris in marine and freshwater environments. Environ Sci: Processes Impacts 17:1363–1369. doi:10.1039/c5em00188a
6. Sadri SS, Thompson RC (2014) On the quantity and composition of floating plastic debris entering and leaving the Tamar Estuary, Southwest England. Mar Poll Bull 81(1):55–60
7. Jambek JR, Geyer R, Wilcox C et al (2015) Plastic waste inputs from land into the ocean. Science 347(6223):768–771
8. van Cauwenberghe L, Devriese L, Galgani F et al (2015) Microplastics in sediments: a review of techniques, occurrence and effects. Mar Environ Res 111:5–17. doi:10.1016/j.marenvres.2015.06.007
9. Rochman CM, Browne MA, Underwood AJ et al (2016) The ecological impacts of marine debris: unraveling the demonstrated evidence from what is perceived. Ecology 97(2):302–312. doi:10.1890/14-2070.1
10. Wagner M, Scherer C, Alvarez-Muñoz D et al (2014) Microplastics in freshwater ecosystems: what we know and what we need to know. Environ Sci Eur 26:12. doi:10.1186/s12302-014-0012-7
11. Harrison JP, Sapp M, Schratzberger M et al (2011) Interactions between microorganisms and marine microplastics: a call for research. Mar Technol Soc J 45(2):12–20. doi:10.4031/MTSJ.45.2.2
12. Oberbeckmann S, Löder MGJ, Labrenz M (2015) Marine microplastic-associated biofilms – a review. Environ Chem 12(5):551–562. doi:10.1071/EN15069
13. Keswani A, Oliver DM, Gutierrez T et al (2016) Microbial hitchhikers on marine plastic debris: human exposure risks at bathing waters and beach environments. Mar Environ Res 118:10–19. doi:10.1016/j.marenvres.2016.04.006
14. Battin TJ, Besemer K, Bengtsson MM et al (2016) The ecology and biogeochemistry of stream biofilms. Nat Rev Microbiol 14:251–263. doi:10.1038/nrmicro.2016.15

15. Widder S, Allen RJ, Pfeiffer T et al (2016) Challenges in microbial ecology: building predictive understanding of community function and dynamics. ISME J 10:2557. doi:10. 1038/ismej.2016.45

16. Hoellein TJ, Tank JL, Rosi-Marshall EJ et al (2009) Temporal variation in the substratum-specific rates of N uptake and metabolism and their contribution at the stream-reach scale. J N Am Benthol Soc 28(2):305–318. doi:10.1899/08-073.1

17. Kominoski JS, Hoellein TJ, Kelly JJ et al (2009) Does mixing litter of different qualities alter stream microbial diversity and functioning on individual litter species? Oikos 118 (3):457–463. doi:10.1111/j.1600-0706.2008.17222.x

18. Yoshida S, Hiraga K, Takehana T et al (2016) A bacterium that degrades and assimilates poly (ethylene terephthalate). Science 351(6278):1196–1199. doi:10.1126/science.aad6359

19. O'Toole G, Kaplan HB, Kolter R (2000) Biofilm formation as microbial development. Annu Rev Microbiol 54:49–79. doi:10.1146/annurev.micro.54.1.49

20. Mincer TJ, Zettler ER, Amaral-Zettler LA (2016) Biofilms on plastic debris and their influence on marine nutrient cycling, productivity, and hazardous chemical mobility. In: Takada H, Karapanagioti HK (eds) Hazardous chemicals associated with plastics in the marine environment. Handbook of environmental chemistry. Springer, Heidelberg. doi:10. 1007/698_2016_12

21. Harrison JP, Schratzberger M, Sapp M, Osborn AM (2014) Rapid bacterial colonization of low-density polyethylene microplastics in coastal sediment microcosms. BMC Microbiol 14:232. doi:10.1186/s12866-014-0232-4

22. Fish KE, Osborn AM, Boxall J (2016) Characterising and understanding the impact of microbial biofilms and the extracellular polymeric substance (EPS) matrix in drinking water distribution systems. Environ Sci Water Res Technol 2:614–630. doi:10.1039/ C6EW00039H

23. Carson HS, Nerheim MS, Carroll KA et al (2013) The plastic-associated microorganisms of the North Pacific Gyre. Mar Poll Bull 75(1–2):126–132

24. Gewert B, Plassmann MM, MacLeod M (2015) Pathways for degradation of plastic polymers floating in the marine environment. Environ Sci Process Impacts 17:1513–1521. doi:10.1039/ C5EM00207A

25. Brandon J, Goldstein M, Ohman MD (2016) Long-term aging and degradation of microplastic particles: comparing in situ oceanic and experimental weathering patterns. Mar Poll Bull 110(1):299–308. doi:10.1016/j.marpolbul.2016.06.048

26. Kowalski N, Reichardt AM, Waniek JJ (2016) Sinking rates of microplastics and potential implications of their alteration by physical, biological, and chemical factors. Mar Poll Bull 109(1):310–319. doi:10.1016/j.marpolbul.2016.05.064

27. Corcoran PL, Norris T, Ceccanese T et al (2015) Hidden plastics of Lake Ontario, Canada and their potential preservation in the sediment record. Environ Poll 204:17–25. doi:10.1016/j. envpol.2015.04.009

28. Free CM, Jensen OP, Mason SA et al (2014) High-levels of microplastic pollution in a large, remote, mountain lake. Mar Poll Bull 85(1):156–163. doi:10.1016/j.marpolbul.2014.06.001

29. Oberbeckmann S, Löder MGJ, Gerdts G et al (2014) Spatial and seasonal variation in diversity and structure of microbial biofilms on marine plastics in Northern European waters. FEMS Microbiol Ecol 90(2):478–492. doi:10.1111/1574-6941.12409

30. Amaral-Zettler LA, Zettler ER, Slikas B et al (2015) The biogeography of the plastisphere: implications for policy. Front Ecol Environ 13(10):541–546. doi:10.1890/150017

31. Hullar MA, Kaplan LA, Stahl DA (2006) Recurring seasonal dynamics of microbial communities in stream habitats. Appl Environ Microbiol 72(1):713–722. doi:10.1128/AEM.72.1. 713-722.2006

32. Newbold JD, Mulholland PJ, Elwood JW et al (1982) Organic carbon spiralling in stream ecosystems. Oikos 38(3):266–272. doi:10.2307/3544663

33. Tank JL, Rosi-Marshall EJ, Griffiths NA et al (2010) A review of allochthonous organic matter dynamics and metabolism in streams. J N Am Benthol Soc 29(1):118–146. doi:10. 1899/08-170.1

34. Webster JR, Benfield EF, Ehrman TP et al (1999) What happens to allochthonous material that falls into streams? A synthesis of new and published information from Coweeta. Freshw Biol 41(4):687–705. doi:10.1046/j.1365-2427.1999.00409.x

35. Long M, Moriceau B, Gallinari M et al (2015) Interactions between microplastics and phytoplankton aggregates: impact on their respective fates. Mar Chem 175:39–46. doi:10. 1016/j.marchem.2015.04.003

36. Nizzetto L, Bussi G, Futter MN et al (2016) A theoretical assessment of microplastic transport in river catchments and their retention by soils and river sediments. Environ Sci Process Impacts 18:1050–1059. doi:10.1039/C6EM00206D

37. Gregory SV, Swanson FJ, McKee WA et al (1991) An ecosystem perspective of riparian zones. Bioscience 41(8):540–551. doi:10.2307/1311607

38. Jones Jr JB, Smock LA (1991) Transport and retention of particulate organic matter in two low-gradient headwater streams. J N Am Benthol Soc 10(2):115–126. doi:10.2307/1467572

39. McCormick AR, Hoellein TJ (2016) Anthropogenic litter is abundant, diverse, and mobile in urban rivers: insights from cross-ecosystem analyses using ecosystem and community ecology tools. Limnol Oceanogr 61:1718–1734. doi:10.1002/lno.10328

40. Osborn AM, Stojkovic S (2014) Marine microbes in the plastic age. Microbiol Aust 35:207–210. doi:10.1071/MA14066

41. Bazire A, Diab F, Jebbar M et al (2007) Influence of high salinity on biofilm formation and benzoate assimilation by *Pseudomonas aeruginosa*. J Ind Microbiol Biotechnol 34(1):5–8. doi:10.1007/s10295-006-0087-2

42. Oberbeckmann S, Osborn AM, Duhaime MB (2016) Microbes on a bottle: substrate, season and geography influence community composition of microbes colonizing marine plastic debris. PLoS One 11(8):e0159289. doi:10.1371/journal. pone.0159289

43. Dris R, Imhof H, Sanchez W et al (2015) Beyond the ocean: contamination of freshwater ecosystems with (micro-)plastic particles. Environ Chem 12(5):539–550. doi:10.1071/ EN14172

44. Hoellein T, Rojas M, Pink A et al (2014) Anthropogenic litter in urban freshwater ecosystems: distribution and microbial interactions. PLoS One 9(6):e98485. doi:10.1371/journal. pone.0098485

45. McCormick A, Hoellein J, Mason SA et al (2014) Microplastic is an abundant and distinct microbial habitat in an urban river. Environ Sci Technol 48(20):11863–11871. doi:10.1021/ es503610r

46. Lagarde F, Olivier O, Zanella M et al (2016) Microplastic interactions with freshwater microalgae: hetero-aggregation and changes in plastic density appear strongly dependent on polymer type. Environ Poll 215:331–339. doi:10.1016/j.envpol.2016.05.006

47. De Tender CA, Devriese LI, Haegeman A et al (2015) Bacterial community profiling of plastic litter in the Belgian part of the North Sea. Environ Sci Technol 49(16):9629–9638. doi:10.1021/acs.est.5b01093

48. Zettler ER, Mincer TJ, Amaral-Zettler LA (2013) Life in the "plastisphere": microbial communities on plastic marine debris. Environ Sci Technol 47(13):7137–7146. doi:10. 1021/es401288x

49. Bryant JA, Clemente TM, Viviani DA et al (2016) Diversity and activity of communities inhabiting plastic debris in the North Pacific Gyre. mSystems 1(3):e00024-16. doi:10.1128/ mSystems.00024-16

50. Kalmbach S, Manz W, Bendinger B et al (2000) *In situ* probing reveals *Aquabacterium commune* as a widespread and highly abundant bacterial species in drinking water biofilms. Water Res 34(2):575–581. doi:10.1016/S0043-1354(99)00179-7

51. Stamps BW, Lyles CN, Suflita JM et al (2016) Municipal solid waste landfills harbor distinct microbiomes. Front Microbiol 7:534. doi:10.3389/fmicb.2016.00534

52. Merga JY, Royden A, Pandey AK et al (2014) *Arcobacter* spp. isolated from untreated domestic effluent. Lett Appl Microbiol 59(1):122–126. doi:10.1111/lam.12256

53. Woo PCY, Lau SKP, Teng JLL et al (2008) Then and now: use of 16S rDNA gene sequencing for bacterial identification and discovery of novel bacteria in clinical microbiology laboratories. Clin Microbiol Infect 14(10):908–934. doi:10.1111/j.1469-0691.2008.02070.x

54. Schmidt VT, Reveillaud J, Zettler E et al (2014) Oligotyping reveals community level habitat selection within the genus *Vibrio*. Front Microbiol 5:563. doi:10.3389/fmicb.2014.00563

55. Ho BT, Dong TG, Mekalanos JJ (2014) A view to a kill: the bacterial type VI secretion system. Cell Host Microbe 15(1):9–21. doi:10.1016/j.chom.2013.11.008

56. Wallden K, Rivera-Calzada A, Waksman G (2010) Type IV secretion systems: versatility and diversity in function. Cell Microbiol 12(9):1203–1212. doi:10.1111/j.1462-5822.2010.01499.x

57. Russell AB, Peterson SB, Mougous JD (2014) Type VI secretion system effectors: poisons with a purpose. Nat Rev Microbiol 12(2):137–148. doi:10.1038/nrmicro3185

58. Nadell CD, Drescher K, Foster KR (2016) Spatial structure, cooperation, and competition in biofilms. Nat Rev Microbiol 14:589–600. doi:10.1038/nrmicro.2016.84

59. Quilliam RS, Jamieson J, Oliver DM (2014) Seaweeds and plastic debris can influence the survival of faecal indicator organisms in beach environments. Mar Poll Bull 84 (1–2):201–207. doi:10.1016/j.marpolbul.2014.05.011

60. Kirstein IV, Kirmizi S, Wichels A et al (2016) Dangerous hitchhikers? Evidence for potentially pathogenic *Vibrio* spp. on microplastic particles. Mar Environ Res 120:1–8. doi:10.1016/j.marenvres.2016.07.004

61. Masó M, Garcés E, Pagès F et al (2003) Drifting plastic debris as a potential vector for dispersing harmful algal bloom (HAB) species. Sci Mar 67(1):107–111. doi:10.3989/scimar.2003.67n1107

62. Carballo J, Seoane RM, Nieto TP (2000) Adhesion of *Aeromonas salmonicida* to materials used in aquaculture. Bull Eur Assoc Fish Pathol 20(2):77–82

63. Müller R-J (2015) Biodegradation behaviour of polymers in liquid environments. In: Bastioli C (ed) Handbook of biodegradable polymers. Rapra Technology Limited, Billingham, pp 33–55

64. Krueger MC, Harms H, Schlosser D (2015) Prospects for microbiological solutions to environmental pollution with plastics. Appl Microbiol Biotechnol 99(21):8857–8874. doi:10.1007/s00253-015-6879-4

65. Lucas N, Bienaime C, Belloy C et al (2008) Polymer biodegradation: mechanisms and estimation techniques – a review. Chemosphere 73(4):429–442. doi:10.1016/j.chemosphere.2008.06.064

66. Ohura T, Aoyagi Y, Takagi K et al (1999) Biodegradation of poly(3-hydroxyalkanoic acids) fibers and isolation of poly(3-hydroxybutyric acid)-degrading microorganisms under aquatic environments. Polym Degrad Stab 63(1):23–29. doi:10.1016/S0141-3910(98)00057-3

67. Manna A, Paul AK (2000) Degradation of microbial polyester poly(3-hydroxybutyrate) in environmental samples and in culture. Biodegradation 11(5):323–329. doi:10.1023/A:1011162624704

68. Brandl H, Püchner P (1991) Biodegradation of plastic bottles made from 'Biopol' in an aquatic ecosystem under *in situ* conditions. Biodegradation 2(4):237–243. doi:10.1007/BF00114555

69. Nauendorf A, Krause S, Bigalke NK et al (2016) Microbial colonization and degradation of polyethylene and biodegradable plastic bags in temperate fine-grained organic-rich marine sediments. Mar Pollut Bull 103(1–2):168–178. doi:10.1016/j.marpolbul.2015.12.024

70. Roy PK, Hakkarainen M, Varma IK et al (2011) Degradable polyethylene: fantasy or reality. Environ Sci Technol 45(10):4217–4227. doi:10.1021/es104042f

71. Headley JV, Gandrass J, Kuballa J et al (1998) Rates of sorption and partitioning of contaminants in river biofilm. Environ Sci Technol 32(24):3968–3973. doi:10.1021/es9804991

72. Tien C-J, Chen CS (2013) Patterns of metal accumulation by natural river biofilms during their growth and seasonal succession. Arch Environ Contam Toxicol 64(4):605–616. doi:10. 1007/s00244-012-9856-2

73. Carr SA, Liu J, Tesoro AG (2016) Transport and fate of microplastic particles in wastewater treatment plants. Water Res 91:174–182. doi:10.1016/j.watres.2016.01.002

74. Mason SA, Garneau D, Sutton R et al (2016) Microplastic pollution is widely detected in US municipal wastewater treatment plant effluent. Environ Pollut 218:1045–1054. doi:10.1016/j. envpol.2016.08.056

75. Murphy F, Ewan C, Carbonnier F et al (2016) Wastewater treatment works (WwTW) as a source of microplastics in the aquatic environment. Environ Sci Technol 50(11):5800–5808. doi:10.1021/acs.est.5b05416

76. Woodall LC, Sanchez-Vidal A, Canals M et al (2014) The deep sea is a major sink for microplastic debris. R Soc Open Sci 1:140317. doi:10.1098/rsos.140317

77. Besseling E, Wang B, Lürling M et al (2014) Nanoplastic affects growth of *S. obliquus* and reproduction of *D. magna*. Environ Sci Technol 48(20):12336–12343. doi:10.1021/ es503001d

78. Rochman CM, Hoh E, Kurobe T et al (2013) Ingested plastic transfers hazardous chemicals to fish and induces hepatic stress. Sci Rep 3:3263. doi:10.1038/srep03263

79. Schratzberger M, Gee JM, Rees HL et al (2000) The structure and taxonomic composition of sublittoral meiofauna assemblages as an indicator of the status of marine environments. J Mar Biol Assoc UK 80(6):969–980. doi:10.1017/S0025315400003039

80. Diaz RJ, Solan M, Valente RM (2004) A review of approaches for classifying benthic habitats and evaluating habitat quality. J Environ Manag 73(3):165–181. doi:10.1016/j. jenvman.2004.06.004

81. Cole M, Lindeque PK, Fileman E et al (2016) Microplastics alter the properties and sinking rates of zooplankton faecal pellets. Environ Sci Technol 50(6):3239–3246. doi:10.1021/acs. est.5b05905

82. Yang Y, Yang J, W-M W et al (2015) Biodegradation and mineralization of polystyrene by plastic-eating mealworms: part 2. Role of gut microorganisms. Environ Sci Technol 49 (20):12087–12093. doi:10.1021/acs.est.5b02663

83. Galloway TS (2015) Micro- and nano-plastics and human health. In: Bergmann M, Gutow L, Klages M (eds) Marine anthropogenic litter. Springer, Berlin, pp 343–366. doi:10.1007/978-3-319-16510-3_13

84. Lambert S, Wagner M (2016) Formation of microscopic particles during the degradation of different polymers. Chemosphere 161:510–517. doi:10.1016/j.chemosphere.2016.07.042

85. Bokulich NA, Mills DA (2013) Improved selection of internal transcribed spacer-specific primers enables quantitative, ultra-high-throughput profiling of fungal communities. Appl Environ Microbiol 79(8):2519–2526. doi:10.1128/AEM.03870-12

86. Visco JA, Apothéloz-Perret-Gentil L, Cordonier A et al (2015) Environmental monitoring: inferring the diatom index from next-generation sequencing data. Environ Sci Technol 49 (13):7597–7605. doi:10.1021/es506158m

87. Grossmann L, Jensen M, Heider D et al (2016) Protistan community analysis: key findings of a large-scale molecular sampling. ISME J 10:2269–2279. doi:10.1038/ismej.2016.10

88. Didelot X, Bowden R, Wilson DJ et al (2012) Transforming clinical microbiology with bacterial genome sequencing. Nat Rev Genet 13:601–612. doi:10.1038/nrg3226

89. Ruan Z, Feng Y (2016) BacWGSTdb, a database for genotyping and source tracking bacterial pathogens. Nucleic Acids Res 44(D1):D682–D687. doi:10.1093/nar/gkv1004

90. Berry D, Mader E, Lee TK et al (2015) Tracking heavy water (D_2O) incorporation for identifying and sorting active microbial cells. Proc Natl Acad Sci U S A 112(2):E194–E203. doi:10.1073/pnas.1420406112

91. Hatzenpichler R, Connon SA, Goudeau D et al (2016) Visualizing *in situ* translational activity for identifying and sorting slow-growing archaeal-bacterial consortia. Proc Natl Acad Sci U S A 113(28):E4069–E4078. doi:10.1073/pnas.1603757113

92. Wang Y, Huang WE, Cui L et al (2016) Single cell stable isotope probing in microbiology using Raman microspectroscopy. Curr Opin Biotechnol 41:34–42. doi:10.1016/j.copbio. 2016.04.018

93. Ojeda JJ, Romero-González ME, Banwart SA (2009) Analysis of bacteria on steel surfaces using reflectance micro-Fourier transform infrared spectroscopy. Anal Chem 81 (15):6467–6473. doi:10.1021/ac900841c

94. Rusconi R, Garren M, Stocker R (2014) Microfluidics expanding the frontiers of microbial ecology. Annu Rev Biophys 43:65–91. doi:10.1146/annurev-biophys-051013-022916

95. Foulon V, Le Roux F, Lambert C et al (2016) Colonization of polystyrene microparticles by *Vibrio crassostreae*: light and electron microscopic investigation. Environ Sci Technol 50:10988. doi:10.1021/acs.est.6b02720

96. Briggs D, Brewis D, Dahm R et al (2003) Analysis of the surface chemistry of oxidized polyethylene: comparison of XPS and ToF-SIMS. Surf Interface Anal 35:156–167. doi:10. 1002/sia.1515

97. Biesinger MC, Corcoran PL, Walzak MJ (2011) Developing ToF-SIMS methods for investigating the degradation of plastic debris on beaches. Surf Interface Anal 43:443–445. doi:10. 1002/sia.3397

98. Jungnickel H, Pund R, Tentschert J et al (2016) Time-of-flight secondary ion mass spectrometry (ToF-SIMS)-based analysis and imaging of polyethylene microplastics formation during sea surf simulation. Sci Total Environ 563-564:261–266. doi:10.1016/j.scitotenv. 2016.04.025

99. Pumphrey GM, Hanson BT, Chandra S et al (2009) Dynamic secondary ion mass spectrometry imaging of microbial populations utilizing ^{13}C-labelled substrates in pure culture and in soil. Environ Microbiol 11(1):220–229. doi:10.1111/j.1462-2920.2008.01757.x

100. Eichorst SA, Strasser F, Woyke T et al (2015) Advancements in the application of NanoSIMS and Raman microspectroscopy to investigate the activity of microbial cells in soils. FEMS Microbiol Ecol 91(10):fiv106. doi:10.1093/femsec/fiv106

101. Roh Y, Gao H, Vali H et al (2006) Metal reduction and iron biomineralization by a psychrotolerant Fe(III)-reducing bacterium, *Shewanella* sp. strain PV-4. Appl Environ Microbiol 72(5):3236–3244. doi:10.1128/AEM.72.5.3236-3244.2006

Risk Perception of Plastic Pollution: Importance of Stakeholder Involvement and Citizen Science

Kristian Syberg, Steffen Foss Hansen, Thomas Budde Christensen, and Farhan R. Khan

Abstract Risk perception has a significant impact on how society reacts to a given risk. There have been cases where a mismatch between the actual risk and the perception of it has led to poor decisions on societal initiatives, such as inappropriate regulatory measures. It is therefore important that the perception of risk is based on an informed foundation acknowledging the biases and drivers that inevitably go with risk perception. Plastic pollution differs in regard to other classical risks, such as those posed by chemicals or genetically modified organisms (GMOs), since the pollution is more visible and already has a significant magnitude. At the same time, everyone is familiar with using plastic, and our daily lives are highly dependent on the use of plastic. This offers some potential to strengthen the societal risk perception and subsequently implement effective measures to address the pollution.

In this chapter, we define eight risk perception drivers (voluntariness, control, knowledge, timing, severity, benefit, novelty, and tangibility) and relate these drivers to plastic pollution. We discuss the process in which plastic pollution has been recognized as an important environmental problem by scientists, the public, and policy makers and elaborate on how the eight risk drivers have influenced this process. Plastic pollution has several of the characteristics that can enhance people's perception of the risk as being important and which has generated great awareness of the problem. The chapter finally discusses how risk perception can

K. Syberg (✉) and F.R. Khan
Department of Science and Environment, Roskilde University, Roskilde, Denmark
e-mail: ksyberg@ruc.dk

S.F. Hansen
Department of Environmental Engineering, Technical University of Denmark, Lyngby, Denmark

T.B. Christensen
Department of Human and Technology, Roskilde University, Roskilde, Denmark

M. Wagner, S. Lambert (eds.), *Freshwater Microplastics*,
Hdb Env Chem 58, DOI 10.1007/978-3-319-61615-5_10,
© The Author(s) 2018

be improved by greater stakeholder involvement and utilization of citizen science and thereby improve the foundation for timely and efficient societal measures.

Keywords Citizen science, Plastic pollution, Public participation, Risk perception, Stakeholder involvement

1 Introduction

Risk is often portrayed as a function of hazard and exposure or in other words as being determined by the probability of an adverse event and the magnitude of this event's consequences [1]. The scientific capabilities for quantifying both probabilities and magnitude related to many risks are often relatively uncertain, which implies that quantification of risk is inherently uncertain [2]. This means that interpretations of risk are very important for human's response to the risk, since the risk perception, rather than an (often unknown) actual estimation of risk, will guide societal response to the risk. Uncertainty furthermore plays a profound role in regard to human's psychological responses to risks [1]. This implies that psychology is important in regard to how we as society react to a given risk, but elements such as communication and social structures also influence risk behavior as they frame the overall social and technical perception of both hazard and exposure.

Risk perception can be explained as the subjective assessment of a negative incident happening together with our concern of the consequences. The term risk perception is perhaps mostly associated with Ulrich Beck's description of the "risk society" in his book of the same name [3]. Beck argues that society must (and will) respond to the growing threat from ecological degradation by acting in a reflexive way [3]. This reflexivity can manifest in different manners, and Beck describes how a public demand for regulation can push a political debate, by drawing upon historical cases regarding oil drilling platforms and nuclear power plants [3]. Since the risk perception is thus often a strong driver for regulation, it has received increasing attention from stakeholders and legislators. In Sweden and Norway, parliamentarians now devote about three times as much attention to risk issues as they did in the first half of the 1960s, as reflected in their submitted private bills [1].

In this chapter we first describe how the historical development of risk perception can be explained within a theoretical framework. After the introduction of these theoretical boundaries, the chapter focuses on risk perception of plastic pollution in a historical perspective, followed by an analysis of stakeholder's role in development of public risk perception and policy measures. The last part of the chapter addresses how citizen science [4] can be an important method to improve societal risk perception of plastic pollution and finally discusses how the concept of citizen science can be expanded to allow for greater stakeholder involvement and better communication between scientist and citizens. Such communication can be vital in regard to informing about plastic pollution and thus improve the foundation for development of risk perception among stakeholders – including citizens and policy makers. For a discussion on the socio-ecological risks of microplastics from a global perspective, see [5].

2 The Theory of Risk Perception

Even though Beck's book on "risk society" might be the best known description of the importance of risk perception, the scientific theory predates Beck's book. One earlier example is that of Slovic et al. [6], who conducted a study where they evaluated several drivers for societal risk perception. One of their conclusions was that the greater a risk is perceived to be, the greater is the public demand for action [6]. The aim of the study was to explain why some hazards were perceived as extreme and others caused less concern, despite inconsistencies in the respective expert opinions [6]. This work built upon earlier studies, where Starr [7] found that risk seemed to be easily accepted if it was associated with benefits and had a voluntary nature. Risk of death in a traffic accident is a classic example of such acceptable risk.

In this paper we will distinguish eight drivers for risk perception (Table 1). The first driver that frames risk perception is voluntariness (driver 1). A person is more likely to accept a given risk if the risk exposure takes place on a voluntary basis compared to an imposed risk. Risks that are perceived to be uncontrollable generally cause greater concern (driver 2: control). The risk associated with flying as a passenger in an airplane, for example, often causes more concern than highway driving in passenger cars. A third driver (driver 3: knowledge) is the degree of familiarity associated with the risk. A known and quantifiable risk (such as the risk of getting cancer from smoking) is often more easily acceptable than the risk posed from an unknown entity. The timing (driver 4) of the risk is also important to its perception. Persons exposed to a given risk are more likely to accept the risk if it is imposed gradually over time than if the risk is imposed instantaneously. Risks with

Table 1 Eight main drivers for risk perception

	Drivers for risk perception	Explanation
1	Voluntariness	If the exposure to the risk factor is voluntary, it is more likely to be accepted compared to a superimposed risk
2	Control	If the risk is perceived to be uncontrollable, it is viewed as more severe
3	Knowledge	A known risk is perceived more acceptable than an unknown and unfamiliar risk
4	Timing	If a hazard has instant and disastrous potential, it is perceived as a higher risk, than hazards, which pose gradual risk over time
5	Severity	Greater perceived risk is correlated with how big a part of the population that is perceived as being at risk
6	Benefits	Risks that are associated with perceived benefits are often deemed more acceptable than risk without any obvious advantage
7	Novelty	Risks from novel entities are generally perceived as more risky than existing risks
8	Tangibility	A risk that is more tangible is perceived as more severe than a risk that is abstract and elusive

Table constructed after [6, 8]

greater potential for immediate disastrous outcome to the individual such as a nuclear power plant meltdown are often perceived as worse than those that inflict slow and gradual damage. A fifth driver is the severity of the risk (driver 5: severity), measured in terms of how many people it might affect, as there seems to be a correlation between number of people potentially affected and the perceived risk. The sixth driver (driver 6: benefit) for risk perception is the degree of benefits that are associated with the risk. People are more likely to accept risks if they believe that taking the risks is associated with high degree of benefit. Driving in cities with intense traffic is an example, where the risk of ending up in a car accident is perceived acceptable due to the benefit of transportation in a car. The seventh driver is the novelty of the risk (driver 7: novelty). Risks associated with new technologies and novel entities are generally perceived as more dangerous than older and more familiar risks, even if the statistical risks are comparable or even lower for the novel risk. The eighth and final driver relates to how tangible the risk is (driver 8: tangibility). It is important to distinguish between risks that by the individual are perceived as tangible and risks that are perceived as abstract and elusive. Abstract and elusive risks, such as those posed by climate change, are typically far more difficult to mobilize political action against, and therefore political action will only take place when the risk has become visible and acute, and by then, it will often be too late to take political action [8]. Giddens himself refers to this phenomenon as the "Giddens paradox" [8].

Before addressing risk perception of plastic pollution in respect to these drivers, it is feasible to explore two historical cases of other yet somewhat related types of risk – i.e., those of hazardous chemicals and genetically modified organisms (GMOs).

2.1 Risk Perception of Hazardous Chemicals and GMOs

In 2009 the European Commission published a study on Europeans' risk perception of potential hazardous chemicals in household products [9]. The results are interesting in the light of the abovementioned framework for "risk" perception, biases, and drivers. The group of chemicals that were associated with the highest perceived risk were pesticides and herbicides used for home use. Of the respondents that answered, 70% said that this group of chemicals posed a risk in their perception. At the other end of the scale were toothpaste (7%) and hair shampoo (11%).

The report concludes that people generally view personal risks lower than risks to the general public. This could be due to a perception of the personal risk being easier to control [9]. The report concludes that if a product is known to be risky, citizens could translate this knowledge into taking precautionary measures, which would again lower the perceived risk. This is in line with the theories about risk perception, i.e., the level of voluntariness as well as the level of control of the risk. Another important aspect for risk perception of household chemicals is the potency of the chemicals [9]. Chemical with high hazardous potential was generally perceived as more risky than less potent chemicals. This could be part of the reason why pesticides and herbicides were perceived as most risky, since these chemicals are designed to kill. The level of control might also be an important driver for the perceived risk in the study (driver 2 in Table 1). Pesticides and herbicides are spread in the environment leading to a loss of control, whereas exposure to toothpaste and hair shampoo is conducted under controlled circumstances (not taking the exposure to the environment from wastewater into consideration). Furthermore, there is a general trust that cosmetic products such as sunscreen are tested and that any potential risk is therefore known to science [9], again in correlation with risk perception driver 3 (whether a risk is understood and quantifiable or unknown and unfamiliar). The report finally concludes that there is a correlation between the educational levels of citizens and their awareness of potential risk but also that the better a risk is understood, the less concerned citizens are about it. These observations are also in accordance with the risk perception, biases, and drivers presented in Table 1.

The use of genetically modified organisms (GMOs) is another controversial topic which has spun an intense debate about risk perception since their introduction in the 1970s [10]. Especially, European citizens have been very reluctant to accept the risk associated with GMOs, not at least due to the high degree of scientific uncertainty associated with the use. Since the 1990s, the debate about the use of GMOs is mainly centered on crops and food products, whereas GMOs in pharmaceuticals have gained broader acceptance [10]. This tendency can possibly be explained by a number of the biases and drivers in Table 1. For instance, citizens will generally view a risk as more tolerable if there is an obvious benefit from taking the risk (driver 6) or if the risk is not directly affecting the individual subject, for example, if the use of a GMO at a farm is affecting the ecosystem on a general level and to a less extent the individual farmer. Production of new pharmaceuticals is often viewed positively, whereas enhanced crop yield might be less related to a consumer benefit and often more related to maximizing the economic outcome to the benefit of the farmer and only very indirectly the consumer.

Another aspect that has had importance for public risk perception of GMOs is the so-called yuck factor [11]. It is a term that was first used to describe citizens' reluctance toward new technologies with unknown consequences: a classic example being the unwillingness toward using purified wastewater as drinking water, regardless of how effective the cleaning is [11]. The "yuck factor" can thus be seen as an emotional response to something that people might find repulsive or in other ways conflicting with their beliefs and values. The emotional attitude toward novel technologies is framed by many factors on the individual level (e.g., age, gender, education, profession, previous experience with technologies, etc.) and on the societal level (e.g., structure and level of educational, media and legal system, norms and values, etc.). The "yuck factor" is therefore only a simplistic explanation to some of the public aversion toward GMOs [11]. The perception that food should be grown in the field and not in the laboratory surely also plays a role for some citizens' reluctance toward accepting this technology. GMOs thereby challenge a public idea of the relation between nature and food as a public set of values, regardless of what might be expressed as objective and scientific truth by the expert community. The "yuck factor" is important in two types of scenarios where a modification to what is perceived as a "natural system" changes this system drastically, whether it is altering the genes in a plant or spreading of artificial objects such as plastic in the environment. First, if the modification is linked with a limited scientific understanding and communication about possible negative consequences, or second if the scientific understanding is conflicting with core values in society and therefore not accepted as trustworthy.

3 Risk Perception of Plastic Pollution and the Role of Stakeholders

After this initial introduction to the field of risk perception, the remaining part of the chapter will focus on how plastic pollution is perceived today and how future efforts with better integration of stakeholders might facilitate a better and more informed risk perception among citizens. However, prior to that we address the historical risk perception of plastic pollution.

3.1 Historical Development

Scientific focus on plastic pollution has increased markedly over the last decades, especially since the turn on the millennium. The first notion of seabirds ingesting plastic debris was published in the 1960s [12]. At this point, research into environmental contamination with plastic debris was a small field, and few papers were published through the 1960s and 1970s (see [13–15]). However, Carpenter and Smith [16] were the first to notice that plastic accumulated in specific oceanic zone, in their Science publication of plastic debris in the Sargasso Sea. It was also in the 1970s that the first reports of beach litter were published [17]. More frequent reports on occurrence were consistently being published from the 1980s (e.g., [18, 19]), and it was in this decade that a systematically growing trend of marine pollution with plastic was first reported [20]. These findings initiated political discussions about the problem and were followed with political initiatives such as the MARPOL Annex V aiming at reducing plastic wastes at sea [21]. However, the Annex was considered optional, and ratification was required by UN member states before it enter into force in 1988 [21] (for a broader discussion on the regulation of microplastics, see [22]). Also in 1988 a report from the US National Oceanic and Atmospheric Administration (NOAA) described the concentration of plastic debris in the North Pacific Gyre. This was later followed by the work of Moore et al. [23], who compared abundance of plastic pellets and planktonic organism in the North Pacific Gyre. They concluded that while planktons were five times as abundant as plastic pellets when measured by number, the mass of the plastic pellets exceeded planktonic mass six times [23]. This "litter artifact" in the middle of the ocean was popularly called the "Great Pacific Garbage Patch," which had a significant impact on the public perception of the problem. The linguistic framing of the plastic pollution repelled the public by playing on the yuck factor, similar to the case of GMOs described above. The pollution was also unknown to many, making the novelty of the problem significant (driver 7). On the other hand, this description did give some backlash since it created an illusion of islands of plastics floating around in the ocean. Since such islands do not exist in reality, some commentators have argued that the environmental problem was exaggerated and that this could erode citizens' trust in institutions [24]. Plastic pollution was not perceived as such a big risk in the decades after the first reports were published. This can be explained using several of the risk perception drivers (Table 1). Since plastic pollution was first reported as a phenomenon on the open ocean and not related directly to severe impacts on marine species and ecosystems, it was not perceived as a risk with "potential for disaster" (driver 4) nor a contamination that impacted a large group of people (driver 5). Debris in the middle of the ocean has no direct link to any human populations per se, which might also have affected the lack of public response (driver 5). Furthermore, oceanic pollution is abstract and not so tangible since it is not easily visible to most people. Therefore, the "Giddens paradox" (driver 8) might also have influenced the lack of perceived risk in these early years. Finally, there was very little information communicated to the public about the problem, for instance, from 2004 to 2010, microplastics were only mentioned a few times in UK newspapers, whereas the number of articles grew markedly in the following years [25]. Since people obviously cannot perceive a risk that they are not aware of, this lack of communication is a final but very important reason for the lack of early alertness to the problem.

4   Risk Perception of Plastic Pollution and Political Actions Since the 2000s

Plastic pollution research declined during the 1990s, only to drastically increase after it was verified during the 2000s that plastic was a ubiquitous marine pollutant [17]. Among several important publications, Thompson et al. [26] published a famous paper in science entitled "Lost at sea: Where is all the plastic?" which is being recognized as a major driver for the elevated scientific interest [17]. The significant increase in scientific publications on the topic was followed with increased international media attention and political measures being enforced. Reports about the plastic pollution problem have thus been broadcasted in international media such as Reuters [27], and political measures have been taken in different regions of the world. In 2008 Rwanda banned the use of non-biodegradable plastic bags throughout the country [28]. This ban followed a national discussion of plastics' negative environmental impacts, especially due to the extensive physical presence of bags in the environment (also discussed by Khan et al. [29] in this volume). This measure was among the first and most comprehensive political acts to control plastic pollution, and it can to a large extent be explained with the risk perception drivers. The spreading of plastic bags was not an environmental risk that the population faced voluntary (driver 1). Since the plastic bags were further spread throughout the environment, it could be viewed as an uncontrollable risk (driver 2), perhaps even with potential for disasters for the ecosystems affected (driver 4). Since it may appear as there is only very limited societal benefit of the pollution to the end consumer (driver 6), there were strong incentives to address the pollution with political measures. Of course, the use of plastic on a societal level includes a vast amount of technical and economic benefits to both producers and consumers, and the current waste management practices where the majority of waste plastics is either landfilled or incinerated may be perceived by some stakeholders as beneficial to the society.

In Europe the debate about the use of resources, waste handling, and the plastic pollution has been ongoing for several years primarily within the context of waste regulation. The first packaging waste directive (Directive 85/339/EEC) was adopted in the mid-1980s aimed at reducing negative environmental aspects of packaging and packaging waste. The Packaging and Packaging Waste Directive has been amended several times since then (1994, 2003, 2004, 2013, and 2015). The 2015 revision resulted in the adoption of Directive (EU) 2015/720 on reducing the consumption of lightweight plastic carrier bags [30]. The overall framework for waste-related regulation is in the EU described in the Waste Framework Directive (Directive 2008/98/EC) that contains the core principles for waste management in Europe. The Waste Framework Directive is related to several directives that target specific waste streams such as batteries, electronic and electrical equipment, end-of-life vehicles, sewage sludge, construction and demolition waste, etc. Many of these waste streams contain plastic, and EU efforts to reduce plastic pollution in the waste sector shall therefore be seen on the background of this wide range of directives. In December 2015, the European Commission launched a Circular Economy (CE) package (also discussed in [31]). The CE package includes proposed revisions to many of the central waste-related directives including the Waste Framework Directive and the Packaging and Packaging Waste Directive. A central element in proposed revisions is common EU-wide 2030 targets for the waste sector. The CE strategy includes five priority areas, one of which is plastic. The commission will in 2017 adopt a strategy on plastic targeting issues such as recyclability, biodegradability, hazardous substances, and marine litter [32].

Microbeads pose a special and interesting case in regard to risk perception of plastic pollution. Microbeads contribute to a relatively small percentage of the total plastic production but have become highly exposed in the media, and risk perception of microplastic is often connected to microbeads. Several campaigns (e.g., Beat the microbead [33]) focus on phasing out microbeads explicitly. Several initiatives have been launched to call for a phaseout of microbeads. In one petition, gathering more than 375,000 signatures called for a ban in the UK [34]. The US state of

California approved Assembly Bill No. 888 banning the use of plastic microbeads in personal care products by 2020 [35], and a new US initiative aims at banning the use of microbeads in personal care products and cosmetics on a national level by mid-2017 [36]. The Canadian House of Commons have proposed a new order which will add microbeads to the national list of toxic substances, as a response to a vote to take immediate measures to phase out microbeads [37]. The environmental presence of microbeads has been documented [38], and the focus on microbeads is therefore scientifically valid. However, the major problem with plastic pollution seems to stem from other sources [39]. The "yuck factor" has played a role in the risk perception and subsequent political action on microbeads. Plastics in products such as toothpaste are viewed as "unnecessary" (driver 6) and "unnatural." Consumers therefore react emotionally negative toward this new, "unnecessary," and "unnatural" use of microbeads in consumer products, and this consumer attitude can be understood as an example of the "yuck factor." Microbeads thus serve as an example of the importance of risk perception to societal action and furthermore how important risk communication and involvement of citizens can be for societal reactions to an environmental problem.

The second part of the chapter addresses how citizen science has improved the risk perception of plastic pollution and finally discusses how it can be further expanded in order to involve citizens and thereby address the pollution better and further enable citizens to obtain informed perceptions of the plastic pollution problem.

5 Citizen Science as Concept

Science as a paid profession started in the later part of the nineteenth century [4]. Up until then scientific data were produced by people who collected the data due to interest. Some famous examples of citizen scientists were Benjamin Franklin and Charles Darwin [4]. Today's citizen science (CS) is most commonly conducted when projects are specifically designed to combine knowledge and expertise from scientists at research institutions with the work of the skilled amateurs, often within conservation biology and monitoring studies [40]. Silvertown [4] proposed that the expanding use of CS is driven by three factors: (1) greater access to the technical tools needed, (2) bringing in additional qualified labor, and (3) a greater demand for outreach within academia. In a historical context, CS has most commonly been used with conservation biology and nature monitoring programs. Examples such as the Atlas Project in Australia, where BirdLife Australia has used CS to obtain more than seven million bird observations for their "Atlas of Australian Bird" [41], and Herbaria@home, where museum collections of wild plants are analyzed by citizens in the UK for more than a decade [42], serve as illustrations of such classic CS projects. CS has however also been used to monitor pollution. The Air Quality Egg Project in the USA and Europe is a CS project that aims at monitoring air quality. It is based on a sensor system designed to allow citizens to collect data on NO_2 and

CO concentrations outside of their home [43]. The IDAH$_2$O Master Water Stewards program, offered by University of Idaho, aims at involving citizens in collection of water quality data in streams in Idaho, USA [44]. In North American citizen scientists have collected data for bird watching programs and have helped scientists develop guidelines for land managers [45]. The increased use of CS can thus be viewed as a way for science to be informed by citizens but at the same time, and very importantly for risk perception, as a means for citizens to obtain a better understanding of the scientific field in focus [46]. As mentioned earlier knowledge is vital for the risk perception. Where risk is perceived as higher by citizens than by experts within the field, it is often the unfamiliarity that is a key psychological driver for risk perception [6]. But there might also be scenarios where citizens are not fully aware of a risk, until they are involved in collecting data for it under a CS program. In these situations, people might underestimate risk due to the lack of knowledge. CS can thus help people to obtain more informed perceptions about a given risk and thereby facilitate a process of transformative learning that can ultimately result in citizens changing the perception of a given problem [47]. Collaboration between citizens and scientists not only influences citizen's risk perception but may also influence the values and beliefs that the scientists possess and ultimately their risk perception as well. This led Gibbons et al. [48] to suggest the distinction between mode 1 and mode 2 researches. Mode 1 research characterizes the traditional disciplinary scientific endeavor in closed scientific communities, and mode 2 research describes a transdisciplinary type of knowledge production where scientists and citizens collaborate to define both problems and solutions. Elements of this way of looking at research can today be found in, for example, the €80 billion European research and innovation program Horizon2020. Horizon2020 is based on three pillars: the excellence pillar that resembles the mode 1 research, the industrial leadership that mainly focuses on innovation in the private sector, and the Societal Challenges pillar that with requirements for multi-actor approach and co-innovation resembles the mode 2 research.

6 Citizen Science and Plastic Pollution

Citizen science has been widely used within the field of plastic pollution [49], often in and around the intertidal zone, e.g., as "beach cleanup" projects. A review conducted by Hidalgo-Ruz and Thiel [49] comparing CS and professional scientist projects concluded that CS can be a useful method for increasing the amount of available information on marine litter. Such events are typically organized by national organizations such as the NOAA in the USA [50] or by private stakeholders such as NGOs. NOAA has developed a mobile application called "Marine Debris Tracker app" (Fig. 1) together with Southeast Atlantic Marine Debris Initiative (SEA-MDI), allowing citizens to report findings of trash from beaches and waterways [51]. The app records the debris location through GPS, and the data can be view directly on the citizens' phone.

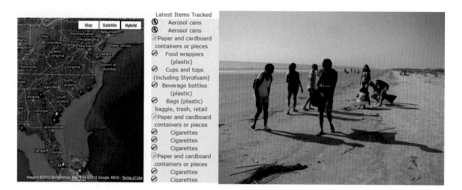

Fig. 1 (*Left*): Picture showing data marine debris collected and reported with the "Marine Debris Tracker app" made by NOAA Marine Debris Program and the Southeast Atlantic Marine Debris Initiative [50]. (*Right*): Citizens using the Marine Debris Tracker app (Picture from SEA-MDI)

Similar initiatives have been developed on the other side of the Atlantic in Europe. The European Environmental Agency, an independent agency financed by the European Union, has developed a mobile application called Marine LitterWatch [52] on the same principles as Marine Debris Tracker. Marine LitterWatch is used by scientists and NGOs in at least ten member states [52].

Apart from these stakeholders (scientist and NGOs), students can also play an active role in collecting and monitoring data using these mobile applications (Fig. 2). In the Roskilde Fjord region in Denmark, students collaborated with scientists to generate data on the occurrence of marine litter at 12 beaches around the fjord [53]. The students analyzed the data using a protocol inspired by the Marine LitterWatch protocol [53]. A similar but much larger project has been conducted in Chile [54]. The "National Sampling of Small Plastic Debris" was a CS project, where schoolchildren from 39 (approximately 1,000 students) from continental Chile and the Easter Island participated in the activity [54]. The project documented the distribution and abundance of small plastic debris on Chilean beaches. Scientist validated the data obtained in the program by recounting all samples in the laboratory. The results showed that the students were able to follow the instructions and generate reliable data [54]. Such involvement of students in collecting data serves as an example of the transformative learning discussed by Ruiz-Mallén et al. [47].

Microplastic is not as visible as meso- and macroplastic and therefore not as easily collected in these CS programs. But since the majority of microplastic pollution is secondary microplastic particles – i.e., breakdown products from meso- and macroplastic – the microplastic pollution is closely interlinked with larger plastic debris. Furthermore the majority of marine plastic debris stems from land-based sources [39], making NOAA arguing that beach cleanups are important contribution to marine protection [50], since they provide additional information for monitoring programs and help protect the environment. The development of Marine Debris Tracker app and the EEA Marine LitterWatch illustrates two aspects

	Boserup	Gundsølille	Jyllinge	Kattinge Vig	Kornerup A	Lejre Vig I	Lejre Vig II	Selsø	Roskilde Vig Syd	Roskilde Vig Nord	Risø bredning	Skomagerkrogen	Sum
Plastic	21	9	9	13	47	26	26	6	29	29	122	39	376
Rubber	1	0	0	0	0	2	0	0	0	3	1	1	8
Sanitary waste	0	0	0	0	0	1	0	0	0	1	10	7	19
Medical waste	0	0	0	1	0	0	0	0	1	1	0	1	4
Cloth	1	0	1	0	0	0	0	1	0	1	3	0	7
Paper/Cardboard	0	0	0	0	3	0	0	1	10	1	7	8	30
Wood (machined)	1	0	1	0	0	0	15	0	0	3	11	5	36
Metal	0	1	3	1	32	0	4	1	2	0	4	0	48
Glass/Ceramics	8	18	92	1	1	9	19	0	7	1	0	0	156
Pollutants	0	0	0	0	0	0	0	0	0	0	0	0	0
Other materials	0	2	2	1	0	15	18	1	7	12	0	0	58
Sum	32	30	108	17	83	53	82	10	56	52	158	61	742

Fig. 2 Data from beach cleanup at Roskilde fjord, Denmark. Litter collected was separated based on type (e.g., plastic, rubber, cloth, etc.) for each station where the cleanup was conducted, following the guideline of the Marine LitterWatch app developed by EEA. Top: pictures of litter on the beach and of the students participating in the beach cleanup [52]

of the CS development within the field of plastic pollution monitoring. Firstly, it serves as an example of how new (mobile) technology enables a systematic gathering of CS-collected data, in accordance with [4] drivers to expansion of CS. Secondly, it illustrates how collaboration between scientist, citizens, and NGOs results in efficient, high-quality data collection and monitoring programs. The quality of data is exactly one of the aspects that have been highlighted as a potential problem with CS-driven data collections [49]. It is therefore important to validate data collected under CS beach cleanup programs, if they are subsequently used in monitoring programs. Lavers et al. [55] found that detection of beach litter varied from 60 to 100% across various types of plastic. The authors further found variation among different observers, depending on observer experience and biological material present on the beach that could be confused with plastic [55]. The authors found that the color of the plastic debris was an important parameter, with blue fragments having the highest detection probability, while white fragments had the lowest.

In 2005 "International Pellet Watch" (IPW) was launched by Prof. Takada from Tokyo University [56]. The aim of the program was to collect monitoring data on POPs adhered to plastic pellets. The program (which can be characterized as a voluntary citizen science program) has participants in more than 50 countries [56]. Yeo et al. [57] describe how the implementation of the IPW program in Australia and New Zealand has been used to collect data. The authors found that the science communication part of the IPW program was so effective that it could be used to generate awareness of both plastic debris and POPs. These two types of pollution are interconnected to some extent, since POPs tend to adsorb strongly to plastic debris, making plastic debris a potential vector for environmental transport of POPs [58] (also discussed by [59, 60]). Since plastic pollution is often visible (especially for meso- and macroplastic), the environmental risk is more readily perceived than risk from hazardous chemicals such as POPs. The visibility generates a higher awareness of the problem than for less visible problems, leading to significant involvement in CS projects, and possibly policy measures as those described above for Rwanda. Furthermore it can also be used to increase the awareness and improve the risk perception or other less visible, but equally problematic, environmental problems, such as the contamination of POPs [57].

Due to the expanding data available from beach cleanup programs, scientists are now using these data to evaluate the ecological importance of plastic debris. Wilcox et al. [61] used expert elicitation to score impact from different types of plastic marine debris. In order to do so, those authors conducted the threat assessment by focusing on the most common types of litter found along the world's coastlines, based on data gathered during three decades of international coastal cleanup efforts [61]. The authors argued that the approach opened opportunities both for policy-based and consumer-driven changes in plastics, by focusing effort on the plastic debris with the highest demonstrable impact on ecologically important taxa serving as indicators of marine ecosystem health [61].

7 Expanding Citizen Science: The Bottom-Up Approach

Thus, citizen science can serve to integrate citizens in scientific projects increasing the scientific data pool and enabling citizens to obtain a more informed foundation for developing risk perception. There are however limitations to the current use of CS. Projects such as Marine LitterWatch and International Pellet Watch can be characterized as top-down CS, where scientists define the problem and ask citizens to help collect data to either illuminate and/or possibly solve the problem. While this is important work, it is dependent on and limited to the problems scientists have identified, and the citizens are primarily seen as "supporters." In this context, we refer to this as a top-down CS approach. An alternative (but not mutually exclusive) approach can be characterized as a bottom-up approach, where citizens are included already in the problem definition phase, potentially transcending a role as "supporters." This can facilitate stronger cooperation between scientist and citizens and lead to a more sustainable development [62]. Such an approach has some advantages that we will address below, before concluding the chapter by evaluating to what extent CS can serve as a valuable tool for increasing and qualifying risk perceptions.

Clausen [62] argues that the current dominating paradigm for inclusion of specific stakeholders in policy and environmental planning processes (i.e., governance) comes at the cost of the participation and influence of citizens in a broader sense and has a tendency to alienate citizens from nature and nature conservation. This is due to the focus on expert elicitation in the governance process (although the current governance paradigm includes more and different stakeholders compared to traditional expert-centered planning and decision processes), which has a tendency to decouple political processes from lay persons' perception of the problem. Clausen [62] further argues that by involving citizens in evolving shared norms and activities within a given topic, it is possible to facilitate the development of a community governance and thereby initiate a continuous sustainable process where citizens gain stronger ownership of (managing) the environment they are a part of. The earlier inclusion of (local) citizens can further strengthen the scientific foundation for addressing an environmental problem. Valinia et al. [63] discussed how the inclusion of local citizens' knowledge about a Swedish lake could improve the scientific foundation for assessing the anthropogenic impact on the water quality. The authors argued that local citizens possessed historical knowledge, which they used to conceptualize reference conditions in regard to the environmental state of the lake [63]. They showed that by comparing local knowledge with data from fish and water chemistry monitoring, as well as paleolimnological reconstructions of water quality, local citizens' knowledge corresponded well with the historical data, helping to deliver a more detailed picture of the present state of the lake. This local knowledge enabled a better assessment of the water quality and could thus contribute to developing a better scientific foundation for regulation [63]. And this is not all. As shown by Nielsen et al. [64], this kind of involvement of local citizens in natural resource management certainly broadened out the total scope of environmental management, integrating biological, cultural, and social dimensions, and through this leads to a growing responsibility within the local community regarding nature protection.

If citizens are included in the problem formulation phase of a risk assessment and meet the scientists in a mutual, not just formal, dialogue, their understanding of the risk will often be more qualified. This again can lead to citizens contributing to determining how risks are best addressed in their local community, which would leave them with higher degree of control over the risk, and if citizens are included in giving different risks priority, this might also ensure that they engage in making the local community a responsible actor in the nature protection. Different kinds of bottom-up CS confirm this, as, for instance, described by [65], discussing experiences from a project in a deprived urban community in London. Since some of the drivers for risk perceptions relate to how known the risk is (driver 3), whether it is faced voluntary (driver 1), and the degree of control over the risk (driver 2), a better integration of citizens via the bottom-up approach might ensure that citizens obtain a more qualified risk perception.

8 Concluding Remarks

The risk perception of plastic pollution has developed markedly since the first discoveries of oceanic plastic debris. In the first part of this chapter, we have illustrated how this development can largely be described with eight common drivers for risk perception. Common themes for many of these drivers are to what degree the problems of character, magnitude, potential impact, and solutions are understood. This implies that greater insight into these aspects by citizens will almost inevitably result in a more informed risk perception. Risk drivers such as how known the risk is (driver 2), possible benefits associated with the risk (driver 3), the novelty of the risk (driver 7), and whether the risk is tangible or abstract (driver 8) are all influenced by the degree of insight people have. In the case of plastic pollution, this greater insight has been built up over the last few decades and has stimulated a development toward a much broader understanding of the problem and a higher degree of perceived risk associated with plastic pollution.

In the second part of the chapter, we address how citizen science generates more awareness of the plastic pollution problem, improves risk perception, evolves science, and even contributes to actively addressing the plastic pollution problem. Classical top-down form of citizen science has had a significant impact of risk perception and subsequent societal measures. However, this type of mode 2 research, where scientists and citizens collaborate more systematically, has the potential to improve this positive process even further. We describe how bottom-up citizen science has been used to improve public participation, increase local ownership, improve scientific understanding, and enhance transparency within several different environmental topics. Within the area of plastic pollution, bottom-up CS has the potential to enable citizens to address the local pollution they are most concerned with, and by involving local citizens, they get higher sense of control over the risk (driver 2). The citizens also generate a better understanding of the consequences of the plastic pollution for the environment (driver 4). Any local conflicts where pollution with plastic is in favor of some citizens or organizations (e.g., cosmetics producers) and not others can further be illuminated in such a process (driver 6), enabling a dialogue about positive and negative consequences of the different behaviors. Finally, the bottom-up citizen science approach has the potential to make the problem even more explicit to the citizens and thereby enhancing the risk perception by reducing the "Giddens paradox" (driver 8). The risk perception of plastic pollution has thus developed markedly over the last decades, due to increased scientific understanding and greater involvement of citizens in collection scientific data.

Acknowledgement This work is supported by the Villum Foundation. VILLUM FONDEN

References

1. Sjøberg L, Moen BE, Rundmo T (2004) Explaining risk perception. An evaluation of the psychometric paradigm in risk perception research. Rotunde publikasjoner, Rotunde no. 84
2. Syberg K, Hansen SF (2016) Environmental risk assessment of chemicals and nanomaterials — the best foundation for regulatory decision-making? Sci Total Environ 541:784–794
3. Beck U (1986) Risk society: towards a new modernity. Sage, London
4. Silvertown J (2009) A new dawn for citizen science. Trends Ecol Evol 24:467–471
5. Kramm J, Völker C (2017) Understanding the risks of microplastics. A social-ecological risk perspective. In: Wagner M, Lambert S (eds) Freshwater microplastics: emerging environmental contaminants? Springer, Heidelberg. doi:10.1007/978-3-319-61615-5_11 (in this volume)
6. Slovic P, Fischhoff B, Lichtenstein S (1981) Facts and fears: societal perception of risk. NA-Adv Consum Res 8:497–502
7. Starr C (1969) Social benefit versus technological risk. Science 165:1232–1238
8. Giddens A (2011) The politics of climate change. Polity Press, Cambridge
9. EC (2009) Europeans' attitudes toward chemicals in consumer products: risk perception of potential health hazards. European Commission, Directorate-General for Communication ("Research and Political Analysis" Unit), Bruxelles, Belgium
10. Torgersen H (2004) The real and perceived risks of genetically modified organisms. EMBO Rep 5:17–21
11. Schmidt CW (2008) The yuck factor when disgust meets discovery. Environ Health Persp 116: A524–A527
12. Kenyon KW, Kridler E (1969) Laysan albatross swallow indigestible matter. Auk 86:339–343
13. Chenxi W, Kai Z, Xiong X (2017) Microplastic pollution in inland waters focusing on Asia. In: Wagner M, Lambert S (eds) Freshwater microplastics: emerging environmental contaminants? Springer, Heidelberg. doi:10.1007/978-3-319-61615-5_5 (in this volume)
14. Khan FR, Mayoma BS, Biginagwa FJ, Syberg K (2017) Microplastics in inland African waters: presence, sources and fate. In: Wagner M, Lambert S (eds) Freshwater microplastics: emerging environmental contaminants? Springer, Heidelberg. doi:10.1007/978-3-319-61615-5_6 (in this volume)
15. Dris R, Gasperi J, Tassin B (2017) Sources and fate of microplastics in urban areas: a focus on Paris megacity. In: Wagner M, Lambert S (eds) Freshwater microplastics: emerging

environmental contaminants? Springer, Heidelberg. doi:10.1007/978-3-319-61615-5_4 (in this volume)

16. Carpenter EJ, Smith KL Jr (1972) Plastic on the Sargasso Sea surface. Science 175:1240–1241

17. Ryan P (2015) A brief history of marine litter research. In: Bergmann M, Gutow L, Klages M (eds) Marine anthropogenic litter. Springer International Publishing, Berlin

18. Van Franeker JA (1985) Plastic ingestion in the North Atlantic fulmar. Mar Pollut Bull 16:367–369

19. Ryan P (1987) The incidence and characteristics of plastic particles ingested by seabirds. Mar Environ Res 23:175–206

20. Harper PC, Fowler JA (1987) Plastic pellets in New Zealand storm-killed prions (*Pachyptila* spp.), 1958–1977. Notornis 34:65–70

21. IMO (2016) Annex V: prevention of pollution by Garbage from Ships. International Maritime Organisation, United Nations. Available at: http://www.imo.org/en/OurWork/environment/pollutionprevention/garbage/Pages/Default.aspx

22. Brennholt N, Heß M, Reifferscheid G (2017) Freshwater microplastics: challenges for regulation and management. In: Wagner M, Lambert S (eds) Freshwater microplastics: emerging environmental contaminants? Springer, Heidelberg. doi:10.1007/978-3-319-61615-5_12 (in this volume)

23. Moore CJ, Moore SL, Leecaster MK, Weisberg SB (2001) A comparison of plastic and plankton in the North Pacific central gyre. Mar Pollut Bull 42:1297–1300

24. Watts A (2015) Are we really "choking the ocean with plastic"? Tracing the creation of an eco-myth. Available at: https://wattsupwiththat.com/2015/12/24/are-we-really-choking-the-ocean-with-plastic-tracing-the-creation-of-an-eco-myth/

25. GESAMP (2015) Sources, fate and effects of microplastics in the marine environment: a global assessment. International Maritime Organisation (IMO), London

26. Thompson RC, Olsen Y, Mitchell RP, Davis A, Rowland SJ, John AWG, McGonigle D, Russell AE (2004) Science 304:838

27. Reuters (2015) Plastic pollution devastating the world's oceans. Reuters International, May 13, 2015. Available at: http://www.reuters.com/video/2015/05/13/plastic-pollution-devastating-the-worlds?videoId=364209898&videoChannel=6&channelName=Technology

28. REMA (2012) Rwanda against plastic bags. Rwanda Environment Management Authority (REMA) Available at: http://www.rema.gov.rw/index.php?id=10&tx_ttnews%5Btt_news%5D=36&cHash=6fee2e51447d80ce6ce1d7d778969aea

29. Khan FR, Mayoma B, Biginagwa FJ, Syberg K Microplastics in inland African waters: presence, sources and fate. In: Warger M, Lambert S (eds) Freshwater microplastics - emerging environmental contaminants? Springer Nature. doi:10.1007/978-3-319-61615-5_6 (in this volume)

30. EC (2015) Directive (EU) 2015/720 of the European Parliament and of the council of 29 April 2015 amending directive 94/62/EC as regards reducing the consumption of lightweight plastic carrier bags. European Commission, Bruxelles

31. Eriksen M, Thiel M, Prindiville M (2017) Microplastic: what are the solutions? In: Wagner M, Lambert S (eds) Freshwater microplastics: emerging environmental contaminants? Springer, Heidelberg. doi:10.1007/978-3-319-61615-5_13 (in this volume)

32. EC (2015) Communication from the Commission to the European Parliament, the council, the European Economic and Social Committee and the Committee of the Regions, closing the loop – an EU action plan for the circular economy, European Commission COM(2015) 614 final, Brussels, Belgium

33. North Sea Foundation and Plastic Soup Foundation (2012) Beat the microbead. International campaign against microbeads in cosmetics. Available at: https://www.beatthemicrobead.org/en/

34. Greenpeace (2016) Petition to phase out microbeads. Available at: https://secure.greenpeace.org.uk/page/s/ban-microbeads

35. State of California (2015) Assembly bill no. 888 waste management: plastic microbeads. CA. Available at: https://leginfo.legislature.ca.gov/faces/billTextClient.xhtml?bill_id=201520160AB888
36. Energy and Commerce Committee (2016) Bipartisan legislation to phase out of plastic microbeads from personal care products. Available at: https://energycommerce.house.gov/news-center/press-releases/breaking-bipartisan-bill-banthebead-now-law#sthash.YJJa0C9E.dpuf
37. Chemical Watch (2016) Canada consults on proposed microbeads regulations. Available at: https://chemicalwatch.com/44980/
38. Eriksen M, Mason S, Wilson S, Box C, Zellers A, Edwards W, Farley H, Amato S (2013) Microplastic pollution in the surface waters of the Laurentian Great Lakes. Mar Pollut Bull 77:177–182
39. Jambeck J, Geyer R, Wilcox C, Siegler TR, Perryman M, Andrady A, Narayan R, Law KL (2015) Plastic waste inputs from land into the ocean. Science 347:768–771
40. Dickinson JL, Zuckerberg B, Bonter DN (2010) Citizen science as an ecological research tool: challenges and benefits. Annu Rev Ecol Evol Syst 41:149–172
41. BirdLife Australia (2016) Atlas & Birdata. Available at: http://birdlife.org.au/projects/atlas-and-birdata
42. BSBI (2016) Herbaria@home, recording historical biodiversity. Botanical Society of Britain & Ireland, Shirehampton. Available at: http://herbariaunited.org/atHome/
43. The Air Quality Egg Project (2016) Available at: http://airqualityegg.com/
44. UoI (2010) IDAH2O master water steward program. University of Idaho, Idaho. Available at: http://www.uidaho.edu/extension/idah2o/about
45. Cohn JP (2008) Citizen science: can volunteers do real research? Bioscience 58:192–197
46. Beecher N, Harrison E, Goldstein N, McDaniel M, Field P, Susskind L (2005) Risk perception, risk communication, and stakeholder involvement for biosolids management and research environ. J Environ Qual 34:122–128
47. Ruiz-Mallén I, Riboli-Sasco L, Ribrault C, Heras M, Laguna D, Perié L (2016) Citizen science: toward transformative learning. Sci Commun 38:523–534
48. Gibbons M, Limoges C, Nowotny H, Schwartsman S, Scott P, Trow M (1993) The new production of knowledge – the dynamics of science and research in contemporary societies. Sage, London
49. Hidalgo-Ruz V, Thiel M (2015) The contribution of citizen scientists to the monitoring of marine litter. In: Bergmann M, Gutow L, Klages M (eds) Marine anthropogenic litter. Springer International Publishing
50. NOAA (2015) How beach cleanups help keep microplastics out of the Garbage Patches. Available at: http://response.restoration.noaa.gov/about/media/how-beach-cleanups-help-keep-microplastics-out-garbage-patches.html
51. NOAA (2016) Marine debris tracker. Available at: https://marinedebris.noaa.gov/partnerships/marine-debris-tracker
52. EEA (2015) Marine litter watch. European Environmental Agency, Copenhagen. Available at: http://www.eea.europa.eu/themes/coast_sea/marine-litterwatch
53. Kallan G, Plasdorf-Vildrik LJ, Christiansen MM, Wagner SB, Søgaard S, Edelfsen SA (2016) Plast Ik. Et studie af plastforurening i Roskilde Fjord. Bachelor project, Roskilde Univerity, Denmark. (in Danish)
54. Hidalgo-Ruz V, Thiel M (2013) Distribution and abundance of small plastic debris on beaches in the SE Pacific (Chile): a study supported by a citizen science project. Mar Env Res 87–88:12–18
55. Lavers JL, Oppel S, Bond AL (2016) Factors influencing the detection of beach plastic debris. Mar Environ Res 119:245–251
56. IPW (2015) International Pellet Watch. Available at: http://www.pelletwatch.org/
57. Yeo BG, Takada H, Taylor H, Ito M, Hosoda J, Allinson M, Connell S, Greaves L, McGrath J (2015) POPs monitoring in Australia and New Zealand using plastic resin pellets, and

international Pellet watch as a tool for education and raising public awareness on plastic debris and POPs. Mar Pollut Bull 101:137–145

58. Syberg K, Khan FR, Selck H, Palmqvist A, Banta GT, Daley J, Sano L, Duhaime MB (2015) Microplastics: addressing ecological risk through lessons learned. Environ Toxicol Chem 34:945–953

59. Scherer C (2017) Interactions of microplastics with freshwater biota. In: Wagner M, Lambert S (eds) Freshwater microplastics: emerging environmental contaminants? Springer Nature, Heidelberg. doi:10.1007/978-3-319-61615-5_8 (in this volume)

60. Lambert S, Wagner M (2017) Microplastics are contaminants of emerging concern in freshwater environments: an overview. In: Wagner M, Lambert S (eds) Freshwater microplastics: emerging environmental contaminants? Springer, Heidelberg. doi:10.1007/978-3-319-61615-5_1 (in this volume)

61. Wilcox C, Mallos N, Leonard GH, Rodriguez A, Hardesty BD (2016) Using expert elicitation to estimate the impacts of plastic pollution on marine wildlife. Mar Policy 65:107–114

62. Clausen LT (2016) Re-inventing the commons: how action research can support the renewal of sustainable communities. Commons, sustainability, democratization: action research and the basic renewal of society. In: Hansen HP, Nielsen BS, Sriskandarajah N, Gunnarsson A (eds) . Taylor & Francis, New York, pp 29–52. (Routledge Advances in Research Methods; No. 19)

63. Valinia S, Hansen HP, Futter MN, Bishop K, Sriskandarajah N, Fölster J (2012) Problems with the reconciliation of good ecological status and public participation in the water framework directive. Sci Total Environ 433:482–490

64. Nielsen H, Hansen HP, Sriskandarajah N (2016) Recovering multiple rationalities for public deliberation within the EU water framework directive. In: Hansen HP, Nielsen BS, Sriskandarajah N, Gunnarsson A (eds) Commons, sustainability, democratization: action research and the basic renewal of society. Taylor & Francis, New York, pp 190–214. (Routledge Advances in Research Methods; No. 19)

65. Egmose J (2016) Organising research institutions through action research. In: Gunnarsson E, Hansen HP, Nielsen BS, Sriskandarajah N (eds) Action research for democracy. Taylor & Francis, New York, pp 182–198. (Routledge Advances in Research Methods; No. 17)

Understanding the Risks of Microplastics: A Social-Ecological Risk Perspective

Johanna Kramm and Carolin Völker

Abstract The diagnosis that we are living in a world risk society formulated by Ulrich Beck 20 years ago (Beck, Kölner Z Soziol Sozialpsychol 36:119–147, 1996) has lost nothing of its power, especially against the background of the Anthropocene debate. "Global risks" have been identified which are caused by human activities, technology, and modernization processes. Microplastics are a by-product of exactly these modernization processes, being distributed globally by physical processes like ocean currents, and causing effects far from their place of origin. In recent years, the topic has gained great prominence, as microplastics have been discovered nearly everywhere in the environment, raising questions about the impacts on food for human consumption. But are microplastics really a new phenomenon or rather a symptom of an old problem? And exactly what risks are involved? It seems that the phenomenon has accelerated political action—the USA has passed the Microbead-Free Waters Act 2015—and industries have pledged to fade out the use of microbeads in their cosmetic products. At first sight, is it a success for environmentalists and the protection of our planet?

This chapter deals with these questions by adopting a social-ecological perspective, discussing microplastics as a global risk. Taking four main characteristics of global risks, we develop four arguments to discuss (a) the everyday production of risk by societies, (b) scientific risk evaluation of microplastics, (c) social responses, and (d) problems of risk management. To illustrate these four issues, we draw on different aspects of the current scientific and public debate. In doing so, we contribute to a comprehensive understanding of the social-ecological implications of microplastics.

J. Kramm (✉) and C. Völker (✉)
ISOE – Institute for Social-Ecological Research, Hamburger Allee 45, 60486 Frankfurt, Germany
e-mail: kramm@isoe.de; voelker@isoe.de

M. Wagner, S. Lambert (eds.), *Freshwater Microplastics*,
Hdb Env Chem 58, DOI 10.1007/978-3-319-61615-5_11,
© The Author(s) 2018

Keywords Global risk, Problem structuring, Risk assessment, Social ecology, Uncertainty

1 The Social-Ecological Risk Perspective: Addressing Global Risks

A common risk definition is that "the term 'risk' denotes the likelihood that an undesirable state of reality (adverse effects) may occur as a result of natural events or human activities" [1]. A classical risk analysis calculates the possibility of an adverse event and the potential damage, for instance, an assessment of ecotoxicity of hazardous substances based on dose-response relationships. For "global risks," also termed systemic risks, classical risk analysis is not so easily applicable, since the characteristics of "global risks" comprise complex cause-effect linkages, which are not fully known, resulting in a high degree of uncertainty and ambiguity in assessing the risk. For this reason, consent to risk management strategies is difficult to obtain [2, 3].

Who or what can be "at risk"? In social-ecological risk research, risks to humans and biophysical entities (e.g., biocoenoses, ecosystems) are considered. The causes of risks mostly lie in human activities, since many natural resources and biophysical processes are influenced by societies [4]. In social-ecological risk research, it has become clear that assessment of the risk alone is not sufficient for management and policy decisions [5]. It is also important to consider the risk perception and concerns of different interest groups [6]. In the case of complex risks which are accompanied by uncertainty, it is important to define the degree of tolerability and acceptability in order to find management strategies acceptable to all interest groups [7]. Therefore, a prerequisite for risk management and related policy-making is not only scientific evidence but also an agreement of the different interest groups on how to understand, interpret, and value the evidence.

Hereafter, we will outline the characteristics of global risks from a social-ecological perspective and present four arguments framing microplastics as a global risk.

(a) Global risks are not produced by an extreme event or a disaster but are created in modern societies as a side effect of an "everyday mode" of system's operation [8, 9] and regulation of the supply system [4]. From this understanding, we derive our first thesis, arguing that the risks of microplastics are produced as an unintended side effect of everyday operations in modern societies.

(b) Global risks are complex; thus, no clear evidence of a cause-effect linkage exists or can be proven, due to "intervening variables," "long delay periods between cause and effect," or "positive and negative feedback loops" [10]. These and the state of "not knowing" [8] contribute to a high degree of uncertainty regarding effects, especially in terms of scope and time. Thus, we argue that the cause-effect linkages of risks associated with microplastics are complex, leading to great uncertainty in their scientific assessment.

(c) Global risks are characterized by a specific vibrancy which affects other linked entities or systems. This can lead to impacts in systems other than the risk-producing system. Such linking may involve natural processes (such as ocean currents, wind) and social processes (like communication, practices). Therefore, we argue, thirdly, that microplastics are vibrant, affecting not only ecosystems but different social, political, and economic spheres.

(d) Global risks are differently perceived, interpreted, and framed, which is an impediment to management strategies. This may be due to the presentation of different kinds of evidence, leading to competing views, or to conflicting interpretations of the same evidence, producing what is referred to as ambiguity [10]. Hence, we argue, fourthly, that microplastics are an example of a complex problem, due not only to uncertainty regarding their negative effects but also to competing views on how to combat the problem.

In the following sections, these four arguments are elaborated by taking into account different aspects of the recent scientific and public debate on microplastics.

2   The Plastic Dilemma and Everyday Modes of Risk Production

Microplastics emerged as a scientific topic about 10 years ago and recently came into public awareness when the debate focused on their release from cosmetic products and potential abundance in human food [11–15]. But are microplastics really a new phenomenon or can we regard them as a newly discovered symptom of an old problem, the problem of plastic pollution? As indicated in the quotation below, microplastics, called "plastic particles," were recognized as part of the problem of plastic pollution in coastal and oceanic waters in the 1970s, though the associated adverse consequences were considered as minor compared to other contaminants:

> At the present levels of abundance of plastic particles in coastal and oceanic waters, adverse biological consequences would appear to be minor compared to the deleterious effect of other contaminants such as petroleum residues and other chemical wastes. Increasing production of plastics, combined with present waste disposal practices, will undoubtedly lead to increases in the concentration of these particles in rivers, estuaries, and the open ocean. [16]

Plastic has been known as a factor in environmental pollution—symbolized by the plastic bag—for a long time. Looking at newspaper headlines dealing with the environment-plastic nexus, it becomes clear that plastic waste in the environment has been perceived as an environmental problem at least since the 1970s (see Table 1 for *The New York Times* headlines).

Table 1 Recurring headlines from *The New York Times*, selected from the period 1970 to 2015

18.02.1973	*Ocean pollution—the very dirty sea around us* Report on scientific surveys discovering pollution at sea. Among other kinds of trash, plastic litter is mentioned as plastic fragments and plastic bottles
25.12.1984	*Deadly tide of plastic waste threatens world's oceans and aquatic life* Report on the first international conference of marine biologists on the issue of "plastic waste in the oceans" held at the University of Hawaii in Honolulu. The article describes plastic waste as a *new and insidious form of pollution*
28.02.1992	*Biologists cite plastic bag in whale death* Article about a humpback whale washed ashore who swallowed a plastic bag, probably the cause of death as stated by researchers
22.06.2008	*Sea of trash—pollution in the world's oceans* Essay on plastic pollution in the oceans describing concrete examples, causes, effects, public perception of the issue, and measures to fight the problem from the 1980s to the 2000s
12.02.2015	*Study finds rising levels of plastics in oceans* Article about a scientific study of the growing amounts and the sources of plastic waste entering the oceans. Nations are urged to *take strong measures to dispose of their trash responsibly*

Although this visible (waste) problem has resulted in a number of new technologies for waste disposal and policies for its regulation, such as the German Law on Circular Economy [17] (see [18] for further discussion), the European Directive on Packaging and Packaging Waste [19], or the European Waste Framework Directive [20], the current debate on the environmental consequences of plastic waste shows that we still have not managed to find effective solutions. But why is it so hard to tackle the problem?

From the social-ecological risk perspective, the environmental implications of plastics can be understood as an unintended side effect produced by modern societies through their normal mode of operation [9]. Plastic products are an integral part of our everyday lives and their consumption is largely inconspicuous. For instance, plastic used in food packaging does not satisfy a demand for plastic but a demand for fresh food. Plastic packaging in the medical sector guarantees aseptic medical products, and plastic bags are an easy way to transport our shopping [21, 22]. These are just a few examples of how plastic products have penetrated our society, contributing to the environmental accumulation of plastic waste. The biggest share of plastic waste is produced by plastic packaging of consumer goods [23]. The environmental risk is thus created in a decentralized way by our everyday lives and not by an extreme event or disaster. To manage the problem, we would need to reconsider our everyday practices and transform our habits and routines in respect of how we produce, use, and dispose of plastic products. Changing everyday habits and routines is certainly challenging. However, it is noteworthy that these routine practices, now referred to as the "throwaway culture," were learned by our society in the not-so-distant past. After Bakelite—the first truly synthetic polymer—was invented as a substitute for natural resources such as horn, ivory, or tortoiseshell in 1907, plastics were soon substituting other materials and used to produce multiple objects. In the first half of the twentieth century, plastic materials enacted a new way of life: first, durable everyday plastic items, like combs, nylon stockings, radios, and telephones, led to "mass culture"—a "democratization of material goods" [24]. Finally, the translation of plastics from the laboratory to the beverage and food packaging industry paved the way for a "throwaway culture." An article published in the late 1950s in the journal *Modern Packaging* captures the shift from a material considered as durable to an ephemeral product:

> The biggest thing that's ever happened in molded plastics so far as packaging is concerned is the acceptance of the idea that packages are made to be thrown away. Plastic molders are no longer thinking in terms of re-use refrigerator jars and trinket boxes made to last a lifetime. Taking a tip from the makers of cartons, cans and bottles, they have come to the realization that volume lies in low-cost, single-use expendability...consumers are learning to throw these containers in the trash as nonchalantly as they would discard a paper cup— and in that psychology lies the future of molded plastic packaging. (n.a. 1957:120 in [25])

The plastic material was coded to be become waste after a short period of use; its use and meaning were changed. This new way of consuming and throwing away metamorphosed into a normal feature of ordinary everyday lives, a practice that is taken for granted nowadays [21]. In the last 50 years, plastics have become *the* workhorse material of the global economy and led to enormous progress for modern societies [23]. And that is the dilemma: society benefits from the attributes of plastic products (they are lightweight, inexpensive, and durable), and at the same time, mass production and durability lead to growing amounts of plastic waste accumulating in the environment [21, 26]. Although plastic has been perceived as a pollutant for a long time, and environmental awareness continues to grow, the per capita consumption of packaging is still increasing [27], so that with the increasing accumulation of (micro)plastics, the associated risks are growing.

3   From Macro to Micro: Unveiling the Complex Side Effects of Plastic Pollution

In recent years, scientific and public debates on plastic pollution have shifted from the visible waste problem to microplastics, an invisible form of plastic pollution. Though already detected in seawater in the 1970s [16, 28–33], it was not until the 2000s that small plastic particles, previously described as pellets, fragments, spherules, granules, etc., were labeled "microplastics" [34], which propelled their scientific career. Since then, the number of studies has grown exponentially (see Fig. 1). With the rising number of studies, microplastics have been discovered in more and more ecosystems, whether deep-sea sediments or freshwater environments [35, 36]. These studies have demonstrated the vast extent of microplastic pollution and its ubiquitous and persistent character and accelerated further research on the sources, environmental fate, and biological effects of microplastics. However, the number of studies is not only the result of a growing scientific interest in a "new"

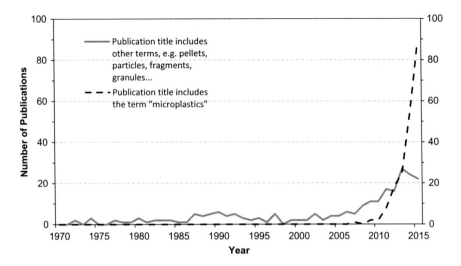

Fig. 1 *Environmental studies on plastic particles from 1970 to today.* The figure shows the rising number of studies in recent years, especially since the introduction of the term "microplastics." Studies were obtained from the search engines "Google Scholar" and "Web of Science." Keywords for the search were: microplastics + environment; plastic particles/fragments/pellets/granules/spheres/fibers + environment

research field—it also indicates the complexity of the problem calling for scientific methods in order to identify and quantify the consequences for the environment and for human health [10].

The traditional approach to environmental risk assessment of chemical substances cannot do justice to the multitude of microplastic particles and intervening variables and, therefore, cannot be applied to determining "safe" or "hazardous" levels of microplastics in natural environments [37]. Microplastics are not a homogenous group of substances, and they stem from various sources. The physicochemical properties of microplastics are as diverse as their sources. They differ in their polymeric composition, their additives, and have various shapes and sizes—all characteristics that can influence their biological effects. Microplastics can be toxic due to associated substances like phthalates and BPA [38], they can result in physical damage due to their shape [39], and they can induce indirect effects after being ingested, such as reduced food consumption due to satiation (malnutrition or even starvation) or intestinal blockage leading to death. Furthermore, biological effects are linked to other environmental contaminants such as persistent organic pollutants (POPs) that are absorbed by microplastic particles [40]. The lack of specific adverse effects leads to great uncertainty regarding predictions of the environmental consequences. These uncertainties were already expressed in early studies of microplastics around 30 years ago. However, despite these knowledge gaps, the problem was addressed pragmatically at that time: microplastics detected in natural waters and proven to be ingested by aquatic organisms were denoted as an "unnecessary contaminant" [33] that is "in all likelihood not beneficial" [41].

Today, pragmatic viewpoints still exist promoting a precautionary strategy with a call for action to reduce the leakage of microplastics into the environment despite evidence on specific adverse effects on ecosystems [42]. Others follow an approach, which Klinke and Renn [43] call "risk based." These studies aim to determine the potential damage of microplastics to provide evidence for the development of risk management strategies. Therefore, they target the existing research gaps in order to reduce uncertainties. But due to the nature of global risks, a broadened knowledge base will reveal even more variables, and it will be hard to achieve clear causality in order to structure the problem. For instance, research on microplastics has discovered even more sources of microplastics [44], and more species that ingest microplastics [45] and exposed methodological questions for assessing the risk, such as adequate detection methods to properly assess and compare the extent of microplastic contamination [46]. This hints at another dilemma: on the one hand, precisely these complexities call for thorough scientific investigation [10], but, on the other hand, exactly these investigations might contribute to higher complexity and greater uncertainty. Finally, the two approaches (risk based vs. precaution based) negotiate the question of how much knowledge is sufficient for action.

4 From Ecosystem Health to Human Health: Vibrancy, Uncertainty, and the Feeling of Insecurity

The impacts of (micro)plastics are not limited to the ecosystems where the plastic materials accumulate; the impacts are vibrant, affecting the political, social, and economic spheres, where they induce secondary and tertiary consequences, a typical characteristic of global risks [10]. For instance, studies point to economic effects, such as income loss among fishermen due to plastic debris [47], damage to marine industries [48], and loss of tourism revenues [49–51], which subsequently have social consequences. Today's discussions center on the impacts of microplastics on food for human consumption [13, 14], with possible but yet unknown threats for food safety and human health. Scientific evidence shows that microplastics are present in organisms, such as shellfish and fish, that play a role in human consumption [14, 52].

Microplastics infiltrating food for human consumption induce social processes. The following case from Germany shows the vibrancy of risk traveling from science into public awareness and how uncertain evidence and risk communication trigger feelings of insecurity.

A study commissioned by the media detected microplastic particles in drinking water, honey, and beer and was covered prominently in the German media [11, 12]. The knowledge produced by this study and the coverage of it in the media were contested by consumer protection agencies and food and beverage industries afraid of reputational effects. The studies were repeated by other scientists who could not verify the results, and some explained the identification of microplastics in German beer as an artifact of laboratory contamination [53].

The German Federal Institute for Risk Assessment (BfR), which deals with risks to human health, published a statement saying on the one hand that they could not detect microplastics in honey and beer in their laboratory studies. On the other hand, they stated that the health risk posed by microplastic-contaminated food and beverages cannot be assessed, due to the lack of reliable data and analytical methods [54]. The European Food Safety Authority (EFSA) started taking first steps toward a future assessment of the potential risks to consumers from microplastics and nanoplastics in food, especially seafood. Uncertainty exists, first, about the scope and quality of the contamination and, second, about the negative health effects for the public.

The media reports led to a raised public awareness of health risks, but the risk management authorities could not clear up the concerns, because despite that their studies had not verified the claim of microplastics in honey and beer, the question remains, if negative effects for human health exist. This feeling of insecurity is also reflected in the consumer survey by the BfR [55], which shows that 63% of the respondents had heard about "microplastics in food" and 52% answered that they were "concerned about microplastics in food". This case shows that there are only single observations of microplastics in food for human consumption and no scientific evidence for negative effects for human health exists. No general statement about risk for human health can be made; nevertheless, people are worried since a hypothetical risk has been communicated. Thus, due to the communication about the hypothetical risk, it becomes symbolically relevant in the first place, and a risk for human health is constructed. Therefore, risk communication is a very important aspect of risk management, with regard to the perception and psychological reactions of people who feel they are at risk. To reduce the social amplification of risk, it is important for laypersons that experts address risks and contextualize them in relations to other risks. Research on risk perception has pointed out that public opinion is steered by media reports scandalizing or exaggerating minor risks, leading to the spending of money to reduce them, while other major risks that failed to attract public attention are insufficiently considered [56, 57]. Risk managers should be sensitive to this and not become misguided by media and public concerns.

5 Risk Decision-Making: From Complex to Structured Problems

In the USA in December 2015, President Obama signed the Microbead-Free Waters Act, banning microbeads from rinse-off cosmetics—a success for microplastic opponents and environmentalists. What led to the quick decision to ban them, despite the complexity of the topic, which impedes risk assessment? A prerequisite for policy decisions is the degree of "consensus on the questions policy is addressing," as well as "certainty about the relevant knowledge" [58]. The degree to which a problem or a risk can be structured depends on consensus, values, views, and secured evidence, which includes knowledge of causes and effects. The continuum ranges from structured problems with common values and consensus on strategies and on the evidence, which comprises secured knowledge including clear causes and effects, to unstructured problems with competing values and no consensus on strategies and on the scientific evidence due to ambiguity and uncertainty [58].

In the case of the adoption of the Microbead-Free Waters Act, different actors were involved in "structuring the problem" [58]. Scientific evidence on the pathways into and the abundance in the environment was provided in strong collaboration with activists. For example, the NGO 5 Gyres Institute published the first microplastic pollution survey of the Great Lakes region in collaboration with the State University of New York in 2013. The concentration of microplastics found in the Great Lakes was higher than that of most samples collected in the oceans [59]. The studies were covered by the media, and the argumentation chain presented was quite clear: the microbeads threaten our lakes and rivers, stem from our cosmetic products, and slip through the sewage plants [60–62]. A clear scientific narrative was established and presented by scientists and activists to big personal care companies. The short "viewpoint" paper by Rochman et al. titled "Scientific Evidence Supports a Ban on Microbeads" [63], comprising a simple calculation of the number of microbeads and their route into the environment, was clearly aimed at strengthening this scientific narrative.

At the same time, environmental and ocean-protection NGOs campaigned for a ban on microplastics in cosmetics. Their campaigning methods included shopping guides that listed all producers using microplastics in their products and the app "Beat the Microbead" which could be used to check whether a product contains plastics. This app was launched by two Dutch NGOs in 2012 and further developed for international use by UNEP and another environmental NGO in 2013 [64]. With the guide and the app, tools were provided which enabled consumers to reduce their use of cosmetic products containing microplastics and to become more aware of the issue.

In the cosmetics industry, the evidence presented by the coalition between scientists and activists was not seriously contested. Global players like Johnson & Johnson, Unilever, and other multinationals announced that their products would be plastic-free within the next few years and that they would use natural substitutes instead. Since then, many more companies have pledged to phase out microplastics, motivated by reputational or environmental concerns.

With the detection of high amounts of microplastics in the Great Lakes, on the doorstep of the USA, the campaign against microplastics was boosted and entered the governmental arena, with several US states passing laws banning microbeads in cosmetics in 2014 and 2015 (e.g., New York, Illinois, California).

In March 2015, legislation to ban microplastics in cosmetics was introduced in the US Congress. How well the problem was structured by then is reflected in the speed with which the bill was passed: In March, it was introduced in the House of Representatives; in December, it was reported on and amended by the Committee on Energy and Commerce, and on the same day it was passed by the House of Representatives. Only 11 days later, it was passed by the Senate unanimously and was signed by the president 10 days later on December 28, 2015 [65]. The "Microbead-Free Waters Act of 2015" (H.R. 1321) prohibits "the manufacture and introduction or delivery for introduction into interstate commerce of rinse-off cosmetics containing intentionally-added plastic microbeads." The law specifies a phase out, starting with a ban on manufacturing the beads from July 2017 on, followed by product-specific manufacturing and sales bans in 2018 and 2019. The law bans only rinse-off and not leave-on products (eye shadow, face powder). Still, the ban can be regarded as a first step toward reducing the emission of microplastics. In Europe, industries have also pledged to phase out the use of microplastics, and Cosmetics Europe, the personal care industry's trade association, though highlighting that the "vast majority" of microplastics come from other sources than personal care products, issued a recommendation to discontinue their use in rinse-off cosmetics, and announced its intention to collaborate closely with regulators. By doing so, they were "addressing public concerns" [66].

At the science-policy interface, interest groups like environmental organizations did play an important role as brokers, but nevertheless further points were also decisive for the structuring of the policy problem. First, clearly structured evidence of cause and effect was presented and was not confused by other conflicting facts (other sources of primary microplastics and secondary microplastics as major sources were almost totally excluded in the US debate). Second, a ban on microbeads in cosmetic products did not constitute a financial risk or any other threat to the personal care sector, since alternatives existed and a change in production was implementable in the set timeframe. In addition, it gave the cosmetic industry the possibility to shape its sustainability profile and to emphasize value sharing with the consumer. This may be a reason why the presented evidence was not contested.

Recently published studies (e.g., [67]) have shifted the focus to land-based sources and the degradation of plastic waste in the oceans and other environments, enhancing the circle of responsibilities from single industries to complex processes of supplying, consuming, and waste management. In this context, it has turned out that cosmetic products as a source of microplastics play a much smaller role than previously thought [68, 69]. In this context, the ban on microbeads is only a tiny drop in the ocean. The complexity of plastics in the environment is becoming more and more obvious and poses a great challenge to risk assessment and management. Against this background, it seems that the Microbead-Free Waters Act was adopted in a window of opportunity in which the problem was perceived as well struc-tured—the scientific evidence was clear to all interest groups, there was consent on the trade-off between the benefits of microbeads in cosmetics and the hazards they pose to ecosystems, and multiple alternatives for microbeads in cosmetics were available (physically and economically). Thus, the case of the USA can be regarded as an example of using a well-structured problem for policy-making, while most of the problems related to plastics are in fact unstructured, e.g., due to competing views of multiple interest groups.

6   Conclusion

Increased research on (micro)plastics has developed the picture that (micro)plastic pollution is ubiquitous. Microplastics have been detected in rivers in Europe (e.g., Danube, [70]), as well as in lakes in Mongolia [50] and the USA [59]. They cross state boarders, passing from rivers into lakes, and finally into the global common, the ocean. They also cross the boundaries between single organisms, accumulating in the food web. From a social-ecological perspective, the risk induces a vibrancy and resonance in socioeconomic, political, and public spheres. Thus, the theses we have presented and their corresponding data clearly identify microplastics as a global risk, leading to the following conclusions regarding further research areas:

Based on an understanding of the risks posed by microplastics as an unintended side effect of the everyday mode of societies, the global dimensions of production and distribution patterns need to be researched in more depth. In many countries of the Global South, a new middle income class with a high demand for plastic products is growing. Relations between the Global North and the Global South need to be addressed more adequately, regarding the production, distribution, consumption, disposal, and leakage into the environment of plastic-packaged products like fast-moving consumer goods.

Due to the complexity of the microplastics phenomenon, its assessment is difficult and requires further scientific investigations to establish the evidence in order to properly address the environmental risk. The same holds true for the assessment of the human health risks. This uncertainty impedes risk management decisions, but nevertheless action is required despite a lack of clear evidence, because microplastics are perceived as a threat by society. Therefore, as the complexity of the phenomenon may never be entirely resolved, future research should also focus on the question of how to handle uncertainty and manage complex global risks.

Although it is common sense that plastics should not be allowed to accumulate in the environment, much less consensus exists regarding the strategies needed to achieve this. As Shaxson [58] points out, the question "How can we make plastics sustainable?" is just too broad and unstructured to enable all the interest groups to speak with one voice. Strategies to combat pollution range from reuse, green chemistry, designs for recycling, improved waste management, standardized labeling, education, cleaning programs, and sustainable consumption. Not a single strategy is required, but each sector needs to be active. However, current debates show that responsibilities are often shifted elsewhere. Thus, identifying the risk producers is not straightforward, as some voices do not regard plastics as the source of the problem but rather their improper disposal; other voices emphasize the design of the plastic material, and yet others target consumer behavior. Risk management is about the negotiation of evidence and values. We should not stop at symbolic goals, like the G7 Action Plan [71], but move on to binding regulations. Research should focus on developing and testing mechanisms to call risk producers to account, for example, with the integration of costs in the benefits, extended producers' responsibility, cost of inaction analysis, etc.

To conclude, we reflect on the risks of microplastics for ecosystems and our health, by drawing on the questions Beck once asked:

How worried should we be? Where is the line between prudent concern and crippling fear and hysteria? [8]

Concerns about microplastics in our food and subsequent health effects, triggered by media reports, lead to social risk amplification, which may be disproportionate to other risks associated with plastics, such as environmental accumulation or the endocrine effects of plasticizers. There is no need for "hysteria" (to quote Beck). Nevertheless, we should take the (micro)plastics issue as a serious symptom of human-made environmental change. Plastic pollution is a visible example of how society and nature interact, and it unveils our relationship with nature. What kind of nature do we want and how do we want to live? We have to explore the intersections between global risks, power relations, and societal relations with nature if we want to bring about their sustainable transformation.

Funding The authors are members of the junior research group "PlastX – Plastics as a systemic risk for social-ecological supply systems" which is funded by the German Federal Ministry for Education and Research (BMBF) as part of the program "Research for sustainable development (FONA)." In FONA, PlastX belongs to the funding priority "SÖF – Social-ecological research" within the funding area "Junior research groups in social-ecological research."

References

1. Renn O (2008) Concepts of risk: an interdisciplinary review. GAIA-Ecol Perspect Sci Soc 17 (1):50–66
2. Wissenschaftlicher Beirat der Bundesregierung Globale Umweltveränderungen (1998) Welt im Wandel: Strategien zur Bewältigung globaler Umweltrisiken: Jahresgutachten 1998. Springer, Berlin
3. Renn O, Keil F (2008) Systemische Risiken: Versuch einer Charakterisierung. GAIA-Ecol Perspect Sci Soc 17(4):349–354
4. Hummel D, Jahn T, Schramm E (2011) Social-ecological analysis of climate induced changes in biodiversity: outline of a research concept. Knowledge Flow Paper. Biodiversität und Klima Forschungszentrum 11: 1–15
5. Renn O, Schweizer P-J, Dreyer M et al (2007) Risiko: Über den gesellschaftlichen Umgang mit Unsicherheit. oekom, München
6. Syberg K, Hansen SF, Christensen TB, Khan FR (2017) Risk perception of plastic pollution: Importance of stakeholder involvement and citizen science. In: Wagner M, Lambert S (eds) Freshwater microplastics: emerging environmental contaminants? Springer, Heidelberg. doi:10.1007/978-3-319-61615-5_10 (in this volume)
7. International Risk Governance Council (2005) Risk governance: towards an integrative approach. White Paper, Genf
8. Beck U (2006) Living in the world risk society. Econ Soc 35(3):329–345
9. Keil F, Bechmann G, Kümmerer K et al (2008) Systemic risk governance for pharmaceutical residues in drinking water. GAIA-Ecol Perspect Sci Soc 17(4):355–361
10. Klinke A, Renn O (2006) Systemic risks as challenge for policy making in risk governance. In: Forum qualitative sozialforschung/forum: qualitative Social Research, Vol 1

11. Der Spiegel (2013) Plastikteilchen verunreinigen Lebensmittel: Unterschätzte Gefahr 17.11.2013. http://www.spiegel.de/wissenschaft/technik/winzige-plastikteile-verunreinigen-lebensmittel-a-934057.html. Accessed 30 Sep 2016
12. NDR (2014) Mikroplastik in Mineralwasser und Bier. http://www.ndr.de/ratgeber/verbraucher/Mikroplastik-in-Mineralwasser-und-Bier,mikroplastik134.html. Accessed 30 Sep 2016
13. Seltenrich N (2015) New link in the food chain? Marine plastic pollution and seafood safety. Environ Health Perspect 123(2):A34–A41
14. van Cauwenberghe L, Janssen CR (2014) Microplastics in bivalves cultured for human consumption. Environ Pollut 193:65–70
15. Lambert S, Wagner M (2017) Microplastics are contaminants of emerging concern in freshwater environments: an overview. In: Wagner M, Lambert S (eds) Freshwater microplastics: emerging environmental contaminants? Springer, Heidelberg. doi:10.1007/978-3-319-61615-5_1 (in this volume)
16. Colton JB, Knapp FD, Burns BR (1974) Plastic particles in surface waters of the northwestern Atlantic. Science 185(4150):491–497
17. Deutscher Bundestag (1994) Gesetz zur Förderung der Kreislaufwirtschaft und Sicherung der umweltverträglichen Beseitigung von Abfällen (Kreislaufwirtschafts- und Abfallgesetz - KrW-/AbfG)
18. Eriksen M, Thiel M, Prindiville M, Kiessling T (2017) Microplastic: What are the solutions? In: Wagner M, Lambert S (eds) Freshwater microplastics: emerging environmental contaminants? Springer, Heidelberg. doi:10.1007/978-3-319-61615-5_13 (in this volume)
19. The European Parliament and the Council of the European Union (1994) European Parliament and Council Directive 94/62/EC of 20 December 1994 on Packaging and Packaging Waste
20. The European Parliament and the Council of the European Union (2008) Directive 2008/98/EC of the European Parliament and of the Council of 19 November 2008 on waste and repealing certain Directives
21. Andrady AL, Bomgardner M, Southerton D et al. (2015) Plastics in a sustainable society, Background paper, Stockholm
22. Andrady AL, Neal MA (2009) Applications and societal benefits of plastics. Philos Trans R Soc, B 364(1526):1977–1984
23. WEF (2016) The new plastics economy: rethinking the future of plastics. http://www3.weforum.org/docs/WEF_The_New_Plastics_Economy.pdf. Accessed 15 Sep 2016
24. Bensaude Vincent B (2013) Plastics, materials and dreams of dematerialization. In: Gabrys J, Hawkins G, Michael M (eds) Accumulation: the material politics of plastic. Routledge, Oxon, pp 17–29
25. Hawkins G (2013) Made to be wasted: PET and topologies of disposability. In: Gabrys J, Hawkins G, Michael M (eds) Accumulation: the material politics of plastic. Routledge, Oxon, pp 49–67
26. Vegter AC, Barletta M, Beck C et al (2014) Global research priorities to mitigate plastic pollution impacts on marine wildlife. Endanger Species Res 25(3):225–247
27. Oregon DEQ (2005) International packaging regulations. State of Oregon Department of Environmental Quality, Portland
28. Carpenter EJ, Anderson SJ, Harvey GR et al (1972) Polystyrene spherules in coastal waters. Science 178(4062):749–750
29. Carpenter EJ, Smith KL (1972) Plastics on the Sargasso Sea surface. Science 175 (4027):1240–1241
30. Gregory MR (1978) Accumulation and distribution of virgin plastic granules on New Zealand beaches. N Z J Mar Freshw Res 12(4):399–414
31. Morris RJ (1980) Floating plastic debris in the Mediterranean. Mar Pollut Bull 11(5):125
32. Bond AL, Provencher JF, Elliot RD et al (2013) Ingestion of plastic marine debris by common and thick-billed murres in the northwestern Atlantic from 1985 to 2012. Mar Pollut Bull 77 (1):192–195
33. Gregory MR (1983) Virgin plastic granules on some beaches of eastern Canada and Bermuda. Mar Environ Res 10(2):73–92

34. Thompson RC, Olsen Y, Mitchell RP et al (2004) Lost at sea: where is all the plastic? Science 304(5672):838
35. van Cauwenberghe L, Vanreusel A, Mees J et al (2013) Microplastic pollution in deep-sea sediments. Environ Pollut 182:495–499
36. Mani T, Hauk A, Walter U et al (2015) Microplastics profile along the Rhine river. Sci Rep 5:17988
37. Brennholt N, Heß M, Reifferscheid G (2017) Freshwater microplastics: challenges for regulation and management. In: Wagner M, Lambert S (eds) Freshwater microplastics: emerging environmental contaminants? Springer, Heidelberg. doi:10.1007/978-3-319-61615-5_12 (in this volume)
38. Oehlmann J, Schulte-Oehlmann U, Kloas W et al (2009) A critical analysis of the biological impacts of plasticizers on wildlife. Philos Trans R Soc B 364(1526):2047–2062
39. Wright SL, Thompson RC, Galloway TS (2013) The physical impacts of microplastics on marine organisms: a review. Environ Pollut 178:483–492
40. Fendall LS, Sewell MA (2009) Contributing to marine pollution by washing your face: microplastics in facial cleansers. Mar Pollut Bull 58(8):1225–1228
41. Zitko V, Hanlon M (1991) Another source of pollution by plastics: skin cleaners with plastic scrubbers. Mar Pollut Bull 22(1):41–42
42. State of New South Wales and the Environment Protection Authority (2016) Plastic microbeads in products and the environment. http://www.epa.nsw.gov.au/resources/waste/plastic-microbeads-160306.pdf. Accessed 20 Dec 2016
43. Klinke A, Renn O (2002) A new approach to risk evaluation and management: risk-based, precaution-based, and discourse-based strategies. Risk Anal 22(6):1071–1094
44. Dris R, Gasperi J, Tassin B (2017) Sources and fate of microplastics in urban areas: a focus on Paris Megacity. In: Wagner M, Lambert S (eds) Freshwater microplastics: emerging environmental contaminants? Springer, Heidelberg. doi:10.1007/978-3-319-61615-5_4 (in this volume)
45. Scherer C, Weber A, Lambert S, Wagner M (2017) Interactions of microplastics with freshwater biota. In: Wagner M, Lambert S (eds) Freshwater microplastics: emerging environmental contaminants? Springer Nature, Heidelberg. doi:10.1007/978-3-319-61615-5_8 (in this volume)
46. Klein S, Dimzon IK, Eubeler J, Knepper TP (2017) Analysis, occurrence, and degradation of microplastics in the aqueous environment. In: Wagner M, Lambert S (eds) Freshwater microplastics: emerging environmental contaminants? Springer, Heidelberg. doi:10.1007/978-3-319-61615-5_3 (in this volume)
47. Nash AD (1992) Impacts of marine debris on subsistence fishermen an exploratory study. Mar Pollut Bull 24(3):150–156
48. McIlgorm A, Campbell HF, Rule MJ (2011) The economic cost and control of marine debris damage in the Asia-Pacific region. Ocean Coast Manag 54(9):643–651
49. Ballance A, Ryan PG, Turpie JK (2000) How much is a clean beach worth? The impact of litter on beach users in the Cape Peninsula, South Africa. S Afr J Sci 96(5):210–230
50. Free CM, Jensen OP, Mason SA et al (2014) High-levels of microplastic pollution in a large, remote, mountain lake. Mar Pollut Bull 85(1):156–163
51. Jang YC, Hong S, Lee J et al (2014) Estimation of lost tourism revenue in Geoje Island from the 2011 marine debris pollution event in South Korea. Mar Pollut Bull 81(1):49–54
52. von Moos N, Burkhardt-Holm P, Köhler A (2012) Uptake and effects of microplastics on cells and tissue of the blue mussel Mytilus edulis L. after an experimental exposure. Environ Sci Technol 46(20):11327–11335
53. Lachenmeier DW, Kocareva J, Noack D et al (2015) Microplastic identification in German beer-an artefact of laboratory contamination? Deutsche Lebensmittel Rundschau 111 (10):437–440
54. Federal Institute for Risk Assessment (2015) Mikroplastikpartikel in Lebensmitteln: Stellungnahme Nr. 013/2015
55. Federal Institute for Risk Assessment (2016) BfR Consumer Monitor 02l2016, Berlin

56. Renn O (1998) The role of risk perception for risk management. Reliab Eng Syst Saf 59 (1):49–62. doi:10.1016/S0951-8320(97)00119-1
57. Okrent D (1996) Risk perception research program and applications: have they received enough peer review? In: Cacciabue C, Papazoglou IA (eds) Probabilistic safety assessment and management. Springer, Berlin, pp 1255–1260
58. Shaxson L (2009) Structuring policy problems for plastics, the environment and human health: reflections from the UK. Philos Trans R Soc B 364(1526):2141–2151
59. Eriksen M, Mason S, Wilson S et al (2013) Microplastic pollution in the surface waters of the Laurentian Great Lakes. Mar Pollut Bull 77(1):177–182
60. The New York Times (2013) Scientists turn their gaze toward tiny threats to great lakes 14.12.2013. http://www.nytimes.com/2013/12/15/us/scientists-turn-their-gaze-toward-tiny-threats-to-great-lakes.html?_r=0. Accessed 19 Aug 2016
61. CBC News (2013) Great Lakes threatened by microplastics: Lakes Erie has highest concentration of small, microscopic plastic 11.10.2013. http://www.cbc.ca/news/canada/windsor/great-lakes-threatened-by-microplastics-1.1990832. Accessed 25 Aug 2016
62. Environmental Leader (2013) Microplastics invade Great Lakes. Environmental Leader. Environmental and Energy Management News 31.10.2013. http://www.environmentalleader.com/2013/10/31/microplastics-invade-great-lakes/#ixzz3wIe3gsLt. Accessed 15 Sep 2016
63. Rochman CM, Kross SM, Armstrong JB et al (2015) Scientific evidence supports a ban on microbeads. Environ Sci Technol 49(18):10759–10761
64. Beat the Microbead (2016) Beat the microbead. International campaign against microbeads in cosmetics. https://www.beatthemicrobead.org/en/in-short. Accessed 28 Sep 2016
65. Bill Information for H.R.1321 – Microbead-Free Waters Act of 2015 (2015). https://www.congress.gov/bill/114th-congress/house-bill/1321/all-info. Accessed 15 Oct 2016
66. Cosmetics Europe (2015) Cosmetics Europe recommendation on solid plastic particles (Plastic micro particles). https://www.cosmeticseurope.eu/news-a-events/news/822-cosmetics-europe-issues-a-recommendation-on-solid-plastic-particles-plastic-micro-particles.html. Accessed 26 Sep 2016
67. Jambeck JR, Geyer R, Wilcox C et al (2015) Plastic waste inputs from land into the ocean. Science 347(6223):768–771
68. Sherrington C, Darrah C, Hann S et al. (2016) Study to support the development of measures to combat a range of marine litter sources. Report for European Commission DG Environment
69. Essel R, Engel L, Carus M et al (2015) Quellen für Mikroplastik mit Relevanz für den Meeresschutz in Deutschland. Umwelt Bundesamt Texte, Dessau-Roßlau
70. Lechner A, Keckeis H, Lumesberger-Loisl F et al (2014) The Danube so colourful: a potpourri of plastic litter outnumbers fish larvae in Europe's second largest river. Environ Pollut 188:177–181
71. G7 Germany (2015) Annex zur Abschlusserklärung. https://www.g7germany.de/Content/DE/_Anlagen/G7_G20/2015-06-08-g7-abschluss-eng.pdf?__blob=publicationFile&v=6. Accessed 10 Sep 2016

Freshwater Microplastics: Challenges for Regulation and Management

Nicole Brennholt, Maren Heß, and Georg Reifferscheid

Abstract The accumulation of plastic debris in aquatic environments is one of the major but least studied human pressures on aquatic ecosystems. Besides the general waste burden in waterbodies, (micro)plastic debris gives rise to ecological and social problems. Related to marine ecosystems, these problems are already in the center of interest of science, policy, and public. The United Nations Environment Programme, for instance, drafted a joint report on "marine plastic debris and microplastics," and the European Community included the issue into the European Marine Strategy Framework Directive, descriptor 10 "marine litter."

However, (micro)plastic litter in freshwater systems is not yet explicitly addressed in the respective regulations, although the issue is relevant for many international and national policy instruments and initiatives. Many conventions, agreements, regulations, strategies, action plans, programs, and guidelines refer to "all wastes" in general. This should also concern (micro)plastic waste.

This chapter provides an overview of the regulatory instruments developed at different levels to address freshwater (micro)plastic litter. Beyond that, specific management options and measures that are either compulsory or voluntary are presented. Nevertheless, only few options have been realized so far. Reasons are numerous, first and foremost the lack of consensus on the definition of microplastics.

Nicole Brennholt and **Maren Heß** contributed equally to this work.

N. Brennholt (✉) and G. Reifferscheid
Department Biochemistry, Ecotoxicology, Federal Institute for Hydrology, Am Mainzer Tor 1, 56068, Koblenz, Germany
e-mail: brennholt@bafg.de

M. Heß (✉)
Department Water Management, Water Protection, North Rhine Westphalia State Agency for Nature, Environment and Consumer Protection, Postfach 101052, 45610, Recklinghausen, Germany

M. Wagner, S. Lambert (eds.), *Freshwater Microplastics*,
Hdb Env Chem 58, DOI 10.1007/978-3-319-61615-5_12,

The complexity of these particulate stressors with very heterogeneous physico-chemical characteristics poses new challenges for regulation and management. We highlight the most important questions from the perspective of freshwater monitoring. Furthermore, we discuss a possible adaption of existing environmental policy instruments and potential management options for single categories of (micro)plastics.

Keywords Environmental plastics, Microplastics definition, National–international, Policy instruments, Science–policy interface

1 Introduction

"Microplastics" (MPs) are a topic of discussion in all types of media and are one of the environmental issues also strongly debated by the public (see [1]). A questionnaire sent to the representatives responsible for water monitoring and management in Europe revealed that around 50% of the European population is discussing about MPs and its potential harm to the environment and human health [2]. Hence, the public expects policy-makers to tackle the problem and to manage it as soon as possible.

In fact, awareness about this issue is increasing in policy. Some of the most important and worldwide acting international and intergovernmental bodies are debating about the global problem of environmental plastics (e.g., United Nations, G7, World Bank, World Economic Forum, etc.). Beyond that, the (micro)plastic issue is already addressed in a few regulations and policy instruments on international and national level (see Sects. 2.1 and 2.2). As most environmental MPs result from incorrect disposed and fragmented plastic litter (see [3]), the management of MPs is closely related to a variety of policy areas. Additionally, regulatory responsibilities can change along the product life of a single plastic product and include plastic production and product design, trade and consumer behavior, recycling and waste management (summarized as "land-based policies"), as well as wastewater management and water protection ("water-based policies"). Hence, the regulation of plastics is already considered in several directives, guidelines, agreements, etc. addressing the application of plastic products, starting with regulations on plastic monomers and additives (e.g., REACH;[1] see Sect. 2.2.3). The use of plastic products is especially regulated in sensible application fields, e.g., food packaging. Recently, management strategies increasingly aim at plastics that either are not needed for the function of a product or do not benefit the user or can easily be replaced by other materials – e.g., carrier bags (see Sect. 2.2.2) or MPs in personal care products (see Sect. 2.3). Various directives address the recycling or disposal of plastics at the end of product life (see Sect. 2.2.2).

Given an efficient plastic management, including waste and wastewater control, plastics should not enter environmental systems. However, they do. Problematically, environmental plastics are outside the intended product life.

---

[1]Regulation concerning the Registration, Evaluation, Authorisation and Restriction of Chemicals.

While regulatory measures can be clearly addressed to one stakeholder at a certain stage of product life (e.g., producer, manufacturer, consumer, waste manager), it is more difficult to identify the correct addressee for plastics already released to the environment. So far, due to the complexity of this issue, it is not clear which (policy) areas have to act first, which concepts would be necessary, and what requirements are needed to promote actions beyond those already initiated.

With regard to aquatic environments, (micro)plastics are mainly considered by marine science and policy and, for instance, implemented into the European Marine Strategy Framework Directive (MSFD, [4], descriptor 10 "marine litter"). A comprehensive overview on regulation and management of marine (plastic) litter is provided by Chen [5]. However, it is assumed that approximately 80% of marine debris is land based [6], even though there is a lack of available quantitative evidence supporting this statement. Rivers are one of the entry pathways for (micro)plastics into marine ecosystems. However, the plastic issue is not explicitly addressed in any regulation regarding freshwater environments so far. In contrast to the MSFD, the 8 years older European Water Framework Directive (WFD, [7]) does not include the issue of plastic pollution.

The management of MPs in aquatic systems is even more complex than the regulation of macroplastic litter. Many questions need to be answered, starting with a commonly accepted definition of MPs. Knowledge gaps about sources, transport pathways, and volumes and the environmental fate of the small particles with their heterogeneous characteristics have to be filled, not at least to define adequate methods for a standardized freshwater monitoring of MPs. The adaption of exposure and hazard assessment to evaluate the risk of freshwater MPs as particulate stressors is one of the major challenges for regulation and management. Currently, essential yet unanswered questions refer to the ecological impacts of plastics on today's environment, let alone their long-term consequences.

Notwithstanding, the issue of (micro)plastic pollution in freshwater environments is one of the major but least studied human pressures on aquatic ecosystems, and further research is required on this issue. Nevertheless, there are many indications for adverse environmental impacts that should lead to preventive measures. As stated in Article 191 of the Lisbon Treaty [8], the European Community policy on the environment "[. . .] shall aim at a high level of protection taking into account the diversity of situations in the various regions of the Union. It shall be based on the **precautionary principle** and on the principles that preventive action should be taken, that environmental damage should as a priority be rectified at source and that the polluter should pay." Therefore, regulation and management should deal with the issue of freshwater (micro)plastics.

This chapter provides a rough overview of the existing regulatory instruments developed at international and national levels which address or at least touch the topic of freshwater (micro)plastics. It does not intend to develop new regulatory approaches dealing with the issue but highlights challenges for regulation and management. Despite the regulation of (micro)plastics being already addressed in a few initiatives, it is still far from a comprehensive management. Reasons might be various as (micro)plastics pose new challenges for freshwater monitoring and regulation. This will be discussed in the third section of this chapter. A compilation of the requirements concerning future standards and guidelines is given from the perspective of specialized authorities conducting monitoring programs on regional, state, and national level (Germany).

2 Regulatory and Policy Instruments

This section provides a brief overview of the current regulatory and policy instruments developed at international, regional, and national levels associated with the issue of (micro)plastics in freshwater systems (Fig. 1). National policy instruments apply only to a particular country, whereas regional instruments tackle certain problems within a specific geographical region, e.g., Europe. International regulation and regional agreements, for instance, are transposed into national legislation, so that similar texts can be found in the instruments at the national level.

The interfaces with the legislation are various: direct links are more likely in marine regulation, whereas freshwater (micro)plastics are not explicitly addressed in regulation so far, although this issue is related, for instance, to many European directives. Within this section, an attempt is being made to demonstrate this link. If possible, regulatory strategies for the integration of (micro)plastics in existing legal instruments are proposed. However, the focus is on European regulatory and policy instruments.

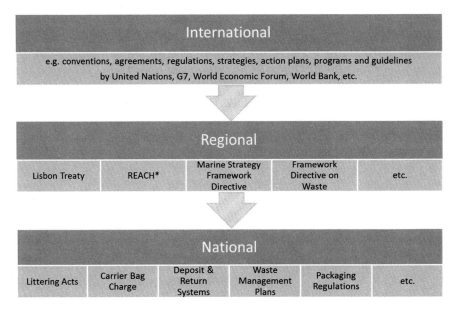

Fig. 1 Levels of regulatory and policy instruments and initiatives for the regulation, assessment, and management of freshwater (micro)plastic litter. * *REACH* Regulation concerning the Registration, Evaluation, Authorisation and Restriction of Chemicals

2.1 International-Level Instruments and Initiatives

International regulation, namely, conventions, agreements, regulations, strategies, action plans, programs, and guidelines, is transposed into regional or national instruments. This is usually done through regional agreements or national legislations so that similar texts can also be found in the instruments at the regional or national level. International instruments apply to the signatory countries and include different geographical regions worldwide.

United Nations (UN)
Based on the Millennium Development Goals, the UN General Assembly adopted the resolution no. A /RES/70/1 "Transforming Our World: The 2030 Agenda for Sustainable Development" on 25 September 2015 [9]. Within this agenda, 17 sustainable development goals with 169 associated targets are announced. Goal 12 "Ensure sustainable consumption and production patterns" includes the following target 12.4: "By 2020, achieve the environmentally sound management of chemicals and all wastes throughout their life cycle, in accordance with agreed international frameworks, and significantly reduce their release to air, water and soil in order to minimize their adverse impacts on human health and the environment" [9, p. 22]. Furthermore, the waste generation shall be substantially reduced by 2030 through prevention, reduction, recycling, and reuse (target 12.5). Although these targets refer to "all wastes" in general, they also cover plastic wastes. The new goals and targets addressed in the agenda have come into effect on 1 January 2016 and will guide the decisions of the member states over the next 15 years. Therefore, the General Assembly "encourages all member states to develop as soon as practicable ambitious national responses to the overall implementation of this Agenda. These can support the transition to the Sustainable Development Goals and build on existing planning instruments, such as national development and sustainable development strategies, as appropriate" [9, p. 33].

The Group of 7 (G7)
The Group of 7 (G7) consists of Canada, France, Germany, Italy, Japan, the UK, and the USA. As environmental issues play an important role alongside economics, foreign policy, and security, topics discussed at G7 summits include climate change, sustainable development, resource efficiency, marine pollution, and nuclear safety. In 2015, the G7 discussed options to address plastic pollution in marine environments and "acknowledge that marine litter, in particular plastic litter, poses a global challenge, directly affecting marine and coastal life and ecosystems and potentially also human health" [10, p. 17]. Among others, the G7 countries are aware of the need for worldwide movement to tackle marine pollution. Therefore, they are developing an action plan to combat marine litter, which provides, inter alia, that practical measures to reduce waste from land- and sea-based sources will be implemented. At this point, it becomes apparent that although marine ecosystems are in the center of interest, freshwater systems cannot be neglected. This is also reflected in the G7 action plan to combat marine litter,

where it is stated that they "support development and implementation of national or regional action plans to reduce waste entering **inland** and coastal waters and ultimately becoming marine litter, as well as to remove existing waste" [11, p. 9]. The G7 action plan lists the following priority actions to address land-based sources:

- Improving countries' systems for waste management, reducing waste generation, and encouraging reuse and recycling;
- Incorporating waste management activities into international development assistance and investments and supporting the implementation of pilot projects where appropriate;
- Investigating sustainable and cost-effective solutions to reduce and prevent sewage and storm water related waste, including micro plastics entering the marine environment;
- Promoting relevant instruments and incentives to reduce the use of disposable single-use and other items, which impact the marine environment;
- Encouraging industry to develop sustainable packaging and remove ingredients from products to gain environmental benefits, such as by a voluntary phase-out of microbeads;
- Promoting best practices along the whole plastics manufacturing, and value chain from production to transport, e. g. aiming for zero pellet loss. [11, p. 10]

The G7 points out that existing platforms and tools for cooperation should be used like the Global Programme of Action for the Protection of the Marine Environment from Land-Based Activities (GPA). The GPA "is the only global intergovernmental mechanism directly addressing the connectivity between terrestrial, freshwater, coastal and marine ecosystems. It aims to be a source of conceptual and practical guidance to be drawn upon by national and/or regional authorities for devising and implementing sustained action to prevent, reduce, control and/or eliminate marine degradation from land-based activities" (retrieved 10.11.2016 from http://www.unep.org/gpa/).

The common understanding of the topic "marine litter" and the most important areas of action and approaches by the G7 can indeed be understood as a step toward an intergovernmental effort against marine litter, but considering that there is a huge potential for reduction regarding litter reaching the sea mainly from land-based sources, litter in freshwater systems should also be in the focus and require concrete measures.

World Economic Forum (WEF)
In January 2016, the World Economic Forum (WEF) published an industry agenda entitled "The New Plastic Economy: Rethinking the Future of Plastics." It states that despite many benefits, the current plastic economy has economic as well as environmental detriments that are becoming more apparent by now: "After a short first-use cycle, 95% of plastic packaging material value, or $80–120 billion annually, is lost to the economy. A staggering 32% of plastic packaging escapes collection systems, generating significant economic costs by reducing the productivity of vital natural systems such as the ocean and clogging urban infrastructure" [12, p. 6]. The agenda assesses the up- and downsides of today's plastic packaging emphasizing the need to rethink the current plastic economy. A new and systemic approach and action plan to achieve a better economic and environmental outcome

are proposed. Thereby, "[t]he circular economy is gaining growing attention as a potential way for our society to increase prosperity, while reducing demands on finite raw materials and minimizing negative externalities" [12, p. 3]. The three main efforts of the New Plastic Economy are to:

1. Create an effective after-use plastics economy by improving the economics and uptake of recycling, reuse and controlled biodegradation for targeted applications. This is the cornerstone of the New Plastics Economy and its first priority, and helps realize the two following ambitions.
2. Drastically reduce leakage of plastics into natural systems (in particular the ocean) and other negative externalities.
3. Decouple plastics from fossil feedstocks by – in addition to reducing cycle losses and dematerializing – exploring and adopting renewably sourced feedstocks. [12, p. 16]

World Bank

The World Bank's Urban Development and Resilience Unit of the Sustainable Development Network produces the Urban Development Series Knowledge Papers to discuss the challenges of urbanization. Within this series, a global review of solid waste management is given by Hoornweg and Bhanda-Tata [13]. This review comprises global management practices, generation, collection, composition, and disposal of waste and compares these across different regions of the world. Furthermore, they did not only estimate global amounts and trends but also make projections on waste generation and composition for the near future in order for decision-makers to prepare accordingly. Further, they describe practical approaches and a range of policy options for governments that could be applied in most cities to encourage waste management practices that will reduce greenhouse gas emissions. They note that "Poorly managed waste has an enormous impact on health, local and global environment, and economy; improperly managed waste usually results in down-stream costs higher than what it would have cost to manage the waste properly in the first place" [13, p. 11].

In 2015, the World Bank established a Pollution Management and Environmental Health (PMEH) program that covers technical assistance and financing for reducing pollution and improving health for all. Three strategic objectives were formulated to progress toward this goal:

- Help selected countries to significantly reduce air, land, and marine pollution levels and thereby improve environmental health outcomes
- Generate new knowledge and improve our understanding of pollution and its health impacts in urban, rural, and marine areas
- Promote increased awareness of environmental health and pollution issues among policy makers, planners, and other relevant stakeholders in low- and middle-income countries (LMICs) through dissemination of scientific evidence in this area, including but not limited to content generated through this program. [14, p. 6]

One component of the PMEH program activities dealing with integrated solid waste management to reduce land-based pollution in marine environments does clearly refer to plastic litter. It addresses issues such as upstream control of solid waste generation to prevent and reduce downstream impacts, focusing on reducing the inflow of plastic litter into marine environments.

2.2 Regional-Level Instruments with Focus on the European Union

Instruments to tackle the problems concerning freshwater (micro)plastics in the European Union are typically regional agreements, regional programs, legislation, or activities dealing with specific problems of freshwater (micro)plastics. The Lisbon Treaty, which aimed at increasing the consistency and coherence of the EU's external actions, stated in Article 191 that the EU policy on the environment shall contribute to preserve, protect, and improve the quality of the environment, protect human health, utilize natural resources in a prudent and rational way, and promote measures at international level to deal with regional or worldwide environmental problems. In this context, the European directives dealing with various aspects of environmental protection can be seen. In the light of the increasing number of scientific publications dealing with the impacts of (micro)plastics on aquatic environments, especially on aquatic organisms, as well as due to the transboundary dimension of plastic pollution, the EU is called upon to develop appropriate policy strategies. As already mentioned, the "water-based policies" such as water protection (MSFD, WFD) and the "land-based policies" such as waste management, plastic production and product design, circular economy, and REACH are affected or likely to be affected by the issue of (micro)plastic pollution of freshwater systems.

2.2.1 Water-Based Policy

The most important directives for the European water policy are the Marine Strategy Framework Directive (MSFD) [4] and the Water Framework Directive (WFD) [7] establishing the legal framework for the protection of European marine and freshwater environments, respectively. Although both aim at implementing a good ecological/environmental status, there is a large discrepancy between them regarding the issue of plastic waste. In the MSFD, waste is defined as one out of 11 qualitative indicators of the good environmental status (descriptor 10 "marine litter"; for further discussion, see [5]), whereas in the WFD, waste is not mentioned. In a possible future revision of the WFD (next review due in 2019), this discrepancy might be clarified.

European Water Framework Directive (WFD)
The WFD has been enacted in October 2000 by the European Commission and focuses on "maintaining and improving the aquatic environment in the Community [. . .] ensuring good [water] quality" [7, p. 2]. Therefore, the amount of pollution entering waterways should be minimized, and the objectives for future water protection should be set. "It does not set exact regulations, but gives each country space to fit the national legislation to put it into practice and arranges and coordinates existing European water legislation" [15, p. 80]. In recital 40, it is noted that

"[w]ith regard to pollution prevention and control, Community water policy should be based on a combined approach using control of pollution at source through the setting of emission limit values and of environmental quality standards" [7, p. 4]. Article 10 describes the combined approach for point and diffuse sources in more detail, "(a) the emission controls based on best available techniques, or (b) the relevant emission limit values, or (c) in the case of diffuse impacts the controls including, as appropriate, best environmental practice" [7, p. 13], and points, in this respect, to further relevant directives.

Even though the 8 years older WFD does not explicitly refer to (micro)plastics or litter in general, Wesch et al. [16] argued that plastic waste is already indirectly integrated in the WFD as it currently stands. In their opinion, litter is broadly associated with relevant quality elements determining the good ecological status of freshwater systems. Consequently, the occurrence of litter, in particular (micro) plastics, could considerably influence the water quality. Furthermore, they point out that a good chemical status of surface waters according to the WFD is achieved when concentrations of listed chemicals (Annex X, WFD) do not exceed the environmental quality standards.

In Article 16, strategies against the pollution of water are mentioned in such a way that "the European Parliament and the Council shall adopt specific measures against pollution of water by individual pollutants or groups of pollutants presenting a significant risk to or via the aquatic environment" [7, p. 17]. Approaches described in Article 16 of the WFD result in a list of priority substances (approved in Annex X). This list registers 45 priority substances or groups of substances, several of which are applied in plastic products such as di (2-ethylhexyl)phthalate, nonylphenol, or octylphenol. As far as priority substances are concerned, the member states are legally obligated to monitor them. However, the measured total concentration of a substance includes all sources of pollution and cannot indicate the plastic-related percentage.

Furthermore, Annex VIII comprised an indicative list of the main pollutants, among others "persistent hydrocarbons and persistent and bioaccumulative organic toxic substances" as well as "substances and preparations, or the breakdown products of such, which have been proved to possess carcinogenic or mutagenic properties or properties which may affect steroidogenic, thyroid, reproduction or other endocrine-related functions in or via the aquatic environment" [7, p. 68]. This might include synthetic polymers and their additives. However, (micro)plastics are not explicitly addressed in the WFD. This discrepancy should be clarified in a possible future revision of the WFD due by 2019, and an assessment system needs to be developed.

Water Protection and Wastewater Treatment Directives

To protect the environment from the adverse effects of urban wastewater discharges and discharges from certain industrial sectors, the European Urban Waste Water Treatment Directive [17] was adopted in 1991. It concerns the collection, treatment, and discharge of domestic effluent or mixture of domestic and certain industrial wastewater (see Annex III of the directive) and/or rainfall water. The issue of (micro)plastic is not included, so that adequate amendments might be needed.

Compared to Europe, the US wastewater regulations established by the Federal Water Pollution Control Act, short Clean Water Act [18], provide the basic structure for regulating discharges of pollutants and regulating quality standards for surface waters. The Clean Water Act refers to regulation of wastewater as well as entry of waste from diffuse sources. Total maximum daily loads of waste are defined aiming at reducing the waste input to freshwater systems. However, it should be noted that, for example, under Californian law, debris less than 5 mm is not considered litter subject to regulation [15, 19, 20]. Accordingly, freshwater MPs, here too, are currently not considered.

2.2.2 Land-Based Policy

Packaging

The Packaging and Packaging Waste Directive [21] calls on the member states to implement return, collection, and recovery systems. The manufacturers, importers, and distributors are directly responsible for reducing packaging waste as well as for developing their own take-back scheme. The "Green Dot Initiative" covering several European countries, for example, collects, sorts, and recycles used packaging. Furthermore, it encourages giving packaging waste a value while being recovered and/or recycled. Hence, the Packaging and Packaging Waste Directive aims not only at "ensur[ing] the functioning of the internal market and to avoid obstacles to trade and distortion and restriction of competition within the Community" *but also at* "prevent[ing] any impact [. . .] on the environment of all Member States as well as of third countries or [. . .] reduc[ing] such impact, thus providing a high level of environmental protection" [21, p. 3].

The Directive 2004/12/EC of the European Parliament and of the Council of 11 February 2004 amending Directive 94/62/EC on packaging and packaging waste [22] aims to ensure that recovery and recycling of packaging waste should be further increased to reduce its environmental impact. With regard to plastics contained in packaging waste, it sets a minimum recycling target of 22.5% by weight no later than 31 December 2008, counting exclusively material that is recycled back into plastics. Annex I gives illustrative examples for criteria of "packaging" referred to in Article 3(1) of [21]. Here, among others, plastic carrier bags are mentioned.

Carrier Bags

Meanwhile, the issue of plastic carrier bags is picked up by another amending directive, known as the Plastic Bags Directive [23]. The Plastic Bags Directive is for the first time considering not only the management of packaging and packaging waste but also its consumption. It was adopted in 2015, and its implementation is currently underway in the member states. It aims at reducing very significantly the use of single-use lightweight plastic carrier bags. The measures to be taken by the member states "may involve the use of economic instruments such as pricing, taxes and levies, which have proved particularly effective in reducing the consumption of plastic carrier bags" [23, p. 2]. The original directive on packaging and packaging waste of 1994 [21] aimed at preventing or reducing the impact of packaging and packaging waste on the environment. Even though plastic carrier bags are included in this directive [21], it does not comprise specific measures on the consumption of such plastic bags.

Waste Legislation
In the present European waste legislation, some strategic elements already exist to tackle the problem of plastic waste in the environment. The Waste Framework Directive [24], for example, relates to issues of product design, life cycle thinking, extended producer responsibility, resource efficiency and conservation, as well as waste prevention through waste operations. This directive aims at "lay[ing] down measures to protect the environment and human health by preventing or reducing the adverse impacts of the generation and management of waste and by reducing overall impacts of resource use and improving the efficiency of such use" [24, p. 6]. It sets general recycling targets for household waste including plastics "[. . .] by 2020, the preparing for re-use and the recycling of waste materials such as at least paper, metal, plastic and glass from households [. . .] shall be increased to a minimum of overall 50 % by weight" [24, p. 11]. Furthermore, in Article 4(1), an explicit waste hierarchy is defined as a priority order in waste prevention and management legislation and policy. It gives precedence to waste prevention; reuse and recycling over recovery, including energy/thermal recovery; and disposal. In addition to the Waste Framework Directive [24], other directives [25–28] also set out recovery and recycling targets.

Another key element in waste management is the extended producer responsibility as described in Article 8 of the Waste Framework Directive [24]. Next to this, it introduces the polluter-pays principle as "guiding principle at European and international levels. The waste producer and the waste holder should manage the waste in a way that guarantees a high level of protection of the environment and human health" [24, p. 4]. Furthermore, "[In] accordance with the polluter-pays principle, the costs of waste management shall be borne by the original waste producer or by the current or previous waste holder" [24, p. 12]. The "polluter-pays principle" is also mentioned in the directive on environmental liability [29] with regard to the prevention and remedying of environmental damage.

According to the waste management hierarchy as laid out in the Waste Framework Directive [24], disposal of waste is the least preferable option and should be limited to the necessary minimum. If disposed waste needs to be landfilled, it has to be sent to landfills, which comply with the requirements of the directive on the landfill of waste [30]. The main objective of this directive is the prevention and reduction of negative effects on the environment, including freshwaters, from the landfilling of waste by introducing strict technical requirements.

In 2014 the European Commission made a legislative proposal [31], which states that "clear environmental, economic and social benefits would be derived from further increasing the targets laid down in Directives 2008/98/EC, 94/62/EC and 1999/31/EC for re-use and recycling of municipal and packaging waste, starting with waste streams which can be easily recycled (e.g. plastics, metals, glass, paper, wood, bio-waste)" [31, p. 9]. With regard to the directive on packaging and packaging waste [21], Article 6 should be amended as follows: "by the end of 2020, the following minimum targets for preparing for re-use and recycling will be met regarding the following specific materials contained in packaging waste: 45% of plastic [...] and by the end of 2025, the following minimum targets for preparing for reuse and recycling will be met regarding the following specific materials contained in packaging waste: 60% of plastic" [31, p. 23]. Concerning the amendment of the directive on the landfill of waste [30], the proposal aims at phasing out landfilling by 2025 for recyclable waste, including plastics, in nonhazardous waste landfills. It is also said that "littering, especially of plastic, has a direct and detrimental impact on the environment and high clean-up costs are an unnecessary economic burden. The introduction of specific measures in waste management plans, financial support from producers within the extended producer responsibility schemes, and proper enforcement from the competent authorities should help eradicate this problem" [31, p. 11].

Circular Economy Package

The discussion on resource efficiency and waste reduction often refers to a systemic change from a linear to a circular economy model (see [32]). In 2015, the European Commission adopted a Circular Economy Package with five priority sectors, among others plastics. It should "stimulate Europe's transition towards a circular economy [...] where resources are used in a more sustainable way" [33, p. 1]. The proposed actions will contribute to "closing the loop" of product life cycles from production and consumption to waste management and the market for secondary raw materials. Concerning the future work on the circular economy, the European Commission schedules a strategy to incentivize plastic recycling ("plastic circular economy strategy") for the following years [34].

Industrial Emissions Directive

In general, the Industrial Emissions Directive (IED) [35] aims at preventing, controlling, and reducing the impact of industrial emissions on the environment (air, water, and land) ensuring a high level of protection for the environment taken as a whole. According to this directive, the guiding principle of sustainable production shall be developed further. For this purpose, an integrative approach takes into account not only pollution emissions but also all production processes to reduce the consumption of resources and energy as well as the environmental damage caused by operation and post-closure of an industrial plant. For this, best available techniques have to be applied. In Annex I, categories of industrial activities giving rise to pollution are listed including the production of organic chemicals, such as plastic materials (polymers, synthetic fibers).

Green Paper on a European Strategy on Plastic Waste in the Environment

In 2013 the European Commission released a Green Paper on a European Strategy on Plastic Waste in the Environment "to launch a broad reflection on possible responses to the public policy challenges posed by plastic waste," because these particular challenges are not specifically addressed in the EU waste legislation at present despite the growing environmental impact of plastic pollution [36, p. 3]. This Green Paper is the first systematic approach to (micro)plastics in the environment at EU level. It explicitly refers to the problem of (micro)plastics and their fate in the environment and the issue of chemicals in and adsorbed to (micro)plastics as well as examines several policy options to improve the management of plastic waste in Europe. The Green Paper addresses the following policy options (as presented by Clayton, 2016[2]):

- *Application of the waste hierarchy to plastic waste management*
- *Achievement of targets, plastic recycling, and voluntary initiatives*
- *Targeting consumer behavior*
- *Toward more sustainable plastics*
- *Durability of plastics and plastic products*
- *Promotion of biodegradable plastics and bio-based plastics*
- *EU initiatives dealing with marine litter including plastic waste*
- *International action*

Thus, in its Green Paper, the European Commission clearly addressed microplastics as part of the waste legislation focusing on mitigation measures.

2.2.3   Chemical Regulation: REACH[3]

For regulating chemical substances, the European REACH regulation [37] has been adopted in 2006. REACH addresses not only the production and use of chemicals but also their potential impacts on both human health and the environment. According to REACH manufacturers, importers and downstream users have to register their chemicals. Furthermore, they are responsible for their safe use. Selected substances are evaluated from public authorities and, if necessary, regulated. Substances of special concern have to go through an authorization procedure. As far as (micro)plastics are concerned, the European REACH Regulation already refers to plastic monomers and additives. The assessment of polymers within REACH is as follows: Because of their high molecular weight, polymer molecules are considered as being of low concern. They are exempted from registration and evaluation, unless the content of (unreacted) monomers exceeds certain limits or they contain certain additives triggering registration and evaluation [38].

---

[2]Presentation by Helen Clayton on the European Conference on plastics in freshwater systems, Federal Press Office, Berlin/Germany, June 21/22, 2016.

[3]Regulation concerning the Registration, Evaluation, Authorisation and Restriction of Chemicals.

2.3   National-Level Instruments

Many national regulations support reducing the amount of (micro)plastic litter in freshwater systems. A selection of these is briefly presented in this section that neither claims to be complete nor to be an assessment. Most of the regulation-based activities aim at reduction actions preventing the environmental plastic pollution.

In the case of preventing littering, there are several regulatory instruments conceivable, for example, the prohibition (of any kind) of littering by prosecuting when disposing litter. Therefore, often a kind of penalty system is established. A non-exclusive list of countries, which have adopted littering acts, is provided in Table 1. As an example, the UK littering act is described below in more detail. In contrast to penalty systems, incentive schemes could be created to encourage a proper return of, e.g., packaging waste.

Table 1   Non-exclusive list of littering acts

Country	Title
Australia:	
• Australian Capital Territory	Litter Regulations 1993
• New South Wales	Protection of the Environment Operations Act 1997
• Northern Territory	Litter Act 1972
• South Australia	Container Deposit Legislation 1977[a]
• Tasmania	Litter Act 2007
• Queensland	Environmental Protection Act 1994
• Victoria	Litter Act 1987
• Western Australia	Litter Act 1979
United States of America (USA):	
• Georgia[b]	Comprehensive Litter Prevention and Abatement Act 2006
• Idaho[b]	Comprehensive Litter Prevention and Abatement Act 2006
• Illinois[b]	Litter Control Act 1974
Ireland	Litter Pollution Act 1997
Jamaica	The Litter Act 1986
Malta	Litter Act 1968
New Zealand	Litter Act 1979
Canada, Saskatchewan	The Litter Control Act 2015
Scotland	Environmental Protection Act 1990 Code of Practice on Litter and Refuse (Scotland) 2006 (COPLAR)
South Africa	White Paper on Integrated Pollution and Waste Management for South Africa 2000
Trinidad and Tobago	Litter Act 1973
United Kingdom (UK)	Environmental Protection Act (EPA) 1990 Clean Neighbourhoods and Environment Act (CNEA) 2005

Most of the acts have been amended since they took effect
[a]Aim of reducing litter by encouraging recycling
[b]http://www.litterbutt.com/stop-litter/litter-laws-by-state.aspx provides a list of US litter laws

In the UK, the Environmental Protection Act (EPA) of 1990 stated that it is an offense to throw down, drop, or otherwise deposit, and then to leave, litter. It enables bans and fines for littering any public places [39]. The Clean Neighbourhoods and Environment Act (CNEA) of 2005 amends the EPA, for example, that a principal litter authority is empowered to specify the amount of a fixed penalty to be applied for a littering offense [40].

In England, the Code of Practice on Litter and Refuse published by the Department for Environment, Food and Rural Affairs in 2006 "applies to all places that are open to the air, including private land and land covered by water. [. . .] There is no restriction on the type of litter for which this may be used, but it is intended primarily to help deal with food and drink packaging and other litter caused by eating 'on-the-go'" [41, p. 42f]. In this code, litter is defined as "materials, often associated with smoking, eating and drinking, that are improperly discarded and left by members of the public; or are spilt during business operations as well as waste management operations" [41, p. 11]. In addition, a law has been passed that requires large shops to charge 5 pence for all single-use plastic carrier bags starting on 5 October 2015. The charge was introduced trying to influence consumer behavior. In the first 6 months since introducing the charge, the plastic bag usage drops to approximately 85% [42]. Wales (started charging in 2011), Northern Ireland (started charging in 2013), and Scotland (started charging in 2014) have also seen a significant drop in plastic bag usage.

In Scotland, The Litter (Fixed Penalties) (Scotland) Order 2013 [43], entering into force in 2014, prescribes fixed penalties for discharging any liability to conviction for the waste (including littering and flytipping) and littering offenses with reference to the EPA. The Scottish Litter Strategy [44], published in 2014 and based on research and extensive consultation, has three main goals to reduce and ultimately prevent litter and flytipping and to encourage personal responsibility and behavior change: "1. Information - improving communications, engagement and education around the issue. 2. Infrastructure - improving the facilities and services provided to reduce litter and promote recycling. 3. Enforcement - strengthening the deterrent effect of legislation and improving enforcement processes" (retrieved 13.11.2016 from http://www.zerowastescotland.org.uk/litter-flytipping/national-strategy).

As already mentioned in the section on regional regulation instruments, the Packaging and Packaging Waste Directive [21] calls on the member states to implement national deposit and return systems, in which disposed plastics are collected and recycled to allow their reuse as new packaging. This should contribute, among others, to a reduction of plastic inputs into freshwater environments. In Denmark, for instance, Dansk Retursystem A/S is such a privately owned nonprofit organization that is regulated by a statutory order (see https://www.dansk-retursystem.dk/). Another example for such a deposit and return system is the Irish company Repak (see https://www.repak.ie/). Deposit and return systems incentivize to correctly dispose (plastic) litter and thus provide the advantage to keep plastics in the economic circle.

Outside Europe, there are further deposit and return systems. In the USA, for instance, container deposit laws, known as bottle bills, are currently implemented in ten states. They require a minimum refundable deposit on beverage containers (usually 5 or 10 cents) in order to promote a high rate of recycling or reuse to reduce waste and prevent littering. By the bottle bills, the refund value of the container provides a monetary incentive to return the container for recycling (see http://www.bottlebill.org/).

Many countries have implemented waste management plans or schemes to prevent and reduce waste production, recover through reuse and recycling, and properly dispose the waste. This helps to prevent environmental pollution including the pollution of freshwater systems.

The Flanders Public Waste Agency (OVAM), developing and monitoring legislation and policies regarding waste management and soil remediation, initiated measures that included promoting source separation, subsidizing the construction of recycling and composting facilities, and discouraging waste. Hence, Flanders, the Flemish region of Belgium, reused, recycled, or composted almost three-fourths of the residential waste produced in this region and has also managed to stabilize waste generation [45]. Furthermore, within the framework of the Flemish Waste Regulation, general regulations that prohibit any kind of littering have been implemented.

The Luxembourgian Waste Management Plan [46] aims at preventing and reducing waste production and pollution from waste; recovering through reuse, recycling, and other environmentally appropriate methods; as well as disposing final waste in an environmentally and economically appropriate way. It set quantitative targets for recovery and recycling including packaging waste. It states that "[o]ther avoided emissions include the benefits of recycling of food and garden waste, paper, glass, metals, plastics, textiles and wood in the municipal solid waste" [47, p. 10].

Regulation instruments do not only address the end (i.e., waste) but also at the beginning of product life or product design. For instance, the UK's The Packaging (Essential Requirements) Regulations 2003 [48] urge the manufacturer to produce the packaging that its "volume and weight be limited to the minimum adequate amount to maintain the necessary level of safety, hygiene and acceptance for the packed product and for the consumer" and to design their products in such a way so as to permit its reuse and recovery and to minimize its environmental impact during the packaging waste disposal. Furthermore, the "Packaging shall be so manufactured that the presence of noxious and other hazardous substances and materials as constituents of the packaging material or of any of the packaging components is minimised with regard to their presence in emissions, ash or leachate when packaging or residues from management operations or packaging waste are incinerated or landfilled" [48, p. 7].

In the USA, for instance, the Microbead-Free Waters Act of 2015 [49], amending the Federal Food, Drug, and Cosmetic Act, prohibits "The manufacture or the introduction or delivery for introduction into interstate commerce of a rinse-off cosmetic that contains intentionally-added plastic microbeads." Here a plastic microbead is defined as "any solid plastic particle that is less than five millimeters in size and is intended to be used to exfoliate or cleanse the human body or any part thereof" (including toothpaste) [49, p. 1]. In this act, the term "microbead" is precisely defined. It refers to primary microplastics and thus provides a direct regulation instrument to tackle the problems related with it, e.g., the pollution of freshwater systems.

3 Challenges of Current Regulation: Reasons and Requirements for Future Management

As shown above, policy-makers are very aware of the problems of environmental plastic waste, and these issues are already considered in several regulatory documents. Nevertheless, most regulations do not clearly refer to microplastics. Therefore, this section aims to highlight the open questions and identify the challenges and the requirements for the future management of MP from the perspective of scientific authorities.

3.1 Do We Need Regulation of Microplastics at All?

Some critics debate the need for a regulation of MP and question whether it is only a "media-made" problem. Indeed, the general public's concern is driven by sensationalized media reports about enormous numbers of MP in the environment. However, there is little scientific data on adverse effects caused by relevant environmental concentrations of MP. Usually, effects were detected in laboratory studies that have tested concentrations far above measured environmental concentrations (see [50] for effects of MP to organisms). So far, only one study reports significant impacts of MP on fish larvae at concentrations found in coastal waters [51].[4]

Nevertheless, MPs occur in almost all types of freshwater environments – ranging from streams in densely populated areas to lakes in almost non-populated areas, e.g., in Mongolia ([52–55], [19, 56, 57]; see [58, 59] for further discussion). Additionally, MPs persist over centuries under common environmental conditions [60]. Thus, in the special case of such extremely persistent pollutants such as MPs, the motivation for any regulatory efforts should not be based solely on the demonstration of adverse effects at current environmental concentrations. If MP input into the environment continues at the current level, environmental concentrations will increase dramatically. Ubiquitous detection, persistency, and continuing release should motivate policy-makers and regulators to act immediately according to the precautionary principle to stop a further plastic accumulation.

---

[4]Note from the editors: The cited publication has been retracted because of scientific misconduct.

Usually, regulation of pollutants in freshwater systems refers to dissolved chemicals, which are different to particulate matter with regard to their environmental fate (e.g., homogeneous versus inhomogeneous distribution). Therefore, we have to critically evaluate the transferability of regulatory options for dissolved chemicals to the issue of MPs. This represents a similar challenge as we know it from engineered nanomaterials. The development of regulation strategies for MP should consider more options than the simple adaptation of the existing regulation strategies for dissolved chemicals or suspended matter. Possibly, entirely new regulation strategies for MP in freshwater need to be developed. To start with, this requires a commonly accepted definition of "microplastics."

3.2 A Precondition for Regulation: The Definition of Microplastics

The term "microplastics" turned into a kind of buzzword in public communication and media, and it is understood as one specific type of pollutant. Hence, expectations rose to find solutions and regulations, which could consider all materials summarized by this single term. In contrast, the term "microplastics" refers to a large group of polymers with various chemical and physical properties, originating from different sources and entering the environment via different pathways (see [3, 59]). Accordingly, these differences among MP particles apply to their environmental fate and persistence and, consequently, also to their bioavailability and potential impacts to organisms.

Verschoor [61] identified five commonly applied criteria to define MP: (1) synthetic materials with high polymer content, (2) solid particles, (3) <5 mm, (4) insoluble in water, and (5) not degradable. However, several points are still under discussion; e.g., some experts are still debating if tire abrasion should be considered as "microplastics" as the monitoring guidance documents for marine litter [62] categorize rubber originating from tires separately from plastics (discussed in more detail in [61]). This decision would significantly influence the measurement results of total environmental MP concentration.

The same applies to the definition of a lower limit for particle size, which is still under discussion. While it is commonly accepted to define all plastic items <5 mm as MPs [e.g., 63], some authors categorize MPs into size-based subgroups. The MSFD Technical Subgroup on Marine Litter [62], for instance, differentiates between larger MPs (1–5 mm) and smaller MPs (20 μm to 1 mm). Various studies set particular methodical limits as a lower size limit – e.g., mesh size of the sampling net or analytical detection limits. As "nano" refers to particles of 1–100 nm [64], the size limit for MPs should consequently start with a lower size limit of 100 nm. Miklos et al. [65] base their size definition on this idea and suggest a size range on "microscale" from 100 nm to 100 μm. Depending on the thresholds defined for these criteria, completely different field concentrations would be obtained. Thus, these standards are fundamental for regulation purposes and should preferably be elaborated scientifically.

The debate about reasonable standards has to face some paradoxical points of discussion: On one hand, a lower limit for particle size would promote the standardization of sampling methods (see Sect. 3.5) and, thus, the elaboration of regulation standards. On the other hand, a lower size limit would exclude small particles from regulation. As MP particles are expected to continuously disintegrate into smaller fragments on sub-micrometer to nanometer scale, present MP particles are future nanoplastic particles, and thus, present regulatory measures on MP will also impact the future concentrations of nanoplastics. Additionally, sources and entry pathways are similar for particles with a wide size range. Therefore, it is questionable to what extent a further differentiation of micro- and nanoplastics is advantageous for the development of regulatory measures. The same plastic item might be documented and assessed as one MP particle in a current monitoring but as many nanoparticles in a future monitoring. From an ecotoxicological perspective, there is also no lower size limit: The smaller the particles are, the more species might potentially ingest those (see [50]). Furthermore, smaller particles can permeate through membranes and, hence, pose a higher risk for adverse effects in organisms. Against this background, it seems unreasonable to exclude small particles from regulation by defining a lower size limit. However, at the same time, general definitions are essential to bring regulatory measures forward. Measures are based on monitoring data, and monitoring again requires standardized and generally accepted methods. Clear guidelines for maximum and minimum particle sizes considered in sampling and analysis are required to generate reliable and legally valid monitoring data. Furthermore, regulation needs to assess the current environmental status with knowledge on the ecotoxicological impacts on organisms. As described, ecotoxicological effects are strongly related to particle size, which determine ingestion, membrane permeation, etc. Apart from size limits, regulators should think about an appropriate categorization of particle size classes. To conclude, final definitions of certain standards are fundamental for regulation purposes and should preferably be elaborated scientifically.

A first attempt to pave the way for future standards has been done by an ad hoc group (AHG) "Microplastics" under the International Organization for Standardization (ISO) Technical Committee (TC) 61 "Plastics." ISO decided to join all forces concerning environmental standards on the plastic issue under this technical committee in order to avoid duplicate work. The scope of this TC is standardization of nomenclature, methods of test, and specifications applicable to materials and products in the field of plastics. The AHG recommended to start a preliminary work item for an ISO technical report "Plastics: Recommendations for the Development of Standards for Investigations of Plastics in the Environment and Biota." It is generally agreed that a global environmental problem needs globally agreed standardization approaches covering the whole range from sampling to effect assessment in order to provide a basis for risk assessment and regulatory options.

3.3 Regulation by Groups?

Besides their size, MP particles vary regarding further physicochemical properties. For regulators, the question follows if a single regulation strategy can address such a comprehensive group of diverse polymers or if it would be more reasonable to tailor regulations specifically to subgroups, especially since (micro)plastics do not solely consist of pure polymers but contain also a number of additives such as plasticizers, UV filters, antioxidants, etc. that alter product properties. Thus, the heterogeneity within the term "microplastics" arises from myriads of combinations of polymers and additives. To make things even more challenging, those additives can change the physicochemical properties and, consequently, also the environmental behavior of particles (for details, see [66]). Therefore, it seems reasonable to develop particular regulatory options focusing on special subgroups. This, in consequence, leads to the question about the main criteria required for a categorization into single groups. Of course, any categorization is depending on the regulatory context and the life stage of a product, as described below. While it can be useful to categorize into very specific subgroups for specific regulatory purposes, in other cases, it might be more efficient to evaluate the whole group of MPs.

As a first approach for MP assessment in freshwater environments, Miklos et al. [65] suggest a modular system starting with the quantification of selected indicator polymers. As soon as the concentration of these polymers exceeds a certain level, more specific analyses should be conducted. These subsequent analyses can take various criteria (such as polymer type, size, shape, additives, etc.) into consideration to further categorize the particles and support the selection of adequate mitigation measures. Here, approaches from chemical regulation might serve as examples.

Chemicals can be categorized based on molecular similarities (e.g., PAHs, PCBs, etc.), by the field of application (e.g., pesticides), or according to their mode of action (e.g., endocrine disruptors). So far, mainly sum parameters for molecularly similar chemicals are implemented to freshwater directives (e.g., dioxin + dlPCB, cyclodiene pesticides, EU WFD). Similarly, MPs could be grouped based on their physicochemical properties (e.g., polymer type, density), by their application fields (e.g., cosmetics, carrier bags, electrical devices), or by (eco)toxicological impacts. The latter might be difficult as little is known about the biological effects of MP, and it will take time to generate comprehensive data (see [50]). In contrast to chemical pollutants, MP can cause both chemical and additionally mechanical effects on organisms. Chemical effects could be caused by the polymers themselves, by their additives, or by a combination of both. Similarly, an occurrence of mechanical effects could depend on particle size, particle shape, or a combination of both. It follows that (eco)toxicologists face the challenge to test the effects of myriads of combinations. Hence, there are efforts to prioritize and start with the presumed most harmful combinations. Ideally, these results will be transferable to a group of similar combinations.

However, ecotoxicity-derived groups are not necessarily suitable for regulation purposes: As each MP particle has unique physicochemical properties (individual polymers, additives, size, shape, etc.), it will induce a unique set of modes of action. Accordingly, one would – in theory – need to perform a multiple stressor assessment of each single particle, which is in itself a complex mixture. From a practical perspective, the integration from multiple stressors in risk assessment is challenging – traditionally each stressor is considered individually. For instance, existing regulations refer to the total concentration of suspended particulate matter (SPM) or for single pollutants adsorbed to SPM. Currently, chemical and particulate parameters are not integrated – as would be required for MP regulation.

Hence, an alternative approach to categorize MP for regulation might by the field of application or by the source for environmental entry. Plastics are used in a wide range of applications including packaging, construction materials, cosmetics, electrical and even medical devices, etc. Obviously, distinct regulatory measures are required to manage the proper recycling of electronic devices compared to throwaway packaging materials or to reduce MP in cosmetics – even though the same polymers might be used in these completely different products. In consequence of their broad use, (micro)plastics enter ecosystems via various pathways. Hence, regulatory measures must not necessarily refer to groups based on MP properties but can also act on groups of sources or entry pathways such as wastewater, incorrect disposal, or agricultural runoff.

As we have seen above, we have different options of grouping MPs. The microbead ban, to name just one example from practice, clearly categorizes MPs by the field of application (cosmetics/personal care products). Which characteristics one select for categorization depends on the regulatory context.

Environmental policy has developed a long list of general and specific management options applicable to a variety of environmental issues (including waste management and water resource management). Some of these might be adopted for the regulation of MPs.

3.4 General Regulation Options by Environmental Policy: Applicable for MP?

Environmental policy aiming at protecting ecosystems and improving the environmental status can be implemented by various regulatory instruments and measures. An intervention can take place on different statutory levels – ranging from voluntary commitments to legally binding bans of certain materials. Furthermore, the interventions can differ regarding the implementation level – including direct regulation of production and application of materials, improvement of waste and water management, and long-term measures aiming at social awareness and changing of behavior. Exhaustive compilations regarding environmental policy instruments are given, e.g., by the OECD [67, 68]. Some of those generally applicable instruments might be transferable to the regulation of MP in freshwater environments. In Fig. 2, some examples are categorized according to the three main types of environmental policy into "regulative," "economic," and "persuasive" instruments. Each category comprises instruments from different statutory and implementation levels. Within each main category, we rated the instruments from "hard" to "soft," depending on how strict measures would affect the application of certain materials. On the horizontal scale, the figure emphasizes the gradient from a "direct" to an "indirect" influence of the regulatory action. The figure does not provide a full compilation of policy tools but rather gives an overview of the range of possible instruments on different implementation levels that could be of interest for the regulation of MP.

In the case of MPs, the choice of regulatory instruments depends on several aspects. It is conceivable that "command and control" (CaC, Fig. 2) instruments such as bans or limitations could either apply to certain polymers, to additives, or to their combination. Similarly, they could be limited to a certain field of application. Measures targeted on certain fields of application could either relate to specific

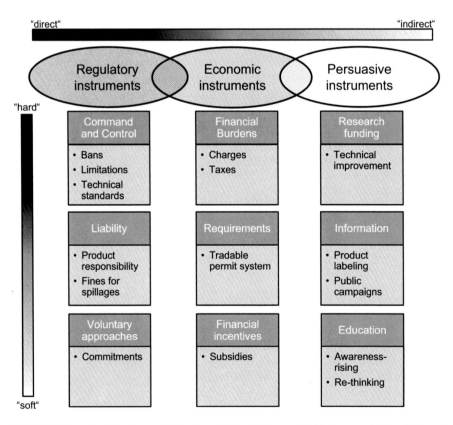

Fig. 2 Instruments of environmental policy, which might be applicable to a regulation of (micro) plastics in freshwater environments (based on [68])

materials or to MPs in general. To give an example, food packaging should not contain any polymers or polymer-additive combinations that could pose any risk to human health by leaching into food products. The same polymer-additive combination might be less harmful in products without direct contact to food and environment. Hence, limitations can be restricted to certain fields of application – provided that proper waste management ensures competent disposal or recycling.

In contrast to the above given example, the Microbead-Free Waters Act 2015 [49] applies not solely to selected materials but bans MPs in general from an application in personal care products (for details, see Sect. 2.3). In rinse-off cosmetic products, where MPs could easily be replaced with natural materials that have similar functions, an entry of MPs into the environment is consciously accepted. "Hard" measures, such as bans and limitations, can potentially be applied to plastic applications that are either not needed for the function of a product or do not benefit the user or can easily be replaced by other materials. In contrast, "softer" instruments need to be applied in areas in which the use of plastics is indisputable (e.g., medical devices).

Regulatory instruments to reduce the emission of MP into the aquatic environment need not necessarily affect production or application but can also be related to an improvement of the management of wastewater and solid waste. Requirements on improved technical standards can be implemented on different levels in a product life cycle. Starting with product design, the range of possibilities includes degradable polymers, polymers with high recycling quotas, or a product design promoting a long and circular product life to reduce waste (see [32]). At the end of product life, enhanced recycling systems can prolong the service time of raw materials to avoid disposal. Any emission of unavoidable waste to the environment needs to be reduced by further regulations. This might be achieved by technical innovations ("CaC") or by a stricter product responsibility from the producer side ("liability").

Economic measures to achieve environmental goals are well known from other fields. They range from imposing financial burden (e.g., taxation) to flexible systems with tradable permits (e.g., CO_2 emission trading) and to financial incentives for increasing recycling rates, to name just a few. With regard to the latter, recycling rates for plastic bottles and further containers for water, soft drinks, milk, etc. increased considerably since the introduction of a container deposit system in several European countries. Deposit systems would be transferable to further plastic-based products (e.g., packaging, carrier bags, etc.)

One of the most sustainable measures would be a social change, with regard to a transformation from a society with linear resource use toward a recycling society, valuing plastics as a precious resource. To achieve such long-term objectives, policy can apply so-called persuasive instruments such as public information, environmental education, and funding of research and development. Compared to the instruments on the left side of Fig. 2, these measures are softer, and effects are less direct. Nevertheless, they might lead to long-lasting input reduction of plastics, MPs, and even further pollutants into the environment.

In case of the heterogeneous group of (micro)plastics, it is important to develop a set of different measures to tackle the problem from different sides. This set might include measures well known from similar environmental issues, ideally complemented with new and specific strategies. The choice of suitable measures depends on the characteristics of the plastics under regulation, on their intended application, and on their current stage of product life cycle. With regard to the latter, various regulatory measures can be applied to one plastic product, as the regulatory responsibilities change during product life. The main stages of product life are schematized in Fig. 3 to emphasize that the management of plastics involves different regulatory authorities as well as various addressees.

As we have seen above, various CaC measures can affect plastic products mainly at the beginning (production, product design) and end (recycling, disposal) of product life. During the actually intended use and function of the product, regulatory measures address consumers and are often realized by financial or persuasive measures. Regulations related to freshwater do usually not concern the intended product life. Plastics usually enter into freshwater systems in consequence of incorrect disposal or insufficient treatment after their intended product life has

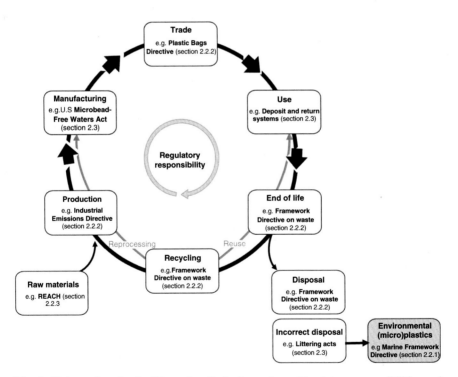

Fig. 3 Main stations in the life cycle of plastic products. Regulatory responsibilities and addressees of regulatory measures change during product life. Examples for regulatory instruments are given for each stage of product life. Environmental plastics mostly occur from incorrect disposal of plastic products after their intended product life has expired

expired. Hence, regulatory measures in early product life stages should minimize plastic waste that could be released into the environment.

However, currently (micro)plastics enter freshwater systems, and monitoring programs need to evaluate the environmental status quo to develop adequate measures. For instance, the type of (micro)plastic and its entry path into the environment should be considered: Depending on their application, plastics enter the environment as macroplastics, secondary microplastics, or primary microplastics (for definition, see [3]). The regulation of primary MP seems to be closer related to the regulation of chemical pollutants: production, application, and entry into the environment are traceable to a certain extent. In freshwater monitoring, source and polluter can potentially be identified, similar to chemical polluters. It has to be noted that the application of some primary MPs even accepts its intended entry into the water cycle, for instance, in the application of MPs in personal care products or as blasting abrasives for cleansing of surfaces (e.g., wheel rims of cars; [69, 70]).

Secondary MPs, in contrast, are not easily traceable due to their various sources and entry pathways. As to their potentially long fate in the environment, the polluter can rarely be identified, and, as a result, it is hardly possible to apply the polluter-pays principle. Secondary MPs usually originate from larger plastic products, which are originally intended to be correctly recycled or disposed. Incorrect disposal (by purpose or because of lacking waste management) leads to fragmentation and distribution of smaller particles in environmental systems. Thus, regulatory measures should intervene before an unintended fragmentation of plastics into MPs can occur. Reasonable strategies should have positive effects on a global level and should be able to prevent the (micro)plastic problems even in regions that lack proper waste management (see [32]).

3.5   Standardization of Sampling and Analysis

The implementation of any regulation measure implies the existence of reliable monitoring data on the status quo and temporal trends in the environment.

In environmental monitoring for regulatory purposes, standardized and harmonized procedures are a prerequisite for reliable, generally accepted, and justiciable data acquisition. On several conferences on plastics in the environment, stakeholders agreed that there is a considerable – not to say a complete – lack of standards for sampling, sample preparation, chemical analysis, and the analysis of biological effects in the field of plastic contamination. A fundamental challenge lies in the fact that the issue of plastic materials in different environmental compartments differs from classical environmental monitoring and assessment issues. Classical monitoring of chemical contamination, e.g., according to the EU WFD, mainly addresses dissolved or particle-bound chemicals in the waterbody or in biota. The plastic contamination issue concerns undissolved material with an extremely inhomogeneous distribution pattern. Up to now, knowledge about representative sampling in wastewater streams and rivers under different hydrological conditions is missing. The picture is similar for sample processing, chemical analysis, sample throughput, and biological effect methods. Crucial criteria like water depth for sampling or mesh size of sampling nets need to be standardized. As mentioned above, such criteria can significantly influence the monitoring results. A marine study conducted in Swedish coastal waters revealed 10^5-fold higher concentrations of MP using 80 µm nets compared to 450 µm nets [71].

Besides a lack of sampling guidelines, no standardized methods for chemical analysis of MP particles in environmental samples exist, so far. The problem starts with sample processing which is a precondition for a precise analysis of plastic particles. Several methods are under development (as summarized in [58]). Some of them are time and work intensive; others are suspected to corrode the plastic items. A generally accepted method for extensive application has still to be developed. However, the required sample cleanup depends on the analysis methods applied. Currently, two main directions of analytical methods are applied to identify MP: spectroscopy and thermogravimetry (see [58]). For both, several techniques exist for the identification of polymer types. However, the choice of a certain technique is determining the outcome less than the choice of the main direction of methods: Spectroscopic methods (e.g., IR microscopy or Raman) can lead to an exact definition of single particles regarding size, shape, color, and main polymer type but are not appropriate for exact mass balancing. In contrast, thermogravimetric methods (e.g., TED-GC-MS or pyrolysis GC-MS) can quantify the exact mass of certain polymers in environmental samples – but thermal degradation of particles does not allow any further characterization of particles. Both directions are appropriate to answer specific questions. (Waste)water management will be more interested in mass balances, while ecology and water conservation will rather ask for an exact description of particle size distribution in order to assess the risk to organisms.

3.6 Mass Balance Versus Particle Characterization as Criterion for Regulation

The choice of mass versus particle concentration depends on the aims of regulation. To give some examples:

The EU MSFD [4] aims to regulate the contribution of plastic waste from single member states to the marine environment via rivers. Here, it seems obvious for the regulation to require information on mass balances instead of particle numbers because (1) the regulation aims to impose financial penalties depending on the contribution of each state to the overall plastic load and (2) plastic particles may disintegrate and break into more pieces on their way through different countries. If regulation should refer to the impact on freshwater ecosystems, it will ask about ecotoxicological effects of MP, for instance. Hence, such regulation requires

Table 2 Overview of the advantages and disadvantages of mass balance versus particle characterization as criteria for regulation

Particle concentration		Mass concentration	
Pro	Contra	Pro	Contra
• Information about size, shape, color → Ecotoxicological relevance → Distinguish primary and secondary MP ➢ Enables source tracking	• No exact mass quantification • Snapshot → Due to continuous fragmentation under environmental conditions	• Exact mass quantification • Fits to conventional regulation options	• No further characterization of particles → Less information
Appropriate for ecotoxicological questions		Appropriate for water management	

information on particle number, shape, and size distribution to assess a potential ingestion by organisms and resulting adverse effects. Table 2 gives an overview on the advantages and disadvantages on both sides.

Apart from any discussion about numbers versus mass balances, further discordances reduce the comparability of monitoring data. Currently, masses or numbers are reported in various units as per m^2 water surface, per m^3 water volume, per m^2 sediment surface, per liter sediment, or per kg sediment, to name just a few. For regulation purpose, comparable units need to be defined and generally applied.

3.7 Adaptation of Ecotoxicological Test Systems

The previous paragraphs discuss the challenges related to the exposure assessment of MPs. Furthermore, comprehensive hazard assessment is required to evaluate the environmental risk of MPs and subsequently formulate reasonable regulation strategies. As described above, particulate pollutants behave differently than dissolved chemicals, and thus, an adaption of test systems of toxicity tests for dissolved chemicals could be required. With regard to this, there is some experience from studies on engineered nanomaterials (see [72]). Researchers can learn from those experiences; however, they need to consider that physicochemical properties of MP might lead to again different behavior in test systems. High-density polymers will sink to sediments. Hence, sediment-living species are required – comparable to the testing of highly lipophilic, sediment-bound chemicals. Vice versa, low-density polymers will float and will only be available for surface-feeding organisms. Those will be at higher risk in environmental systems as they might feed selectively on floating materials and accumulate them from the water phase. Chemical testing is usually not focused on this feeding type.

Furthermore, MP particles can impact organisms in various manners – chemically and mechanically (see [50]). One particle can be seen as a multiple stressor itself. Hence, ecotoxicologists have to face different challenges and to adapt common test systems and endpoints. Potentially, the chosen endpoints of mono-substance test designs simply overlook the effects of MP. Fundamental research would be required to formulate the right questions and to adapt or develop suitable test designs with adequate endpoints. This also applies to biomarkers used in field studies.

Above all, scientists should consider that organisms are adapted to natural particles of different materials (sand, clay, or similar), but with similar properties as MP, in their natural habitats. It is crucial to perform tests on MP particles in comparison with such natural particles. This applies especially to tests regarding the "Trojan horse effect" – the transport of hydrophobic substances via MP into organisms (see [50]). Studies need to address whether there are differences in the sorption of chemicals to MP versus natural particles and, consequently, in the impact to organisms. In fact, there are regulations for priority substances bound to suspended matter – and MP particles are included in the suspended matter. Unfortunately, most ecotoxicological studies lack a direct comparison of suspended matter spiked with chemicals toward spiked MP particles. Knowledge about those aspects could help to prioritize regulatory questions.

3.8 The Information Base for Regulation: Too Scarce? Too Much? Inapplicable?

Even though scientists continuously provide new findings about MPs, it still provides a huge challenge for numerous scientific fields. The group of MPs comprises particles with countless physicochemical properties determining their environmental fate and risk to organisms. It seems to be a playground for researchers to investigate open questions ranging from degradation process, uptake by organisms, and interaction with chemicals to special surface properties of aged plastics. Knowledge about those aspects is fundamental but in sum too complex to be considered for regulation.

We have to formulate what information is needed for regulation and what kind exceeds the scope of generalized regulation instruments. While the current lack of knowledge is obvious, this should not serve as a general excuse for delaying an implementation of regulation instruments for those persistent materials. To refer to the initial statements of this chapter, we have to ask ourselves which kind of knowledge is required to justify the need for regulation of highly persistent MP?

While we can clearly state that regulatory strategies for a reduction of environmental (micro)plastics are urgently needed, many questions about the implementation of monitoring and regulatory strategies are still open:

- How to define microplastics?
- Can we adopt existing regulative options or do we need to develop new strategies?
- Which criteria can categorize MPs for regulatory purpose?
- Which particles have to be regulated with priority?
- Which monitoring methods can adequately answer regulatory questions?

- How should standardization of monitoring methods look like?
- How can ecotoxicological test designs be adopted for an assessment of combined chemical and mechanical effects?
- What are the relevant sources and pathways into the environment?
- Which measures can reduce MP entry into the environment?

However, the list of questions could be continued far beyond the one above – especially with regard to different perspectives as human toxicology, drinking water supply, etc. exceeding the scope of this freshwater-related chapter. Nevertheless, it emphasizes the need of interdisciplinary cooperation to address the issue of freshwater (micro)plastics.

4 Conclusions

This chapter provided a rough overview of the existing regulation instruments developed at international, regional, and national levels to address freshwater (micro)plastics. While several regulations address plastics, concrete regulations on microplastics – especially with regard to freshwater systems – are rare. Hence, we discussed possible reasons for that and formulated a list of questions to be answered with priority.

Despite many open questions, we want to conclude that:

- In our point of view, regulation of freshwater (micro)plastics is urgently required.
- An important step toward the management of environmental (micro)plastics has been accomplished by awareness raising in society and policy.
- International and intergovernmental bodies already discuss measures to reduce environmental plastics (e.g., UN, G7, World Bank, World Economic Forum)
- So far, policy-makers integrated the (micro)plastic issue into a few regulatory directives on international and national level.
- These regulations concern diverse fields of policy (e.g., chemical regulation, waste management, water resource management).
- Environmental policy provides a long list of instruments, which might be adopted to develop further management options for the issue of MP.

Nevertheless, further research should be promoted to fill current knowledge gaps. The compilation of challenges for regulation and management as presented in Sect. 3 highlights the most important needs from the perspective of freshwater monitoring. Thus, the key issues to be tackled in a systematic approach are:

- Microplastics are a heterogeneous group of pollutants.
- Hence, a definition of regulatory (sub)groups is as important as the definition of MPs itself in order to define management options more precisely.
- As MPs are mainly derived from larger plastic items, its management needs to be closely linked to the regulation of plastic production, consumption, and litter.

- MPs in the freshwater environment are already outside their intended product cycle. Hence, it is difficult to address responsible stakeholders (producer, consumer, waste manager, etc.).
- Thus, it is even more important to clarify entry pathways into aquatic environment, to define standardized methods for exposure and hazard assessment, and to work in an integrated approach.

An adequate regulation of environmental (micro)plastics is a huge challenge for research and policy. As plastics influence all parts of society, single fields of science or policy cannot tackle this issue individually. During the lifetime of a single plastic product – from design and production to trade and consumption to the correct recycling or disposal at the end of its functional product life – regulatory responsibilities change. This provides various possibilities for regulators to intervene before plastics enter the environment. However, it requires an interdisciplinary coordination of measures on different statutory, political, economic, and social levels.

Only the interplay between all stakeholders from all countries results in success. (Micro)plastic particles do not respect political frontiers and, thus, accumulate in interregional waterbodies. For this reason, the need to treat this emerging environmental issue in an international context is increasing. Although – or precisely because – we currently know little about the consequences of MPs in aquatic systems, we should develop and implement measures to reduce further emissions. This is especially true regarding the high persistence and accumulation of these materials in the environment and in accordance with the precautionary principle.

References

1. Kramm J, Völker C (2017) Understanding the risks of microplastics. A social-ecological risk perspective. In: Wagner M, Lambert S (eds) Freshwater microplastics: emerging environmental contaminants? Springer, Heidelberg. doi:10.1007/978-3-319-61615-5_11 (in this volume)
2. Breuninger E, Bänsch-Baltruschat B, Brennholt N, Hatzky S, Reifferscheid G, Koschorreck J (2016) Plastics in European freshwater environments. Issue paper (final version). In: Bänsch-Baltruschat B, Brennholt N, Kochleus C, Reifferscheid G, Koschorreck J Conference on plastics in freshwater environments. pp 16–71. UBA Dokumentationen 05/2017. ISSN 2199-6571
3. Lambert S, Wagner M (2017) Microplastics are contaminants of emerging concern in freshwater environments: an overview. In: Wagner M, Lambert S (eds) Freshwater microplastics: emerging environmental contaminants? Springer, Heidelberg. doi:10.1007/978-3-319-61615-5_1 (in this volume)
4. EU (2008) Directive 2008/56/EC of the European Parliament and of the Council of 17 June 2008 establishing a framework for community action in the field of marine environmental policy. Marine Strategy Framework Directive (MSFD). http://eur-lex.europa.eu. Retrieved 4 May 2009
5. Chen C-L (2015) Regulation and management of marine litter. In: Bergmann M, Gutow L, Klages M (eds) Marine anthropogenic litter. Springer, Berlin, pp 185–200

6. Gordon M (2006) Eliminating Land-based discharges of Marine Debris. In: California State Water Resources Control Board. California: A Plan of Action from The Plastic Debris Project. Sacramento
7. EU (2000) Directive 2000/60/EC of the European Parliament and of the Council of 23 October 2000 establishing a framework for Community action in the field of water policy. Water Framework Directive (WFD). http://eur-lex.europa.eu. Retrieved 4 May 2009
8. EU (2007) The Lisbon Treaty. http://www.lisbon-treaty.org/wcm/. Retrieved 09 Nov 2016
9. UN (2015) Transforming our world: the 2030 Agenda for Sustainable Development. Resolution No. A /RES/70/1 adopted by the General Assembly of the United Nations on 25 Sep 2015
10. G7 (2015) Leaders' Declaration G7 Summit, 7–8 June 2015. Schloss Elmau, Germany, pp 17–18
11. G7 (2015) Annex to the Leaders' Declaration G7 Summit, 7–8 June 2015. Schloss Elmau, Germany, pp 9–11
12. WEF (2016) The New Plastics Economy: Rethinking the future of plastics. Industry Agenda by the World Economic Forum. p 36. https://www.weforum.org/reports/the-new-plastics-economy-rethinking-the-future-of-plastics/
13. Hoornweg D, Bhada-Tata P (2012) What a Waste - A Global Review of Solid Waste Management. Urban Development Series Knowledge Papers No. 15. p 116. http://siteresources.worldbank.org/INTURBANDEVELOPMENT/Resources/336387-1334852610766/What_a_Waste2012_Final.pdf
14. PMEH (2016) Supporting Pollution Action for Health. Pollution Management & Environmental Health Program Annual Report 2016. p 56. http://documents.worldbank.org/curated/en/905491479734253523/pdf/110353-AR-PMEHAnnualRprtFINALWEBHI-PUBLIC.pdfREACH. Regulation (EC) No 1907/2006 of the European Parliament and of the Council of 18 December 2006 concerning the Registration, Evaluation, Authorisation and Restriction of Chemicals
15. Gorycka M (2009) Environmental risks of microplastics. Faculteit der Aard- en Levenswetenschappen, Vrije Universiteit Amsterdam, Amsterdam, p 171
16. Wesch C, Stöfen A, Klein R, Paulus M (2014) Microplastics in freshwater environments: a need for scientific research and legal regulation in the context of the European water framework directive. Zeitschrift für Europäisches Umwelt- und Planungsrecht 12:258–274
17. EU (1991) Council Directive of 21 May 1991 concerning urban waste water treatment (91/271/EEC) (Urban Waste Water Treatment Directive). http://eur-ex.europa.eu. Retrieved 4 May 2009
18. U.S. (2002) Federal Water Pollution Control Act [As Amended Through P.L. 107–303, November 27, 2002]. Clean Water Act (CWA). https://www.epw.senate.gov/water.pdf. Retrieved 12 Nov 2016
19. Moore CJ, Lattin GL, Zellers AF (2011) Quantity and type of plastic debris flowing from two urban rivers to coastal waters and beaches of Southern California. J Integr Coast Zone Manag 11:65–73
20. Moore CJ (2008) Synthetic polymers in the marine environment: a rapidly increasing, long-term threat. Environ Res 108:131–139
21. EU (1994) European Parliament and Council Directive 94/62/EC of 20 December 1994 on packaging and packaging waste. http://eurlex.europa.eu. Retrieved 12 Nov 2016
22. EU (2004) Directive 2004/12/EC of the European Parliament and of the Council of 11 February 2004 amending Directive 94/62/EC on packaging and packaging waste. http://eurlex.europa.eu. Retrieved 12 May 2009
23. EU (2015) Directive (EU) 2015/720 of the European Parliament and of the Council of 29 April 2015 amending Directive 94/62/EC as regards reducing the consumption of lightweight plastic carrier bags. http://eurlex.europa.eu. Retrieved 11 Nov 2016
24. EU (2008) Directive 2008/98/EC of the European Parliament and of the Council of 19 November 2008 on waste and repealing certain Directives. http://eur-lex.europa.eu. Retrieved 12 Nov 2016

25. EU (1994) European Parliament and Council Directive 94/62/EC of 20 December 1994 on packaging and packaging waste. http://eur-lex.europa.eu. Retrieved 12 Nov 2016
26. EU (2000) Directive 2000/53/EC of the European Parliament and of the Council of 18 September 2000 on end-of life vehicles. http://eur-lex.europa.eu. Retrieved 12 Nov 2016
27. EU (2002) Directive 2002/96/EC of the European Parliament and of the Council of 27 January 2003 on waste electrical and electronic equipment (WEEE). http://eur-lex.europa.eu. Retrieved 12 Nov 2016
28. EU (2006) Directive 2006/66/EC of the European Parliament and of the Council of 6 September 2006 on batteries and accumulators and waste batteries and accumulators and repealing Directive 91/157/EEC. http://eur-lex.europa.eu. Retrieved 12 Nov 2016
29. EU (2004) Directive 2004/35/CE of the European Parliament and of the Council of 21 April 2004 on environmental liability with regard to the prevention and remedying of environmental damage. http://eur-lex.europa.eu. Retrieved 08 Nov 2016
30. EU (1999) Council Directive 1999/31/EC of 26 April 1999 on the landfill of waste. http://eur-lex.europa.eu. Retrieved 12 Nov 2016
31. European Commission (2014) COM/2014/0397. Proposal for a Directive of the European Parliament and of the Council amending Directives 2008/98/EC on waste, 94/62/EC on packaging and packaging waste, 1999/31/EC on the landfill of waste, 2000/53/EC on end-of-life vehicles, 2006/66/EC on batteries and accumulators and waste batteries and accumulators, and 2012/19/EU on waste electrical and electronic equipment. http://eur-lex.europa.eu. Retrieved 12 Nov 2016
32. Eriksen M, Thiel M, Prindiville M, Kiessling T (2017) Microplastic: what are the solutions? In: Wagner M, Lambert S (eds) Freshwater microplastics: emerging environmental contaminants? Springer, Heidelberg. doi:10.1007/978-3-319-61615-5_13 (in this volume)
33. European Commission (2015) Closing the loop: Commission adopts ambitious new Circular Economy Package to boost competitiveness, create jobs and generate sustainable growth. Press release, Brussels, 2 December 2015, IP/15/6203. http://europa.eu/rapid/press-release_IP-15-6203_en.htm. Retrieved 13 Nov 2016
34. Plastics circular economy strategy not due until end of 2017. ENDS Europe DAILY. ISSN 1463–1776, www.endseurope.com/article/47616/
35. EU (2010) Directive 2010/75/EU of the European Parliament and of the Council of 24 November 2010 on industrial emissions (integrated pollution prevention and control). http://eur-lex.europa.eu. Retrieved 13 Nov 2016
36. European Commission (2013) Green Paper on a European Strategy on Plastic Waste in the Environment. /* COM/2013/0123 final */. Document 52013DC0123. http://eur-lex.europa.eu/legal-content/EN/TXT/?uri=CELEX:52013DC0123. Retrieved 10 Nov 2016
37. EU (2006) Regulation (EC) No 1907/2006 of the European Parliament and of the Council of 18 December 2006 concerning the Registration, Evaluation, Authorisation and Restriction of Chemicals (REACH), establishing a European Chemicals Agency, amending Directive 1999/45/EC and repealing Council Regulation (EEC) No 793/93 and Commission Regulation (EC) No 1488/94 as well as Council Directive 76/769/EEC and Commission Directives 91/155/EEC, 93/67/EEC, 93/105/EC and 2000/21/EC. http://eur-lex.europa.eu. Retrieved 11 Nov 2016
38. ECHA (2012) Guidance for monomers and polymers. ECHA-12-G-02-EN. European Chemicals Agency. http://echa.europa.eu/. Retrieved 11 Nov 2016
39. UK (1990) Environmental Protection Act 1990. http://www.legislation.gov.uk. Retrieved 13 Nov 2016
40. UK (2005) Clean Neighbourhoods and Environment Act 2005. http://www.legislation.gov.uk. Retrieved 13 Nov 2016
41. DEFRA (2006) Code of Practice on Litter and Refuse 2006. https://www.gov.uk/government/uploads/system/uploads/attachment_data/file/221087/pb11577b-cop-litter.pdf. Retrieved 13 Nov 2016

42. https://www.gov.uk/government/publications/carrier-bag-charge-summary-of-data-in-england-for-2015-to-2016/single-use-plastic-carrier-bags-charge-data-in-england-for-2015-to-2016. Data Retrieved 13 Nov 2016
43. UK (2013) The Litter (Fixed Penalties) (Scotland) Order 2013. http://www.legislation.gov.uk. Retrieved 13 Nov 2016
44. Scottish Government (2014) Zero Waste - Towards A Litter-Free Scotland: A Strategic Approach To Higher Quality Local Environments. http://www.gov.scot/Resource/0045/00452542.pdf. Retrieved 13 Nov 2016
45. Allen C (2012) Flanders, Belgium: Europe's Best Recycling and Prevention Program. Global Alliance for Incinerator Alternatives (GAIA)
46. PGGD (2010) Plan général de gestion des déchets. Published by the Government of Luxembourg. Ministry of Sustainable Development and Infrastructures. Environment Administration. Luxembourg, 404 pp
47. European Environment Agency (2013) Municipal waste management in Luxembourg. ETC/SCP working paper prepared by Emmanuel C. Gentil, Copenhagen Resource Institute
48. UK (2003) The Packaging (Essential Requirements) Regulations 2003. http://www.legislation.gov.uk. Retrieved 13 Nov 2016
49. U.S. (2015) Microbead-Free Waters Act of 2015. https://www.congress.gov. Retrieved 13 Nov 2016
50. Scherer C, Weber A, Lambert S, Wagner M (2017) Interactions of microplastics with freshwater biota. In: Wagner M, Lambert S (eds) Freshwater microplastics: emerging environmental contaminants? Springer Nature, Heidelberg. doi:10.1007/978-3-319-61615-5_8 (in this volume)
51. Lönnstedt OM, Eklöv P (2016) Environmentally relevant concentrations of microplastic particles influence larval fish ecology. Science 352:1213–1216
52. Dris R, Gasperi J, Rocher V, Saad M, Renault N, Tassin B (2015) Microplastic contamination in an urban area: a case study in Greater Paris. Environ Chem 12:592–599
53. Mani T, Hauk A, Walter U, Burkhardt-Holm P (2015) Microplastics profile along the Rhine river. Sci Rep 5:17988
54. Free CM, Jensen OP, Mason SA, Eriksen M, Williamson NJ, Boldgiv B (2014) High-levels of microplastic pollution in a large, remote, mountain lake. Mar Pollut Bull 85:156–163
55. McCormick A, Hoellein TJ, Mason SA, Schluep J, Kelly JJ (2014) Microplastic is an abundant and distinct microbial habitat in an urban river. Environ Sci Technol 48:11863–11871
56. Zbyszewski M, Corcoran P (2011) Distribution and degradation of fresh water plastic particles along the beaches of Lake Huron, Canada. Water Air Soil Pollut 220:365–372
57. Eriksen M, Mason S, Wilson S, Box C, Zellers A, Edwards W, Farley H, Amato S (2013) Microplastic pollution in the surface waters of the Laurentian Great Lakes. Mar Pollut Bull 77:177–182
58. Klein S, Dimzon IK, Eubeler J, Knepper TP (2017) Analysis, occurrence, and degradation of microplastics in the aqueous environment. In: Wagner M, Lambert S (eds) Freshwater microplastics: emerging environmental contaminants? Springer, Heidelberg. doi:10.1007/978-3-319-61615-5_3 (in this volume)
59. Dris R, Gasperi J, Tassin B (2017) Sources and fate of microplastics in urban areas: a focus on Paris Megacity. In: Wagner M, Lambert S (eds) Freshwater microplastics: emerging environmental contaminants? Springer, Heidelberg. doi:10.1007/978-3-319-61615-5_4 (in this volume)
60. Barnes DKA, Galgani F, Thompson RC, Barlaz M (2009) Accumulation and fragmentation of plastic debris in global environments. Philos Trans R Soc B 364:1985–1998
61. Verschoor AJ (2015) Towards a definition of microplastics. Considerations for the specification of physico-chemical properties. RIVM Letter report 2015–0116
62. JRC – Joint Research Centre of the European Commission (2013) MSDF Technical Subgroup on Marine Litter. Guidance on Monitoring of Marine Litter in European Seas. A guidance

document within the Common Implementation Strategy for the Marine Strategy Framework Directive. Report No. EUR 26113 EN

63. Arthur C, Baker J and Bamford H 2009. Proceedings of the International Research Workshop on the Occurrence, Effects and Fate of Microplastic Marine Debris. 9–11 Sep 2008. NOAA Technical Memorandum NOS-OR&R-30

64. EU (2011) Commission Recommendation of 18 October 2011 on the definition of nanomaterial (2011/696/EU). http://eur-lex.europa.eu. Retrieved 13 Nov 2016

65. Miklos D, Obermaier N, Jekel M (2016) Mikroplastik: Entwicklung eines Umweltbewertungskonzepts – Erste Überlegungen zur Relevanz von synthetischen Polymeren in der Umwelt. UBA Texte 32/2016. ISSN 1862–4804

66. Lambert S, Scherer C, Wagner M (2017) Ecotoxicity testing of microplastics: considering the heterogeneity of physicochemical properties. Integr Environ Assess Manag 13:470. doi:10.1002/ieam.1901

67. OECD (2016) Environmental policy tools and evaluation. http://www.oecd.org/env/tools-evaluation/. Retrieved 28 Nov 2016

68. OECD (2001) Sustainable Development: Critical Issues. Paris, France

69. Essel, R., Engel, L., Carus, M., Ahrens, R.H. (2015) Sources of microplastics relevant to marine protection in Germany. UBA Texte 64/2015. Umweltbundesamt. Dessau-Roßlau, Germany. http://www.umweltbundesamt.de/publikationen/sources-of-microplastics-relevant-to-marine

70. Lassen C, Hansen SF, Magnusson K, Hartmann NB, Rehne Jensen P, Nielsen TG, Brinch A (2015) Microplastics: occurrence, effects and sources of releases to the environment in Denmark. Danish Environmental Protection Agency, Copenhagen K

71. Norén F (2007) Small plastic particles in Coastal Swedish waters. KIMO, Sweden

72. Rist SE, Hartmann NB (2017) Aquatic ecotoxicity of microplastics and nanoplastics: lessons learned from engineered nanomaterials. In: Wagner M, Lambert S (eds) Freshwater microplastics: emerging environmental contaminants? Springer, Heidelberg. doi:10.1007/978-3-319-61615-5_2 (in this volume)

Microplastic: What Are the Solutions?

Marcus Eriksen, Martin Thiel, Matt Prindiville, and Tim Kiessling

Abstract The plastic that pollutes our waterways and the ocean gyres is a symptom of upstream material mismanagement, resulting in its ubiquity throughout the biosphere in both aquatic and terrestrial environments. While environmental contamination is widespread, there are several reasonable intervention points present as the material flows through society and the environment, from initial production to deep-sea microplastic sedimentation. Plastic passes through the hands of many stakeholders, with responsibility for environmental contamination owned, shared, or rejected by plastic producers, product/packaging manufacturers, government, consumers, and waste handlers.

The contemporary debate about solutions, in a broad sense, largely contrasts the circular economy with the current linear economic model. While there is a wide agreement that improved waste recovery is essential, how that waste is managed is a different story. The subjective positions of stakeholders illuminate their economic philosophy, whether it is to maintain demand for new plastic by incinerating

M. Eriksen (✉)
5 Gyres Institute, Los Angeles, CA, USA
e-mail: marcuseriksen@gmail.com

M. Thiel
Facultad Ciencias del Mar, Universidad Católica del Norte, Larrondo 1281, Coquimbo, Chile

Millennium Nucleus Ecology and Sustainable Management of Oceanic Island (ESMOI), Coquimbo, Chile

Centro de Estudios Avanzados en Zonas Áridas (CEAZA), Coquimbo, Chile
e-mail: thiel@ucn.cl

M. Prindiville
Upstream Policy Project, Damariscotta, ME, USA

T. Kiessling
Facultad Ciencias del Mar, Universidad Católica del Norte, Larrondo 1281, Coquimbo, Chile

M. Wagner, S. Lambert (eds.), *Freshwater Microplastics*,
Hdb Env Chem 58, DOI 10.1007/978-3-319-61615-5_13,
© The Author(s) 2018

postconsumer material or maintain material efficacy through recycling, regulated design, and producer responsibility; many proposed solutions fall under linear or circular economic models. Recent efforts to bring often unheard stakeholders to the table, including waste pickers in developing countries, have shed new light on the life cycle of plastic in a social justice context, in response to the growing economic and human health concerns.

In this chapter we discuss the main solutions, stakeholder costs, and benefits. We emphasize the role of the "honest broker" in science, to present the best analysis possible to create the most viable solutions to plastic pollution for public and private leadership to utilize.

Keywords Extended producer responsibility, Marine debris solutions, Microplastic, Plastic marine pollution, Recycling, Reuse

1 Research Conclusions Guide Solutions

Since 2010 there have been more research publications about plastic marine pollution than in the previous four decades, bringing the issue mainstream as a robust field of science and in public discourse. Much of what we know can be summarized in three conclusions: fragmented plastic is globally distributed, it is associated with a cocktail of hazardous chemicals and thus is another source of hazardous chemicals to aquatic habitats and animals, and it entangles and is ingested by hundreds of species of wildlife at every level of the food chain including animals we consider seafood [1].

Global Distribution of Microplastics The global distribution of plastics is a result of the fragmentation and transportation by wind and currents to the aquatic environment, from inland lakes and rivers to the open ocean and likely deposition to coastlines or the seafloor [2]. New studies are showing increasing abundances of microplastic upstream, showing that microplastic formation is not limited to the sea, though it was discovered there.

The first observations of plastic in the ocean were made in 1972 in the western North Atlantic consisting of preproduction pellets and degraded fragments found in plankton tows [3]. Studies in the North Pacific [4, 5], and South Atlantic followed [6]. Scientists were beginning to understand the global implications of fragmented plastics traveling long distances. "Data from our oceanic survey suggests that plastic from both intra- and extra-gyral sources becomes concentrated in the center of the gyre, in much the same fashion that *Sargassum* does [7]."

In 2001 Captain Charles Moore published his discovery of an accumulation of microplastics in the North Pacific Subtropical Gyre [8]. This finding might have joined the trickle of research that had been published in the previous quarter century, but sensationalized media stories reported fictional islands of trash converging in the ocean that were forming garbage patches twice the size of Texas.

This subsequently catalyzed the attention, interest, and concern of the public, policymakers, industry, and science.

Regional and global estimates of floating debris have come forth [9, 10]. Estimates of environmental concentrations have ranged from 8 million tons of plastic leaving shorelines globally each year [11], compared to one estimate of a quarter million tons drifting at sea [12]. This represents a huge disparity suggesting that plastics sink, wash back ashore, or fragment long before they arrive in the subtropical gyres. Analysis of the size distribution of plastic in the oceans has found hundred times less microplastics than expected [10, 12], supporting the suggestion that fragmented microplastics do not survive at the sea surface indefinitely and likely invade marine food chains before moving subsurface to be captured by deeper circulating currents and ultimately deposited as sediment. Recent studies have unveiled microplastics frozen in sea ice [13] and deposited on shorelines worldwide [14] and across the sea floor [15, 16], even the precipitation of synthetic fibers as fallout from the skies [17]. Collectively, these observations suggest widespread contamination in all environments.

Inherent Toxicity and the Sorption of Pollutants While plastic products entering the ocean represent a range of varied polymers and plasticizers, many absorb (taking in) and adsorb (sticking to) other persistent organic pollutants and metals lost to the environment, resulting in a long list of toxicants associated with plastic debris [18]. Gas stations will sometimes use giant mesh socks full of polyethylene pellets draped around storm drains to absorb hazardous chemicals before they reach the watershed. In the aquatic environment, plastic behaves similarly, mopping up chemicals in surrounding water. Several persistent organic pollutants (POPs) bind to plastic as it is transported through the watershed, buried in sediment, or floating in the ocean [19, 20]. A single pellet may attract up to one million times the concentration of some pollutants in ambient seawater [21], and these chemicals may be available to marine life upon ingestion.

The chemistry of plastic in consumer products raises human health as well as ecological concerns. For example, they include polyfluorinated compounds ("PFCs") [22–24] and the pesticide/sanitizer triclosan [25, 26], also used in over-the-counter drugs, antimicrobial hand soaps and some toothpaste brands, flame retardants, particularly PBDEs [27, 28], and nonylphenols. Bisphenol A (BPA), the building block of polycarbonates, and phthalates – the plastic additive that turns hardened PVC into pliable vinyl – are both known endocrine disruptors [29, 30].

This is not surprising in the case of BPA, which was invented as a synthetic estrogen [31], yet proved to be a usable form of plastic. Exposure may come from the lining of metal cans for food storage [32], CDs, DVDs, polycarbonate dishware, and receipt paper from cash registers. BPA has been linked to many developmental disruptions, including early puberty, increased prostate size, obesity, insulin inhibition, hyperactivity, and learning disabilities [33]. Phthalates are similarly problematic as endocrine disruptors [34], with effects including early puberty in females, feminization in males, and insulin resistance [35]. Different phthalates are found in paints, toys, cosmetics and food packaging, added for the purpose of increasing durability, elasticity, and pliability. In medical applications, such as IV bags and tubes, phthalates are prone to leaching after long storage, exposure to elevated temperatures, and as a result of the high concentration present – up to 40% by weight [36]. Although phthalates metabolize quickly, in a week or less, we are exposed continuously through contact with associated products.

Widespread Effects on Marine Life Of the 557 species documented to ingest or entangle in our trash, at least 203 [1] of them are also ingesting microplastic in the wild, of which many are fish [37] and other vertebrates [38, 39]. In addition, laboratory data suggest a growing list of zooplankton [40], arthropods [41], mollusks [42], and sediment worms [43] is also susceptible, along with phytoplankton interactions that may affect sedimentation rates [44]. In addition, examples of clams [45] and fish [46] recovered from fish markets have been found with abundant microplastics in the gut. A study of mussels in the lab demonstrated that 10 μm microplastics were translocated to the circulatory system [47], leading to studies that now demonstrate evidence that micro- and nanoplastics can bridge trophic levels into crustaceans and other secondary consumers [48, 49]. Ingested microplastic laden with polybrominated diphenyls (PBDEs) may transfer to birds [50, 51] and to lugworms [52]. The evidence is growing that there are impacts on individual animals including cancers in fish [53] and lower reproductive success and shorter lifespan in marine worms [43]. Some studies even show impacts to laboratory populations: one study of oysters concludes that there is "evidence that micro-PS (polystyrene) cause feeding modifications and reproductive disruption [. . .] with significant impacts on offspring" [54].

While some research shows that plastic can be a vector, or entry point, for these toxicants to enter food webs, others do not. Some studies of microplastic ingestion have shown that complete egestion follows, as in the marine isopod *Idotea emarginata* [55], or ingestion of non-buoyant microplastics by the mud snail *Potampoyrgus antipodarum*, which showed no deleterious effects in development during the entire larval stage [56]. A recent review concluded that hydrophobic chemicals bioaccumulated from natural prey overwhelm the flux from ingested microplastic for most habitats, implying that microplastic in the environment is not likely to increase exposure [57].

Section Summary These three themes dominate the literature today, with an increasing resolution on ecotoxicology and human health. Understanding the fate of micro- and nanoplastics is necessary for a better understanding of the distribution and disposition of plastic pollution. These themes collectively imply microplastic is hazardous to the aquatic environment in the broadest sense. As the literature expands, these themes become benchmarks, tools for policymakers, to mitigate foreseen problems of microplastic contamination of all environments and the social impacts they have on communities worldwide.

2 Mitigation Where There Is Harm

Demonstrated harm to wildlife from plastic is documented from entanglement and macrodebris ingestion, and ingestion of microplastics have shown negative impacts on individual organisms, but demonstrating that microplastics cause harm to the whole ecosystems is unclear [58]. In a recent meta-analysis of available research demonstrating impacts on wildlife from marine debris, 82% of 296 demonstrated impacts were caused by plastic [59]. Interestingly, the vast majority of those (89%) were impacts at suborganismal levels from micro- and nanoplastics, including damages to tissues or organ function, with only 11% due to impacts from large debris, such as entanglement in ropes and netting or death from ingestion of larger items.

According to Rochman et al. [59] there are many cases of suborganismal level impacts, like the ingestion of 20 µm microplastic particles by the copepod *Calanus helgolandicus* affecting survival and fecundity [60], toxic effects on the embryonic development of the sea urchin *Lytechinus variegatus* [61], and reduced feeding in the annelid worm *Arenicola marina* after ingesting 400 µm particles [43]. What these studies and others have in common is that they are limited to laboratory settings, often using PS microspheres only, and use a narrow scale of particle size, shape, and duration of exposure [62]. This criticism was also pointed out in a recent study of the freshwater mud snail *Potamopyrgus antipodarum*, whereby five common and environmentally relevant non-buoyant polymers were introduced in a range of sizes and high concentrations in their food, resulting in no observed effects [56], suggesting that more work in real settings with environmentally relevant microplastic particle size, shape, and polymer type is needed to better understand ecological harm.

Can we say ecological harm exists without the weight of evidence in the literature to say so? One could argue that the volume of research published lately, especially the proposal from Rochman and others to classify plastic marine debris as a hazardous substance [63], indicates substantial concern from the scientific community. That classification would meet criteria for mitigation from policymakers in terms of shifting the burden of proof that plastic is safe to the producer [64]. While further studies of ecological impacts are needed, it is reasonable to employ the precautionary principle considering the risk of widespread and irreversible harm.

Equally, we must not forget the harm to society from plastic pollution. The flow of the material from plastic production to waste management and environmental pollution affects societies in ways that are often difficult to quantify and are often ignored. For example, plastic waste has been shown to incubate water-borne insects and act as a vector for dengue fever in the Philippines [65]. The industry of waste-picking in developing countries is plagued with substantial human health costs from illness and injury from collecting and handling plastics. Open-pit and low-tech incineration is correlated with respiratory illness and cancer clusters among the populations that live near them [66]. While this book aims to understand the impacts of freshwater microplastics, in this chapter we aim to understand and include the upstream social costs in our assessment of the sources and true costs associated with micro- and nanoplastics.

3 Downstream (Ocean Recovery) Versus Upstream Intervention

Then where do our actions to prevent the potential of irreversible harm begin? The three research themes (global distribution, toxicity, marine life impacts) guide mitigation upstream, but it did not begin that way.

The sensationalized mythology of trash islands and garbage patches that had dominated the public conversation about plastic marine pollution in the mid-2000s invoked well-intentioned schemes to recover plastic from the ocean gyres, like giant floating nets to capture debris and plastic-to-fuel pyrolysis machines on ocean-going barges, to seeding the seas with bacteria that consume PET, polyethylene, and polypropylene (which, if this could work, would have the unintended consequence of consuming fishing nets, buoys, docks, and boat hulls). All of these schemes fail on several fronts: economics of cost-benefit, minimizing ecological impacts, and design and testing in real ocean conditions [67]. Recent analysis of debris hot spots and current modeling support the case for nearshore and riverine collection rather than mid-ocean cleanup [68].

This begs the question, "What should be done about what is out there now?" If we do nothing, the likely endgame for microplastic is sedimentation on shore [14] or the seafloor [16], as a dynamic ocean ejects floating debris. Consider the precedent of how tar balls plagued the open ocean and shorelines until MARPOL Annex V stopped oil tankers from rinsing their ship hulls of petroleum residue to the sea in the mid-1980s. A relatively rapid reduction in tar ball observations soon followed [69]. Though we will live with a defining stratigraphy of micro- and nanoplastic in sediments worldwide [70], the ocean can recover if we stop doing more harm.

Still, what can be done about macrodebris? In the 2015 G7 meeting in Germany, Fishing for Litter was presented as the only viable ocean cleanup program, and described as "a useful last option in the hierarchy, but can only address certain types of marine litter" [71]. While Fishing for Litter campaigns can be effective at capturing large persistent debris, like fishing nets, buoys, buckets, and crates before they fragment further, like the KIMO International efforts in North Sea and around Scotland [72], they do not address the source.

4 Upstream Interventions at the Sources of Freshwater Microplastic

Doing no more harm requires upstream intervention. The further upstream mitigation occurs, the greater the opportunity to collect more plastic with less degradation and fragmentation and identifying sources before environmental impacts occur. For most scientists and policymakers, ocean cleanup is not economically or logistically feasible, moving the debate to upstream efforts, like zero waste strategies, improving waste recovery, and management and mitigating point and nonpoint sources of microplastic creation and loss to the environment.

Measuring Microplastic Sources There is wide agreement that microplastic at sea is a case of the tragedy of the commons, whereby its abundance in international waters and untraceability makes it nearly impossible to source to the company or country of origin. In terrestrial environments, identification to source is easier due to less degradation, but capturing and quantifying microplastics in any environment is difficult and can easily be contaminated or misidentified [73], and in inland waterways there is the challenge of sorting debris from large amounts of biomass. In the United States provisions under the Clean Water Act and state TMDLs (Total Max Daily Loads) direct environmental agencies to regulate plastic waste in waterways, like California's TMDLs, though they are often limited to >5 mm and miss microplastic entirely.

While there are processes in the environment that degrade plastic into smaller particles (UV degradation, oxidation, embrittlement and breakage, biodegradation), there are other terrestrial activities and product/packaging designs that create microplastic (Table 1). These may include the mishandling of preproduction pellets at production and distribution sites, industrial abrasives, synthetic grass in sports arenas, torn corners of sauce packets, vehicle tire dust, tooled shavings from plastic product manufacture, road abrasion of plastic waste on roadsides, unfiltered dryer exhaust at laundry facilities losing microfibers to the air [17], or combined sewage overflow that discharges plastics from residential sewer lines, like personal care products, fibers from textiles, and cosmetics, into the aquatic environment. These many sources lack specific methods of measurement.

There are examples of observed microplastic abundance in terrestrial and freshwater environments leading to mitigations, such as the US Microbead-Free Waters Act of 2015 [74] and state laws on the best management practices on preproduction pellet loss [75]. Interestingly, these two examples share three common characteristics: (a) they are quantified by standard methods using nets to measure discharges in waterways, (b) they are found in high abundance, and (c) they are primary microplastics, making it easier to identify responsible sources. Considering the many terrestrial activities that create small amounts of difficult to quantify micro- and nanoplastics, often called secondary microplastics, there is a need for new methods to measure their significance.

Table 1 Sources, measurements, and strategies for upstream mitigation of microplastics

Tackling upstream microplastics		
Category	Source	Potential mitigation
Production	Microplastics in cosmetics	Removing them from products. Replace with benign alternatives
	Mismanaged preproduction pellets	Regulate pellet handling. Operation clean sweep
Commerce	Industrial abrasives	Improve containment and recovery and require alternatives
	Laundromat exhaust	Improved filtration
	Agriculture – degraded film, pots, and pipes	Improve recovery, biodegradable plastics
Consumer	Tire dust	Technological advances, road surface
	Littering of small plastic items (cigarette filters, torn corners of packaging, small film wrappers, etc.)	Enforcement of fines for littering, consumer education, EPR on design
	Domestic laundry. Waste water effluent	Wash with top-load machines. Wastewater containment, single-fiber woven textiles. Textile coatings
Waste management	Fragmentation by vehicles driving over unrecovered waste	Improved waste management
	UV and chemically degraded terrestrial plastic waste	Improved waste management
	Sewage effluent (synthetic fibers)	Laundry filtration, textile industry innovation
	Combined sewage overflow (large items)	Infrastructure improvement
	Mechanical shredding of roadside waste during regular cutting of vegetation (mostly grass)	Better legislation and law enforcement; valorization of waste products

Why wait until microplastic reaches water to quantify its existence? The current methods of storm drain catchment and waste characterization measure macroplastic only. Microplastics, such as synthetic grass, tooled shavings, road abrasion, etc., are sources of microplastic with unknown abundances, which could be measured by sampling surface areas on the ground nearby the activities that create them. Methodologies might include square meter sweeping of sidewalks and roadsides to quantify abundances. A recent study of microfiber fallout used containers on rooftops in Paris to capture airborne particles [17]. These micro- and nanoplastic fibers can be measured closer to the source. Surveying the surface of foliage near laundromats (Eriksen, unpublished data) recently discovered abundant microfibers. Other methods might employ footbaths outside hotels or shops with carpeted floors to measure the transport of fibers due to foot traffic. The production of household microplastics could be estimated from dust particles accumulated in the filter bags of vacuum cleaners. Quantifying the significance of these point and nonpoint sources might assist efforts to mitigate their contributions.

5 Competing Economic Models Impact Microplastic Generation

The contemporary debate about solutions largely contrasts the circular economy with the current linear economic model. These competing economic models reveal subjective stakeholder motives, whether it is a fiduciary responsibility to shareholders, an environmental or social justice mission, or an entrepreneurial opportunity. These economic models influence the design and utility of plastic and therefore the abundance and exposure of plastic waste to the environment, thus influencing the formation of microplastic.

Material Loss Along the Value Chain in the Linear and Circular Economic Models Given the many sources of microplastic, the different sectors of economy and society producing these and the relatively limited knowledge about them (Table 1), it becomes apparent how difficult it would be trying to "plug" leaks of microplastics to the environment. Some of the sources could be stopped by effective legislation (e.g., banning microbeads in cosmetic products), education and regulation enforcement (litter laws), and technological advancements (effluent filters, biodegradable polymers).

However, in the end it becomes increasingly difficult to mitigate these leak points the further from the source intervention begins. The closest point to the source is the choice of polymer and how it is managed throughout the supply chain and once it becomes waste. Some efforts have included an upfront tax to fund cleanup efforts or mitigate environmental impacts, but those appear impractical due to the diffusion and difficulty in collecting small microplastics. Given the low value of most postconsumer plastic products and lack of recovery incentives, the chances of downstream mitigation are extremely low.

Consequently, leaks of microplastics to the terrestrial and ultimately aquatic environment (primary or secondary by input in form of large objects which later degrade into microplastics) occur throughout the supply chain, e.g., in form of loss of preproduction pellets, littering, or irresponsible waste management (Fig. 1). Little material remains in the system, and most would not be fit for effective recycling (i.e., reusing) because of contamination or expensive recuperation schemes. Deposition in landfills or energy recovery through incineration therefore appears as the ultimate strategy to remove almost all material from the system, effectively creating a linear economic model. Energy recovery is not a form of recycling and does not break this linearity, because it essentially removes used plastics from the economic system through destruction, converting them into ashes and atmospheric CO_2 (Fig. 1).

A circular economic model on the other hand could address leaks of plastics at all life cycle stages. The reduction of leakage to the environment requires adaptation and consensus of all stakeholders, e.g., designing for reuse; discouraging littering, for example, by introducing deposit return schemes; and ensuring a high recycling quota during the waste stage (Fig. 2). Most likely one key to the

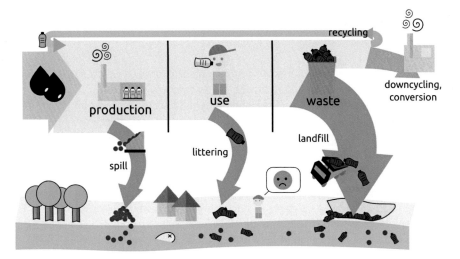

Fig. 1 Linear economy model for plastic products and packaging and system leaks. Product is manufactured using principally new resources, largely petroleum based. Most of the product's value is lost during its life cycle because of leakage along the entire value chain (*red arrows*), including pellet loss, littering, combined sewage overflow, loss during transport and improper storage of waste, and poorly designed products that are easily lost to the environment and difficult to recover (microbeads, small wrappers, torn corners of packaging). This leads to a contamination of the environment, affecting wildlife and human well-being. A small proportion is recycled (*green arrow*) for remanufacture, with the remainder utilized for energy recovery

implementation of this circular economic model is to modify the value chain of plastics throughout all phases of its functional life. A number of economic alternatives are already being implemented as will be described below. This model also puts emphasis on preventive measures when accounting for environmental problems caused by excessive leakage. Prevention is also much more cost-effective and environmentally friendly than postconsumer cleanup schemes, many of which are economically or technologically unfeasible.

Most stakeholders agree waste management must improve globally to prevent pollution of the aquatic environment, and that landfilling waste is not a viable strategy in the future. What some have called "uncontrolled biochemical reactors" [76] are landfills which are increasingly losing popularity as the costs and hazards outweigh the benefits. In "Zero Plastics to Landfill by 2020" [77], the European Union, and the trade organizations Plastics Europe and the American Chemistry Council [78], advocates ending landfill reliance. Where the circular and linear economies largely differ is the role of policy to drive design, and the end-of-life plan for recovered plastic.

Zero Waste vs. Waste-to-Energy This division could be considered the frontline where sharp divisions exist. Whether plastics are incinerated for energy recovery or sorted for recycling and remanufacture reflects stakeholder positions and influences

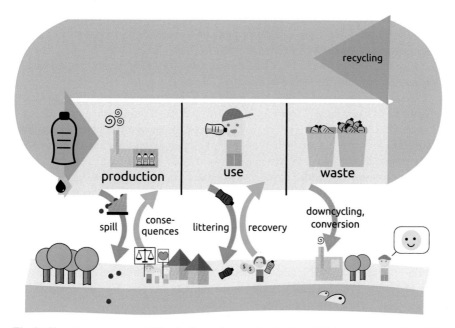

Fig. 2 Circular economy model for plastic products and packaging. A high percentage of recycled content is required as feedstock for new products, and the remainder from sustainable sources (potentially biopolymers). Poor practices (*red arrows*) throughout the life cycle are mitigated, for example, by proper legislative policy, public awareness that leads to proper consumer waste handling, and incentivized recovery systems (e.g., returnable bottles). Recovery is further improved by regulating end-of-life design in products and packaging. This leads to reduced leakage of plastic to the environment from all sectors of society, and significant improvements are social justice concerns for communities that manage waste. The small amount of residual plastic is then disposed of responsibly

decisions about product and packaging design and regulation far upstream. The end-of-life plan for plastic affects the entire value chain.

A recent document produced by the Ocean Conservancy (2015) titled "Stemming the Tide," with strong industry support, called for a $5 billion investment in waste management, with large-scale waste-to-energy incinerator plants targeting SE Asia, specifically China, Taiwan, Philippines, Indonesia, and Vietnam, based on a study reporting 4–12 million tons of waste entering the oceans annually, primarily from that region [11]. It was released 1 week prior to the October 2015 Our Ocean Conference. Within days, the organization Global Alliance for Incinerator Alternatives (GAIA) submitted a letter in response with 218 signatories, mostly environmental and social justice NGOs, arguing that incinerators historically exceed regulatory standards for emissions and subsequently cause harm to the environment and human health and that the financial cost to build infrastructure, maintenance, and management are typically underestimated [79]. In many cases, the financial structure includes long-term waste quotas that lock communities into mandatory waste generation [66]. For example, the $150 million cost to build the H-Power incinerator in Oahu, Hawaii, also comes with an 800,000 ton per year "put or pay" trash obligation. If they don't get their quota of waste, the city pays a portion of the revenue they would have earned burning the trash they didn't get. The public calls it "feeding the beast" [80], which had undermined recycling, waste diversion, and composting programs, for fear of fines.

Two earlier documents, "On the Road to Zero Waste" from GAIA [79] and "Waste and Opportunity," from As You Sow and the National Resources Defense Council (NRDC) [81], both lay out a framework for sustainable material management from resource extraction to recovery and remanufacture, without the need for incineration, or the legacy of associated toxicity and human health effects.

In the developing world, circular economic systems are expanding. There are material recovery facilities, or MRFs, sprouting up everywhere. Waste sorting and collection happens door to door, with the collector keeping the value of recyclables after delivering all materials to the local MRF. Organics are composted, recyclables are cashed in, and the rest is put on public display to show product/packaging design challenges. According to the Mother Earth Foundation, 279 communities in the Philippines have MRFs, and waste diversion from landfills and open-pit burning now exceeds 80%. The template for the community MRF is proving its scalability across Asia, India, Africa and South America.

Rationale of the Linear Economy In 2014 Plastics Europe released an annual report titled "Plastics – the facts 2013: An analysis of European latest plastics production, demand and waste data" [82], outlining the forecast for plastic demand and challenges in the years ahead. Worldwide, there has been a historical trend of a 4% increase of annual plastic production since the 1950s, with slight dips during the OPEC embargo in the 1970s and the 2008 economic downturn, but otherwise it's been steady growth from almost no domestic plastic produced post-WWII to 311 million tons of new plastic produced in 2013 alone. If this growth rate continues as anticipated worldwide, there will be close to 600 million tons produced annually by 2030 and over a billion tons a year by 2050.

This trajectory is partially based on rising demand from a growing global middle class and is coupled with the rising population. Yet, these demands will stabilize, leaving waste-to-energy through incineration a key driver in the security of demand for new plastic production. Recycled plastic is a direct competitor with new plastic production, being inversely proportional to the available supply. This has been largely acknowledged and has kept recycling rates generally very low worldwide. Consider recycle rates in the United States alone, with the highest recovery per product in 2013 won by PET bottles (31.3%) seconded by HDPE milk containers (28.2%), and national average for all plastic combined was 9.2% after 53 years of keeping score [83].

The industry transition in light of these trends is to advocate energy recovery after maximizing the utility of plastic, arguing that the cost vs. benefit of plastic favors unregulated design and improved waste management. A careful look at the life cycle of alternative materials (paper, metals, glass), from extraction to manufacture, transportation, and waste management, must be weighed against the benefits of plastic. Plastics make food last longer [84], offer more durable and lightweight packaging for transportation of goods, maintain clean pipes for drinking water distribution, and facilitate low-cost sterile supplies for hospitals, each having degrees of efficiency over alternative materials in terms of waste generation, water usage, and CO_2 emissions, like lightweighting cars with plastic resulting in lower fuel consumption [85].

For example, an industry analysis comparing the impacts of transportation, production, waste management, and material/energy recovery on the environment concluded that the upstream production and transportation phases of the value chain for plastics accounted for 87% of total costs [78], leaving 13% of the impacts on the environment caused downstream by how waste is managed. Plastic producers have suggested that some of these upstream production impacts could be further mitigated by sourcing low-carbon electricity that by doubling the current use of alternate energy for production could cut the plastics sector's own greenhouse gas emissions by 15% [78]. Mitigating the problems of microplastics requires understanding not only where waste is generated but also where other environmental harms can be avoided at all points along the value chain.

The Case for Bridge Technologies While large-scale incinerators are criticized for cost, waste quotas, emissions, and the effect of undermining zero waste strategies, is there a case for the temporary use of small-scale waste to energy until more efficient systems of material management evolve?

While the H-Power plant in Oahu, Hawaii has been criticized, alternatives have been proposed. One firm recently proposed gasification (high heat conversion of waste to a synthetic gas), submitting evidence that the initial cost of infrastructure is far less than the H-Power plant, pays for itself in 1.4 years with current waste input, is three times more efficient than incineration in terms of energy conversion, and has no long-term waste quota, allowing zero waste strategies to alleviate existing waste streams. The system could then be relocated to other waste hot spots to manage waste or reduce waste volumes in exposed landfills (Sierra Energy, personal communication).

Although volumes of waste reduced on land become volume of waste increased in the air (conservation of mass), any form of combustion (pyrolysis, gasification, incineration) to create energy results in greenhouse gas (GHG) emissions, a principle concern of any form of waste incineration.

A study of waste incineration and greenhouse gas (GHG) emissions found that once it came to energy recovery, "the content of fossil carbon in the input waste, for example, as plastic, was found to be critical for the overall level of the GHG emissions, but also the energy conversion efficiencies were essential" [86]. Increased plastic in the waste stream meant increased overall GHG emissions. Reliance on energy recovery from waste in the linear economic model will have a net balance of more GHG than upstream mitigation strategies in the circular economic model, though the linear vs. circular economy may not be so black and white. A combination of multiple end-of-life strategies could collectively manage the diversity of waste in both efficiency and economy.

Another analysis of GHG emissions compared the current strategy in Los Angeles of landfilling the vast majority of waste to a combination of three strategies in a modern MRF, namely, (a) anaerobic digestion of wet waste, (b) thermal gasification of dry waste, and (c) landfilling residuals [87]. Their analysis did not consider economic, environmental, or social parameters, only GHG emissions, and was based on an assumption of 1,000 ton of waste per day entering each scenario for 25 years; then they modeled the GHG emissions for the century that followed. In each scenario, the GHG emissions from transportation, operation, and avoided emissions by replacing fossil fuels were factored in. Results showed that continued landfilling resulted in a net increase of approximately 1.64 million metric tons of carbon dioxide equivalent (MTCO2E), while the MRF scenario results in a net avoided GHG emissions of (0.67) million MTCO2E, showing that a shift to a MRF where multiple waste management strategies are employed resulted in a total GHG reduction of approximately 2.31 million MTCO2E.

Those residuals that exist after diversion of waste to recycling and anaerobic digestion could be landfilled, and in some cases waste-to-energy could have a role. This would be appropriate only after diversion efforts of recyclables and compostables have been maximized. Also, building incinerator infrastructure could create tremendous debt or include a demand for large volumes of waste, also called a "waste quota" that could undermine local efforts to eliminate products and packaging that generate microplastics. Simultaneously, a market for recycled materials must be encouraged, while all environmental and worker health concerns are prioritized. Waste-to-energy could have a role, but long after all other efforts to manage waste have been employed.

Section Summary In the linear economy contrasted with the circular economy, we see two world views on how to solve the plastic pollution problem. While the linear economic system benefits production by eliminating competition from recycled material, it is more polluting than the circular system because of multiple points of leakage along the supply chain. Plastic pollution is lost at production as pellet spills, lost by the consumer as litter with no inherent value, and lost at collection and disposal as waste is transported. In the circular system these are mitigated when systems to focus on material control and capture are implemented. Zero waste is the ideal of the circular economy, where the need for destruction through energy capture, or landfill, are increasingly unnecessary.

6 Microplastic Mitigation Through a Circular Economy

In the emerging circular economy, the flow of technical materials through society returns to remanufacture, with products and packaging designed for material recovery, low toxicity, ease of dismantling, repair and reuse, and where this doesn't work, a biological material may substitute so circularity in a natural system can prevail. Shifting to a circular economy has prompted interest in a range of interventions, including bioplastics, extended producer responsibility, and novel business approaches.

Green Chemistry as a Biological Material Bioplastic has been in production since Henry Ford's soybean car in the 1930s, made from soy-based phenolic resin, which he bashed with a sledgehammer to demonstrate its resilience, but the WWII demand for a cheap, better-performing material induced him to chose petroleum-based plastic. Today, bioplastics are viewed with new interest. These plant-based plastics are considered a means to create a more reliable and consistently valued resource, decoupled from fossil fuels. The Bioplastic Feedstock Alliance, created with wide industry alliance and support from the World Wildlife Fund (WWF), intends to replace fossil fuels with renewable carbon from plants, representing no net increase in GHG emissions. Referred to as [the] "bioeconomy," these companies envision bioplastics as "reducing the carbon intensity of materials such as those used in packaging, textiles, automotive, sports equipment, and other industrial and consumer goods" [88].

It is important to distinguish biodegradable from bio-based plastics. Bioplastic is the loosely defined catch-all phrase that describes plastic from recent biological materials, which includes true biodegradable materials and nonbiodegradable polymers that are plant based. While the label "biodegradable" has a strict ASTM standard and strict guidelines for usage in advertising, the terms bioplastic, plant based, and bio based do not. Despite all of the leafy greenery in labeling for these bioplastics, it is still the same polymer that would otherwise have come from fossil fuels.

The biodegradability of bio-based and biodegradable plastics will vary widely based on the biological environment where degradation may occur. Poly-lactic acid (PLA) is a compostable consumer bio-based plastic requiring a large industrial composting facility that's hot, wet, and full of compost-eating microbes, unlike a backyard composting bin. Poly-hydroxy-alkanoate (PHA), made from the off-gassing of bacteria, is a marine-degradable polymer (ASTM 7081), but rates of degradation vary with temperature, depth, and available microbial communities [89].

PHA and PLA are both recyclable and compostable, but how these materials are managed depends on available infrastructure. While recycling could be energetically more favorable than composting, it may not be practical because of sorting and cleaning requirements. Kale et al. point out the lack of formal agreement between stakeholders (industry, waste management, government) about the utility of biodegradable plastics and their disposal [90], but the compostability of bioplastic packaging materials could become a viable alternative if society as a whole would be willing to address the challenges of cradle-to-grave life of compostable polymers in food, manure, or yard waste composting facilities. The industries that make bioplastic polymers recognize these challenges and therefore their limited applications. PHA is ideal to be used where you need functional biodegradation, such as some agriculture and aquaculture applications, where a part has a job to do in the environment but it would be either impractical or very costly to recover (Metabolix, personal communication). Also, many single-use throwaway applications may be replaced by PHA, including straws or the polyethylene lining on paper cups (Mango Materials, personal communication). Without the infrastructure widely available to recycle bio-based and biodegradable plastics, manufacturers are aiming for compostability in compliance with organic waste diversion initiatives.

Extended Producer Responsibility (EPR) There is a wide agreement that waste management must be improved, including public access to recycling, composting, and waste handling facilities. Equally, there is a need to improve the design of products and packaging to facilitate recovery in the first place. Regulating primary microplastics has been successful with microbeads and preproduction pellets, yet there are many characteristics of product and packaging design that could be improved to minimize the trickle of irrecoverable microplastics from terrestrial to aquatic environments.

Product and packaging design must move "beyond the baseline engineering quality and safety specifications to consider the environmental, economic and social factors," as explained in "Design through the 12 Principles of Green Engineering" [91]. When designing for the full life cycle of a product, manufactures and designers talk with recyclers to reduce environmental impacts by improving recovery, which may include avoiding mixed materials or laminates, reduced toxicity, and ease of repair, reuse, and disassembly, as well as the systems that move materials between consumer and the end-of-life plan. Reducing microplastic formation by design might also include eliminating tearaway packaging (opening chip/candy wrappers, individual straw/toothpick covers), small detached components (bottle caps and safety rings), or small single-use throw-away products (coffee stirrers, straws, bullets in toy air rifles). These mitigations can be voluntary, but are often policy-driven through fees or bans [92].

Extended producer responsibility is a public policy tool whereby producers are made legally and financially responsible for mitigating the environmental impacts of their products. When adopted through legislation, it codifies the requirement that the producer's responsibility for their product extends to postconsumer management of that product and its packaging. With EPR, the responsible legal party is usually the brand owner of the product.

EPR is closely related to the concept of "product stewardship," whereby producers take action to minimize the health, safety, environmental, and social impacts of a product throughout its life cycle stages. Producers' being required to take back and recycle electronic equipment through the EU's Waste of Electrical and Electronic Equipment (WEEE) Directive is an example of EPR. The Closed Loop Fund – which accepts corporate money to loan to US municipalities to boost packaging recycling – is an example of voluntary product stewardship [93]. Different schemes of EPR have been implemented [94], and even though some first success is achieved in recycling of plastics and other packaging products [95], these systems still require many improvements ranging from economic models [96] to logistic aspects [97].

While EPR has primarily been applied as a materials management strategy, the concept can also be applied to plastic pollution prevention and mitigation. In 2013, the Natural Resources Defense Council helped advance how EPR can more directly impact plastic pollution beyond boosting the collection and recycling of packaging [98]. NRDC developed policy concepts and legislation to make the producers of products which have a high tendency to end up as plastic pollution, responsible not just for recycling, but for litter prevention and mitigation as well. Legislation introduced in California would have (a) had State Agencies identify the major sources of plastic pollution in the environment and (b) required the producers of those products to reduce the total amount in the environment by 75% in 6 years and 95% in 11 years. While the legislation did not advance far in California, this was a significant development and provides an example of how to incorporate litter prevention and pollution mitigation in future EPR policy.

Section Summary The utility of green chemistry has led to public confusion over the biodegradability of polymers, stemming from an important differentiation between biopolymers and biodegradable polymers, as well as the true conditions where biodegradability occurs. While biopolymers offer a promising divestment from fossil fuel feedstocks, biodegradable plastics are challenged by the infrastructure requirements for identification, sorting, and degradability. In a circular economy, biopolymers and biodegradable polymers must exist in a system, either manufactured or natural, where the material is recovered and reprocessed. Extended producer responsibility is the policy mechanism that creates those systems, with the intention to mitigate the true economic, social, and environmental costs associated with waste.

7 Business Transformation Through Novel Policy and Design

The status quo for much of product and packaging manufacture is planned obsolescence, which drives cheap-as-possible chemistry and design and has been largely subsidized by municipalities that agree to manage all that waste at a limited cost to the manufacture and principal cost to the tax payer. With an abundance in the waste stream of plastics embedded in difficult-to-recover products and packaging (electronics, laminates, food-soiled packaging), energy recovery becomes a more attractive alternative.

The effort to rely on energy recovery through incineration is largely a perpetuation of the "planned obsolescence" strategy of securing demand for new products, employed historically since post-WWII manufacture. Planned obsolescence encourages material consumption in several ways: technological (software and upgrades overwhelming old hardware), psychological (fashion), and conventional (designed weakness and impractical repair).

The Ellen MacArthur Foundation [99] published in February 2016 "The New Plastics Economy" proposed business solutions that manage materials through the consumer, beyond planned obsolescence, where product designers talk to recyclers to create an end-of-life design, systems of "leasing" products over ownership, allowing product upgrades over planned obsolescence. By making a business case for managing the circular flow of technical materials, the status quo of cradle to grave can be put to rest.

The market dominance of poorly designed products will likely not self-regulate a transformation, requiring policy tools. EPR in some ways can be facilitated by novel policy tools. In London in 2015 a 5p fee on plastic bags, rather than a ban, resulted in an 85% reduction in their consumption. In areas where citizens "pay to pitch" the waste they generate, consumers commonly strip packaging at the point of purchase, which in turn is communicated to the distributor of goods to redesign the delivery of goods. This system of pay to pitch has been applied to some remote communities, such as islands, to require importers to export postconsumer materials.

Andrew Winston, author of *The Big Pivot*, suggests an alternate model of doing business, the Benefit Corporation, or "B-Corp," whereby corporations take on a mission statement of social or environmental justice that is on equal par with the profit motive. A rapidly changing consumer base that is more connected through communication is forcing corporations to be transparent, accountable, and behave ethically. The B-Corp is the bridge across the divide.

8   Reducing and Reusing Plastic Waste

Avoiding the production of new plastics altogether whenever possible is the most reliable way to avoid the generation of microplastics, whether primary microplastics (needed for the production of new plastic articles) or secondary (resulting during breakdown of larger plastic items).

As the market for ethically produced products is growing worldwide (e.g., Fairtrade [100], organic food produce [101, 102]), and consumers become aware of the possible impacts of marine pollution [103], several examples are demonstrating a successful reduction of plastic waste or the reuse of discarded plastics in order to create other products (upcycling), thereby saving natural resources and, in some cases, even removing ocean plastic pollution.

Among popular recent innovations are the production of clothes, shoes, skateboards, sun-glasses, and swimming gear from derelict fishing gear [104, 105]. Such lines of products, making a pro-environmental statement, are likely to be especially appealing to customers of the Generation Y/Millenials (see references in [106]). Another example for a consumer-driven desire to combat excessive plastic litter, this time in the form of packaging waste, is the recent development of zero waste stores, sprouting up in Europe and the United States (Fig. 3a) [107, 108]. Many of these stores are crowd funded [107] and require customers to bring their own food container which also avoids food waste by allowing customers to buy the quantities they consume. Many of those shops do not offer products from large brands to distance themselves from supermarket chains and emphasize a community-based economy model.

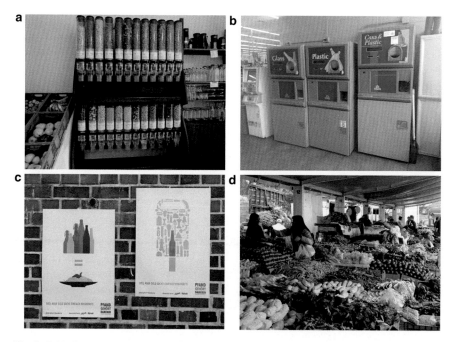

Fig. 3 Initiatives to reduce or recuperate packaging waste. (**a**) = "Unverpackt" store in Germany where customers can buy food in bulk, bringing their own containers. ©Martin Thiel. (**b**) = Reverse vending machines accepting glass and plastic bottles and aluminum cans in a supermarket in the United States. ©Alex Kirsch. (**c**) = Advertisement of the "Pfand gehört daneben"-campaign in Germany, advocating to leave deposit return bottles in Germany next to the garbage bin in order for easy pick up ©Pfand gehört daneben 2016. (**d**) = "Feria libre" in Chile, allowing customers to buy vegetables and fruits in bulk (public domain, Jorge Valdés R. Joval)

An example of a large retail store taking up waste reduction strategies is the Amazon.com, Inc., with its program "Frustration-Free Packaging," which aims to reduce packaging volume and complexity. The company claims to have saved 11,000 tons of packaging during 5 years, including reductions of styrofoam and thin plastic films [109].

Possibly the most established way of avoiding excessive waste and saving valuable resources is in the form of container deposit fees, especially for beverages (Fig. 3b). This has been shown as highly effective to reduce the amount of waste in the environment with return rates as high as 90% and higher in Sweden and Germany for several materials commonly used in beverage production (metal, glass, plastic) [110, 111]. Deposit return strategies are more efficient than curbside recycling programs [112], largely because of the monetary incentive for recovery ("One man's trash is another man's treasure"). For example, the "Pfand gehört daneben" campaign in Germany ("Deposit bottles belong next to it [the garbage bin]") encourages the public to leave unwanted deposit return bottles accessible for easy pick up by private waste collectors and not trashing them in a garbage bin

(Fig. 3c). However, a return deposit fee on food containers does not ensure that the container is reused as the large and growing proportion of returnable but single-use plastic bottles in Germany illustrate [113]; therefore, further incentives are necessary.

Another way to reduce plastics is prohibition or taxing of plastic products that can be easily replaced, such as microbeads in cosmetic and daily care products and plastic bags for groceries. A survey conducted in Ireland revealed that fees/taxes on plastic bags seem to be well received among customers [114].

Buying from local farmers' markets is another way for a customer to procure less packaging (Fig. 3d). While farmers' markets were replaced in most of Europe and North America by large supermarket chains, they are celebrating a comeback over the last two decades [115]. In other countries it is still normal to procure the majority of fresh foods from farmers' markets, despite the introduction of large supermarket chains. This is the case in Chile where "Ferias libres" (neighborhood outdoor markets) supply the population with 70% of its demand for fruit and vegetables and 30% of seafood products [116].

Collectively, all these strategies help reduce the leakage of low-value/single-use plastics into terrestrial and aquatic environments and subsequent formation of microplastics from their degradation. Regardless of the most modern waste management systems available, leakage of single-use throwaway products and packaging occurs. Their reduction is the most efficient mitigation effort to reduce microplastics in the environment.

9   Conclusion

An environmental movement may be defined as a loose, noninstitutionalized network of organizations of varying degrees of formality, as well as individuals and groups with no organizational affiliation, who are engaged in collective action motivated by shared identity or concern about environmental issues [117].

In July of 2016, the American Chemistry Council published "Plastics and Sustainability: A valuation of environmental benefits, costs and opportunities for continuous improvement," largely a comparison of life cycle analyses putting plastic in a positive light against alternative materials (glass, metal, paper). At the same time, the Plastic Pollution Policy Project convened 18 organizations focused on zero waste initiatives to align on policy and campaigns and to create common messaging to counter industry-dominated narratives. A movement has emerged, while stakeholder positions have dug in their heels.

Here we have discussed solutions to microplastics in freshwater ecosystems, which largely form in terrestrial environments from primary or secondary microplastics. We know that microplastics are global, increasingly toxic over time, and impacts to wildlife are pervasive, leading to the collective conclusion that plastic in the environment causes harm. We also know that capturing microplastic downstream is extremely difficult and requires upstream intervention.

Once in natural water bodies (rivers, lakes, oceans), recovery of microplastics is impossible. Therefore, one challenge is to identify and quantify the upstream sources – a prerequisite to mitigation. In the cases of microbeads and preproduction pellets, we witnessed the role of science to present observations of microplastic pollution, followed by a movement to pressure policymakers to regulate industry. The work of scientists continues to illuminate microplastic impacts, such as recent reports from the Group of Experts on the Scientific Aspects of Marine Environmental Protection (GESAMP) [118], a working group gathered by UNEP to synthesize and report on the state of the scientific evidence regarding the plastic pollution issue and distribute the information to the United Nations Environment Assembly.

There are four principal solutions that will have high impact on preventing terrestrial and freshwater microplastics from forming. They are: (1) identify and quantify terrestrial microplastic sources, (2) scale zero waste strategies, (3) pursue policy-driven EPR, and (4) develop novel business solutions. These solutions will bring greater alignment between stakeholders on the utility of plastic in society and a more equitable end-of-life, where environmental and social justice are integrated in the full cost of plastic. The bridge between the linear and circular economy is about material circularity coupled with a sincere investment in common decency and democracy, and corporate responsibility toward those ends, what Severyn Bruyn calls a Civil Economy, whereby government, business, nonprofits and civic groups "can develop an accountable, self-regulating, profitable, humane, and competitive system of markets" [119] (Bruyn 2000).

This a thoughtful approach that considers the chemistry of materials, the design of products, the processes required to make things, and finally the systems that manage how materials flow back into the production chain, all in the context of causing no harm to people and the environment, benign by design in its totality.

Acknowledgments MT was supported by the Chilean Millennium Initiative (grant NC120030).

References

1. Kühn S et al (2015) Deleterious effects of litter on marine life. In: Bergmann M, Gutow L, Klages M (eds) Marine anthropogenic litter. Springer, New York, pp 75–116
2. Shimanaga M, Yanagi K (2016) The Ryukyu trench may function as a "depocenter" for anthropogenic marine litter. J Oceanogr 72(6):895–903. doi:10.1007/s10872-016-0388-7
3. Carpenter EJ, Smith KL (1972) Plastics on the Sargasso sea surface. Science 175:1240–1241
4. Wong CS et al (1974) Quantitative tar and plastic waste distributions in the Pacific Ocean. Nature 247:30–32
5. Shaw DG, Mapes GA (1979) Surface circulation and the distribution of pelagic tar and plastic. Mar Pollut Bull 10:160–162
6. Morris RJ (1980) Floating plastic debris in the Mediterranean. Mar Pollut Bull 11:125
7. Wilber R (1987) Plastic in the North Atlantic. Oceanus 30:61–68
8. Moore CJ et al (2001) A comparison of plastic and plankton in the North Pacific central gyre. Mar Pollut Bull 42:1297–1300

9. van Sebille E et al (2015) A global inventory of small floating plastic debris. Environ Res Lett 10:124006

10. Cózar A et al (2014) Plastic debris in the open ocean. Proc Natl Acad Sci USA 111:10239–10244

11. Jambeck J et al (2015) Plastic waste inputs from land into the ocean. Science 347:768–771

12. Eriksen M et al (2014) Plastic pollution in the world's oceans: more than 5 trillion plastic pieces weighing over 250,000 tons afloat at sea. PLoS One 9:e111913

13. Obbard RW et al (2014) Global warming releases microplastic legacy frozen in Arctic Sea ice. Earth's Future 2:315–320

14. Browne MA et al (2011) Accumulation of microplastic on shorelines worldwide: sources and sinks. Environ Sci Technol 45:9175–9179

15. Woodall LC et al (2014) The deep sea is a major sink for microplastic debris. R Soc Open Sci 1:140317

16. van Cauwenberghe L et al (2013) Microplastic pollution in deep-sea sediments. Environ Pollut 182:495–499

17. Dris R et al (2016) Synthetic fibers in atmospheric fallout: a source of microplastics in the environment? Mar Pollut Bull 104(1–2):290–293. doi:10.1016/j.marpolbul.2016.01.006

18. Gauquie J et al (2015) A qualitative screening and quantitative measurement of organic contaminants on different types of marine plastic debris. Chemosphere 138:348–356

19. Teuten EL et al (2007) Potential for plastics to transport hydrophobic contaminants. Environ Sci Technol 41:7759–7764

20. Rios LM et al (2010) Quantitation of persistent organic pollutants adsorbed on plastic debris from the Northern Pacific Gyre's "eastern garbage patch". J Environ Monit 12:2226–2236

21. Mato Y et al (2001) Plastic resin pellets as a transport medium for toxic chemicals in the marine environment. Environ Sci Technol 35:318–324

22. Fromme H et al (2009) Perfluorinated compounds – exposure assessment for the general population in western countries. Int J Hyg Environ Health 212:239–270

23. Giesy JP et al (2001) Global biomonitoring of perfluorinated organics. Sci World J 1:627–629

24. Kovarova J, Svobodova Z (2008) Perfluorinated compounds: occurrence and risk profile. Neuroendocrinol Lett 29:599–608

25. Ahn KC et al (2008) In vitro biologic activities of the antimicrobials triclocarban, its analogs, and triclosan in bioassay screens: receptor-based bioassay screens. Environ Health Perspect 116:1203–1210

26. Chalew TE, Halden RU (2009) Environmental exposure of aquatic and terrestrial biota to triclosan and triclocarban. J Am Water Works Assoc 45:4–13

27. Costa LG et al (2008) Polybrominated diphenyl ether (PBDE) flame retardants: environmental contamination, human body burden and potential adverse health effects. Acta Biomed 79:172–183

28. Yogui GT, Sericano JL (2009) Polybrominated diphenyl ether flame retardants in the US marine environment: a review. Environ Int 35:655–666

29. Bonefeld-Jørgensen EC et al (2007) Endocrine-disrupting potential of bisphenol A, bisphenol A dimethacrylate, 4-n-nonylphenol, and 4-n-octylphenol in vitro: new data and a brief review. Environ Health Perspect 115:69–76

30. U.S. EPA (2016) Bisphenol A (BPA) action plan. https://www.epa.gov/assessing-and-man aging-chemicals-under-tsca/bisphenol-bpa-action-plan. Accessed 5 Oct 2016

31. Dodds EC, Lawson W (1936) Synthetic estrogenic agents without the phenanthrene nucleus. Nature 137:996

32. Brotons JA et al (1995) Xenoestrogens released from lacquer coatings in food cans. Environ Health Perspect 103:608–612

33. vom Saal FS, Hughes C (2005) An extensive new literature concerning low-dose effects of bisphenol A shows the need for a new risk assessment. Environ Health Perspect 113:926–933

34. Meeker JD et al (2009) Phthalates and other additives in plastics: human exposure and associated health outcomes. Philos Trans R Soc Lond B Biol Sci 364:2097–2113

35. Stahlhut RW et al (2007) Concentrations of urinary phthalate metabolites are associated with increased waist circumference and insulin resistance in adult US males. Environ Health Perspect 115:876–882
36. Buchta C et al (2005) Transfusion-related exposure to the plasticizer di(2-ethylhexyl) phthalate in patients receiving plateletpheresis concentrates. Transfusion 45:798–802
37. Lusher A et al (2015) Microplastic interactions with North Atlantic mesopelagic fish. ICES J Mar Sci 73(4):1214–1225. doi:10.1093/icesjms/fsv241
38. Anastasopoulou A et al (2013) Plastic debris ingested by deep-water fish of the Ionian Sea (Eastern Mediterranean). Deep-Sea Res I Oceanogr Res Pap 74:11–13
39. de Stephanis R et al (2013) As main meal for sperm whales: plastics debris. Mar Pollut Bull 69:206–214
40. Cole M et al (2013) Microplastic ingestion by zooplankton. Environ Sci Technol 47:6646–6655
41. Ugolini A et al (2013) Microplastic debris in sandhoppers. Estuar Coast Shelf Sci 129:19–22
42. Von Moos N et al (2012) Uptake and effects of microplastics on cells and tissue of the blue mussel *Mytilus edulis* L. after an experimental exposure. Environ Sci Technol 46:11327–11335
43. Wright SL et al (2013) Microplastic ingestion decreases energy reserves in marine worms. Curr Biol 23:1031–1033
44. Long M et al (2015) Interactions between microplastics and phytoplankton aggregates: impact on their respective fates. Mar Chem 175:39–46
45. Li J et al (2015) Microplastics in commercial bivalves from China. Environ Pollut 207:190–195
46. Seltenrich N (2015) New link in the food chain? Marine plastic pollution and seafood safety. Environ Health Perspect 123:34–41
47. Browne MA et al (2008) Ingested microscopic plastic translocates to the circulatory system of the mussel *Mytilus edulis* (L.) Environ Sci Technol 42:5026–5031
48. Farrell P, Nelson K (2013) Trophic level transfer of microplastic: *Mytilus edulis* to *Carcinus maenas*. Environ Pollut 177:1–3
49. Setälä O et al (2014) Ingestion and transfer of microplastics in the planktonic food web. Environ Pollut 185:77–83
50. Yamashita R et al (2011) Physical and chemical effects of ingested plastic debris on short-tailed shearwaters, *Puffinus tenuirostris*, in the North Pacific Ocean. Mar Pollut Bull 62:2845–2849
51. Tanaka K et al (2013) Accumulation of plastic-derived chemicals in tissues of seabirds ingesting marine plastics. Mar Pollut Bull 69:219–222
52. Besseling E et al (2013) Effects of microplastic on fitness and PCB bioaccumulation by the lugworm *Arenicola marina* (L.) Environ Sci Technol 47:593–600
53. Rochman CM et al (2013) Ingested plastic transfers hazardous chemicals to fish and induces hepatic stress. Sci Rep 3:3263
54. Sussarellu R et al (2015) Oyster reproduction is affected by exposure to polystyrene microplastics. Proc Natl Acad Sci USA 113:2430–2435
55. Hämer J et al (2014) Fate of microplastics in the marine isopod Idotea emarginata. Environ Sci Technol 48:13451–13458
56. Imhof HK, Laforsch C (2016) Hazardous or not – are adult and juvenile individuals of *Potamopyrgus antipodarum* affected by non-buoyant microplastic particles? Environ Pollut 218:383–391
57. Koelmans AA et al (2016) Microplastic as a vector for chemicals in the aquatic environment: critical review and model-supported reinterpretation of empirical studies. Environ Sci Technol 50:3315–3326
58. Syberg K et al (2015) Microplastics: addressing ecological risk through lessons learned. Environ Toxicol Chem 34:945–953
59. Rochman C et al (2016) The ecological impacts of marine debris: unraveling the demonstrated evidence from what is perceived. Ecology 97:302–312

60. Cole M et al (2015) The impact of polystyrene microplastics on feeding, function and fecundity in the marine copepod *Calanus helgolandicus*. Environ Sci Technol 49:1130–1137
61. Nobre CR et al (2015) Assessment of microplastic toxicity to embryonic development of the sea urchin *Lytechinus variegatus* (Echinodermata: Echinoidea). Mar Pollut Bull 92:99–104
62. Phuong NN et al (2016) Is there any consistency between the microplastics found in the field and those used in laboratory experiments? Environ Pollut 211:111–123
63. Rochman CM et al (2013) Policy: classify plastic waste as hazardous. Nature 494:169–171
64. Inter-Departmental Liaison Group of Risk Assessment (2002) The precautionary principle: policy and application. http://www.hse.gov.uk/aboutus/meetings/committees/ilgra/pppa.pdf. Accessed 5 Oct 2016
65. Gubler D (2012) The economic burden of dengue. Am J Trop Med Hyg 86:743–744
66. Global Alliance for Incinerator Alternatives (2012) On the road to zero waste: successes and lessons from around the world. http://www.no-burn.org/on-the-road-to-zero-waste-suc cesses-and-lessons-from-around-the-world. Accessed 5 Oct 2016
67. Deep Sea News (2014) The ocean cleanup, part 2: technical review of the feasibility study. http://www.deepseanews.com/2014/07/the-ocean-cleanup-part-2-technical-review-of-the-feasibility-study/. Accessed 5 Oct 2016
68. Sherman P, van Sebille E (2016) Modeling marine surface microplastic transport to assess optimal removal locations. Environ Res Lett 11:1
69. Peters AJ, Siuda ANS (2014) A review of observations of floating tar in the Sargasso Sea. Oceanography 27:217–221
70. Corcoran P et al (2014) An anthropogenic marker horizon in the future rock record. GSA Today 24(6):4–8. doi:10.1130/GSAT-G198A.1
71. Watkins E et al (2015) Marine litter: socio-economic study. Scoping Report, London
72. KIMO (2016) What is fishing for litter?. http://www.fishingforlitter.org.uk/what-is-fishing-for-litter. Accessed 6 Oct 2016
73. Song YK et al (2015) A comparison of microscopic and spectroscopic identification methods for analysis of microplastics in environmental samples. Mar Pollut Bull 93:202–209
74. U.S. Congress (2015) H.R. 1321 – Microbead-Free Waters Act of 2015. https://www.con gress.gov/bill/114th-congress/house-bill/1321. Accessed 6 Oct 2016
75. California Environmental Protection Agency (2014) Preproduction plastic debris program. State Water Resources Control Board, Sacramento. http://www.waterboards.ca.gov/water_ issues/programs/stormwater/plasticdebris.shtml. Accessed 5 Oct 2016
76. Krohn W, van den Daele W (1998) Science as an agent of change: finalization and experimental implementation. Soc Sci Inf 37:191–222
77. Plastics Europe (2013) Zero plastics to landfill by 2020. http://www.plasticseurope.org/documents/document/20131017112406-03_zero_plastics_to_landfill_by_2020_sept_2013.pdf. Accessed 6 Oct 2016
78. Trucost (2016) Plastics and sustainability: a valuation of environmental benefits, costs and opportunities for continuous improvement. https://plastics.americanchemistry.com/Plastics-and-Sustainability.pdf. Accessed 5 Oct 2016
79. Global Alliance for Incinerator Alternatives (2015) Open letter to ocean conservancy regarding the report "Stemming the Tide". http://www.no-burn.org/downloads/Open_Letter_Stem ming_the_Tide_Report_2_Oct_15.pdf. (Accessed 15th May 2016).
80. Green Magazine Hawaii (2012) Feeding the beast – Hawai'i's waste-to-energy plant isn't killing two birds with one stone – it's burning the candle on both ends. http://greenmagazinehawaii.com/feeding-the-beast/. Accessed 6 Oct 2016
81. MacKerron CB, Hoover D (2015) Waste and opportunity 2015: Environmental progress and challenges in food, beverage and consumer goods packaging. As You Sow. Natural Resource Defense Council, New York
82. Plastics Europe 2013 Plastics – the facts 2013: an analysis of European latest plastics production, demand and waste data

83. U.S. EPA (2015) Advancing sustainable materials management: facts and figures 2013 – assessing trends in materials generation, recycling and disposal in the United States. https://www.epa.gov/sites/production/files/2015-09/documents/stanislaus.pdf. Accessed 6 Oct 2016

84. Denkstatt (2015) How packaging contributes to food waste prevention. http://denkstatt.at/files/How_Packaging_Contributes_to_Food_Waste_Prevention_V1.2.pdf. Accessed 28 Aug 2016

85. Andrady A, Neal M (2009) Applications and societal benefits of plastics. Philos Trans R Soc Lond B Biol Sci 364:1977–1984

86. Astrup T et al (2009) Incineration and co-combustion of waste: accounting of greenhouse gases and global warming contributions. Waste Manag Res 27:789–799

87. Los Angeles Department of Public Works 2016 Comparative greenhouse gas emissions analysis of alternative scenarios for waste treatment and/or disposal. Los Angeles, California

88. Bioplastic Feedstock Alliance (2016) Who we are. http://bioplasticfeedstockalliance.org/who-we-are/. Accessed 6 Oct 2016

89. Gross RA, Kalra B (2002) Biodegradable polymers for the environment. Science 297:803–807

90. Kale G et al (2007) Compostability of bioplastic packaging materials: an overview. Macromol Biosci 7:255–277

91. Anastas P, Zimmerman J (2003) Design through the 12 principles of green engineering. Environ Sci Technol 37:94–101

92. Oosterhuis F et al (2014) Economic instruments and marine litter control. Ocean Coast Manag 102:47–54

93. Closed Loop Fund (2016) Closed loop fund. http://www.closedloopfund.com/. Accessed 5 Oct 2016

94. Lifset R et al (2013) Extended producer responsibility. J Ind Ecol 17:162–166

95. Niza S et al (2014) Extended producer responsibility policy in Portugal: a strategy towards improving waste management performance. J Clean Prod 64:277–287

96. Pires A et al (2015) Extended producer responsibility: a differential fee model for promoting sustainable packaging. J Clean Prod 108:343–353

97. Wagner TP (2013) Examining the concept of convenient collection: an application to extended producer responsibility and product stewardship frameworks. Waste Manag 33:499–507

98. Monroe L (2013) Tailoring product stewardship and extended producer responsibility to prevent marine plastic pollution. Tulane Environ Law J 27:219

99. EMF (2016) The new plastics economy: rethinking the future of plastics. https://www.ellenmacarthurfoundation.org/publications/the-new-plastics-economy-rethinking-the-future-of-plastics. Accessed 6 Oct 2016

100. Fairtrade (2015) Fairtrade by the numbers. http://www.fairtrade.org.za/uploads/files/Business/Fairtrade_by_the_Numbers_2015.pdf. Accessed 16 Sep 2016

101. The Co-operative Group (2012) The ethical consumerism report 2012. http://www.coop.co.uk/PageFiles/416561607/Ethical-Consumer-Markets-Report-2012.pdf. Accessed 16 Sep 2016

102. Organic Trade Association (2016) U.S. organic state of the industry. http://ota.com/sites/default/files/indexed_files/OTA_StateofIndustry_2016.pdf. Accessed 16 Sep 2016

103. Gelcich S et al (2014) Public awareness, concerns, and priorities about anthropogenic impacts on marine environments. Proc Natl Acad Sci USA 111:15042–15047

104. Bureo.co (2016) Bureo skateboards, net positiva. http://bureo.co/net-positiva.php. Accessed 16 Sep 2016

105. The Verge (2016) Adidas' limited edition sneakers are made from recycled ocean waste. http://www.theverge.com/2016/6/8/11881670/adidas-3d-printed-sneaker-competition. Accessed 16 Sep 2016

106. Smith K (2010) An examination of marketing techniques that influence Millennials' perceptions of whether a product is environmentally friendly. J Strateg Mark 18:437–450

107. Bepakt.com (2016) List of packaging-free shops. http://bepakt.com/packaging-free-shops/. Accessed 5 Oct 2016
108. NABU (2016) Unverpackt einkaufen. https://www.nabu.de/umwelt-und-ressourcen/ressourcenschonung/einzelhandel-und-umwelt/nachhaltigkeit/19107.html. Accessed 16 Sep 2016
109. Jeff Bezos (2013) Amazon frustration-free packaging letter to customers. https://www.amazon.com/gp/feature.html/ref=amb_link_84595831_1?ie=UTF8&docId=1001920911&pf_rd_m=ATVPDKIKX0DER&pf_rd_s=merchandised-search-5&pf_rd_r=99H1ZVH9M6C C8V6HM5YS&pf_rd_t=101&pf_rd_p=2104054182&pf_rd_i=5521637011. Accessed 16 Sep 2016
110. ECOTEC Research & Consulting (2001) Study on the economic and environmental implications of the use of environmental taxes and charges in the European Union and its member states. http://ec.europa.eu/environment/enveco/taxation/environmental_taxes.htm. Accessed 16 Sep 2016
111. Schüler K (2015) Aufkommen und Verwertung von Verpackungsabfällen in Deutschland im Jahr 2012. Umweltbundesamt, Dessau-Roßlau
112. Ashenmiller B (2009) Cash recycling, waste disposal costs, and the incomes of the working poor: evidence from California. Land Econ 85:539–551
113. BMUB – Bundesministerium für Umwelt (2016) Naturschutz, Bau und Reaktorsicherheit, Anteile der in Mehrweg-Getränkeverpackungen sowie in ökologisch vorteilhaften Einweg-Getränkeverpackungen abgefüllten Getränke in den Jahren 2004 bis 2014. http://www.bmub.bund.de/fileadmin/Daten_BMU/Bilder_Infografiken/verpackungen_mehrweganteile_oeko.png. Accessed 16 Sep 2016
114. Convey F et al (2007) The most popular tax in Europe? Lessons from the Irish plastic bags levy. Environ Resour Econ 38:1–11
115. U.S. Department of Agriculture (2016) Number of U.S. farmers' markets continues to rise. http://www.ers.usda.gov/data-products/chart-gallery/detail.aspx?chartId=48561&ref=collection&embed=True. Accessed 16 Sep 2016
116. Observatorio Feria Libre (2013) Características económicas y sociales de ferias libres de Chile – encuesta nacional de ferias libres. Proyecto de Cooperación Técnica FAO – ODEPA – ASOF TCP CHI/3303: fortalecimiento de las ferias libres para la comercialización agroalimentaria
117. Rootes C (2004) Environmental movements. In: Snow D, Soule S, Kriesi H (eds) The Blackwell companion to social movements. Blackwell, Oxford, pp 608–640
118. GESAMP 2016 Sources, fate and effects of microplastics in the marine environment – part two of a global assessment.
119. Bruyn S (2000) A civil economy: transforming the marketplace in the twenty-first century. University of Michigan Press, Ann Arbor

Index

A

Adsorption, 9, 25, 32–36, 57, 153, 169–172
Adverse outcome pathways (AOPs), 145
Africa, 89, 101–120
  Great Lakes, 59, 101
African catfish (*Clarias gariepinus*), 168
Agriculture, 5, 87, 114, 126, 185, 259, 280
Algae, 11, 36, 128, 155, 164, 183, 189
Amphibians, 158
Anodonta cygnea, 155, 157
Antimicrobial agents, 9, 275
Antioxidants, 4, 9, 13, 61, 258
Anuraeopsis fissa, 155
Aquabacterium commune, 187
Arenicola marina, 118, 277
Asian green mussel (*Perna viridis*), 167
Atmospheric fallout, 72, 76
Attenuated total reflectance (ATR)
  micro-FT-IR, 31
Aubonne, 59
Autecology, 153

B

Bacterivores, 155, 160
Bags, 29, 91, 103–116, 131, 209, 226, 248, 253, 258, 276, 280
Beach cleanup programs, 117, 212, 215
Bioaccumulation, 2, 12, 169, 247
Bioavailability, 9, 12, 118, 169, 256
Biodegradation, 60, 181, 190
Biofilms, 7, 57, 62, 170, 181–195
Biological effects, 25
Biomagnification, 12
Bioorthogonal noncanonical amino acid
  tagging (BONCAT), 194

Bisphenol A (BPA), 169, 275
Bivalves, 35, 155, 166, 167, 189
Black Sea, 59, 126
Bosmina coregoni, 11, 156
Bottles, 29, 103, 108, 128, 131, 186, 190, 226, 254, 261, 284, 291
Brachionus calyciflorus, 156, 160
Brienz Lake, 59

C

Cairo, 103, 105, 118
Calanus helgolandicus, 166, 277
Carbon nanotubes (CNTs), 4, 29, 132
Carcinus maenas, 11, 166
Carrier bags, 248
Cellophane, 89, 94
Ceriodaphnia quadrangula, 156, 158
Cerium dioxide, CeO_2, 29, 36
Chaoborus flavicans, 157
China, 6, 59, 87–96, 283
Chlamydomonas reinhardtii, 171
Ciliates, 155, 158, 160
Citizen science, 203, 211
Clams, 90, 155, 276
Clarias gariepinus, 168
Coastal cleanup efforts, 215
Collectors, 155, 291
Combined sewer overflows (CSOs), 72
Composting, 254, 284–288
Congo River, 103
Conochilus unicornis, 156, 159
Copepods, 13, 155–160, 166, 277
Co-pollutants, 32
Cosmetic products, 5, 27, 29, 107, 136, 182, 207, 211, 223–232, 254, 261, 276, 292

M. Wagner, S. Lambert (eds.), *Freshwater Microplastics*,
Hdb Env Chem 58, DOI 10.1007/978-3-319-61615-5,
© The Author(s) 2018

Printed in the United States
By Bookmasters